A·N·N·U·A·L E·D·I·T·I·O·N·S

Environment

Twenty-fifth Edition

W9-DIE-526

06/07

EDITOR

John L. Allen

University of Wyoming

John L. Allen is professor and chair of geography at the University of Wyoming. He received his bachlor's degree in 1963 and his M.A. in 1964 from the University of Wyoming, and in 1969 he received his Ph.D. from Clark University. His special area of interest is the impact of contemporary human societies on environmental systems.

McGraw Hill **Contemporary Learning Series**

2460 Kerper Blvd., Dubuque, IA 52001

Visit us on the Internet
http://www.mhcls.com

Credits

1. **The Global Environment: An Emerging World View**
   Unit photo—U.S. Air Force photo by: Staff Sgt. Dennis J. Henry Jr.
2. **Population, Policy, and Economy**
   Unit photo—© Superstock/Thinkstock
3. **Energy: Present and Future Problems**
   Unit photo—© Getty Images/Russell Illig
4. **Biosphere: Endangered Species**
   Unit photo—Getty Images/PhotoLink/C. McIntyre
5. **Resources: Land and Water**
   Unit photo—© CORBIS/Royalty-Free
6. **The Hazards of Growth: Pollution and Climate Change**
   Unit photo—© PunchStock/Creatas

Copyright

Cataloging in Publication Data
Main entry under title: Annual Editions: Environment. 2006/2007.
1. Environment—Periodicals. I. Allen, John L., *comp.* II. Title: Environment.
ISBN-13: 978–0–07–351542–7   ISBN-10: 0–07–351542–6   658'.05   ISSN 0272–9008

Twenty-fifth Edition

Cover image © Digital Vision/PunchStock and Photos.com
Printed in the United States of America   1234567890QPDQPD9876   Printed on Recycled Paper

Editors/Advisory Board

Members of the Advisory Board are instrumental in the final selection of articles for each edition of ANNUAL EDITIONS. Their review of articles for content, level, currentness, and appropriateness provides critical direction to the editor and staff. We think that you will find their careful consideration well reflected in this volume.

Preface

In publishing ANNUAL EDITIONS we recognize the enormous role played by the magazines, newspapers, and journals of the public press in providing current, first-rate educational information in a broad spectrum of interest areas. Many of these articles are appropriate for students, researchers, and professionals seeking accurate, current material to help bridge the gap between principles and theories and the real world. These articles, however, become more useful for study when those of lasting value are carefully collected, organized, indexed, and reproduced in a low-cost format, which provides easy and permanent access when the material is needed. That is the role played by ANNUAL EDITIONS.

At the beginning of our new millennium, environmental dilemmas long foreseen by natural and social scientists began to emerge in a number of guises: regional imbalances in numbers of people and the food required to feed them, international environmental crime, energy scarcity, acid rain, build-up of toxic and hazardous wastes, ozone depletion, water shortages, massive soil erosion, global atmospheric pollution, climate change, forest dieback and tropical deforestation, and the highest rates of plant and animal extinction the world has known in 65 million years.

These and other environmental problems continue to worsen in spite of an increasing amount of national and international attention to the issues surrounding them and increased environmental awareness and legislation at both global and national levels. The problems have resulted from centuries of exploitation and unwise use of resources, accelerated recently by the shortsighted public policies that have favored the short-term, expedient approach to problem-solving over longer-term economic and ecological good sense. In Africa, for example, the drive to produce enough food to support a growing population has caused the use of increasingly fragile and marginal resources, resulting in the dryland deterioration that brings famine to that troubled continent. Similar social and economic problems have contributed to massive deforestation in middle and South America and in Southeast Asia.

Part of the problem is that efforts to deal with environmental issues have been intermittent. During the decade of the 1980s, economic problems generated by resource scarcity caused the relaxation of environmental quality standards and contributed to the refusal of many of the world's governments and international organizations to develop environmentally sound protective measures, which were viewed as too costly. More recently, in the late 20th and early 21st century, as environmental protection policies were adopted, they were often cosmetic, designed for good press and TV sound bites, and—even worse—seemingly designed to benefit large corporations rather than to protect environmental systems. Even with these policies based more in public relations than in environmental ones, governments often lacked either the will or the means to implement them properly. The absence of effective environmental policy has been particularly apparent in those countries that are striving to become economically developed. But even in the more highly developed nations, economic concerns tend to favor a loosening of environmental controls. In the United States, for example, the interests of maintaining jobs for the timber industry imperil many of the last areas of old-growth forests, and the desire to maintain agricultural productivity at all costs causes the continued use of destructive and toxic chemicals on the nation's farmlands. In addition, concerns over energy availability have created the need for foreign policy and military action to protect the developed nations' access to cheap oil and have prompted increasing reliance on technological quick fixes, as well as the development of environmentally sensitive areas to new energy resource exploration and exploitation. Yet the simpler measures of energy conservation do not seem to be important to policy makers.

Despite the recent tendency of the U.S. government to turn its back on environmental issues and refuse to participate in important international environmental accords, particularly those related to global warming, there is some reason to hope that a new environmental consciousness is awakening with the new global economic system. Unfortunately, increasing globalization of the economy has meant the increased spread of other things as well, such as conflict and infectious diseases. The emergence of international terrorism as an instrument of national or quasi-national policy—particularly where terrorism may employ environmental contamination or disease as a weapon—has the potential to produce future environmental problems that are almost too frightening to think about. It has long been an accepted doctrine that we would all be better off when economic and political barriers dropped. We are now learning that is not always the case.

In *Annual Editions: Environment 06/07* every effort has been made to choose articles that encourage an understanding of the nature of the environmental problems that beset us and how, with wisdom and knowledge and the proper perspective, they can be solved or at least mitigated. Accordingly, the selections in this book have been chosen more for their intellectual content than for their emotional tone. They have been arranged into an order of topics—the global environment; population, policy, and economy; energy; the biosphere; land and water resources; and the hazards of growth—that lends itself to a progressive understanding of the causes and effects of human modifications of Earth's environmental systems. We will not be protected against the ecological consequences of human actions by remaining ignorant of them.

Readers can have input into the next edition of *Annual Editions: Environment* by completing and returning the post-paid *article rating form* at the back of the book.

John L. Allen
Editor

Contents

UNIT 1
The Global Environment: An Emerging World View

UNIT 2
Population, Policy, and Economy

The concepts in bold italics are developed in the article. For further expansion, please refer to the Topic Guide and the Index.

UNIT 3
Energy: Present and Future Problems

The concepts in bold italics are developed in the article. For further expansion, please refer to the Topic Guide and the Index.

UNIT 4
Biosphere: Endangered Species

UNIT 5
Resources: Land and Water

The concepts in bold italics are developed in the article. For further expansion, please refer to the Topic Guide and the Index.

UNIT 6
The Hazards of Growth: Pollution and Climate Change

The concepts in bold italics are developed in the article. For further expansion, please refer to the Topic Guide and the Index.

Topic Guide

This topic guide suggests how the selections in this book relate to the subjects covered in your course. You may want to use the topics listed on these pages to search the Web more easily.

On the following pages a number of Web sites have been gathered specifically for this book. They are arranged to reflect the units of this *Annual Edition*. You can link to these sites by going to the student online support site at *http://www.mhcls.com/online/*.

ALL THE ARTICLES THAT RELATE TO EACH TOPIC ARE LISTED BELOW THE BOLD-FACED TERM.

Internet References

The following internet sites have been carefully researched and selected to support the articles found in this reader. The easiest way to access these selected sites is to go to our student online support site at *http://www.mhcls.com/online/*.

AE: Environment 06/07

The following sites were available at the time of publication. Visit our Web site—we update our student online support site regularly to reflect any changes.

General Sources

Britannica's Internet Guide
http://www.britannica.com
This site presents extensive links to material on world geography and culture, encompassing material on wildlife, human lifestyles, and the environment.

CIA Factbook
http://www.cia.gov/cia/publications/factbook/
This site is the United States government's official source for data on the population, production, resources, geography, political systems, and other important characteristics of each of the world's countries.

EnviroLink
http://www.envirolink.org/
One of the world's largest environmental information clearinghouses, EnviroLink is a grassroots nonprofit organization that unites organizations and volunteers around the world and provides up-to-date information and resources.

Library of Congress
http://www.loc.gov
Examine this extensive Web site to learn about resource tools, library services/resources, exhibitions, and databases in many different subfields of environmental studies.

The New York Times
http://www.nytimes.com
Browsing through the archives of the *New York Times* will provide a wide array of articles and information related to the different subfields of the environment.

SocioSite: Sociological Subject Areas
http://www.pscw.uva.nl/sociosite/TOPICS/
This huge sociological site from the University of Amsterdam provides many discussions and references of interest to students of the environment, such as the links to information on ecology and consumerism.

U.S. Geological Survey
http://www.usgs.gov
This site and its many links are replete with information and resources in environmental studies, from explanations of El Niño to discussion of concerns about water resources.

UNIT 1: The Global Environment: An Emerging World View

Alternative Energy Institute (AEI)
http://www.altenergy.org
The AEI will continue to monitor the transition from today's energy forms to the future in a "surprising journey of twists and turns." This site is the beginning of an incredible journey.

Earth Science Enterprise
http://www.earth.nasa.gov
Information about NASA's Mission to Planet Earth program and its Science of the Earth System can be found here. Surf to learn about satellites, El Niño, and even "strategic visions" of interest to environmentalists.

IISDnet
http://www.iisd.org
The International Institute for Sustainable Development, a Canadian organization, presents information through gateways entitled Business, Climate Change, Measurement and Assessment, and Natural Resources. IISD Linkages is its multimedia resource for environment and development policy makers.

National Geographic Society
http://www.nationalgeographic.com
Links to *National Geographic*'s huge archive are provided here. There is a great deal of material related to the atmosphere, the oceans, and other environmental topics.

Research and Reference (Library of Congress)
http://lcweb.loc.gov/rr/
This research and reference site of the Library of Congress will lead to invaluable information on different countries. It provides links to numerous publications, bibliographies, and guides in area studies that can be of great help to environmentalists.

Santa Fe Institute
http://acoma.santafe.edu
This home page of the Santa Fe Institute—a nonprofit, multidisciplinary research and education center—will lead to many interesting links related to its primary goal: to create a new kind of scientific research community, pursuing emerging science.

Solstice: Documents and Databases
http://solstice.crest.org/index.html
In this online source for sustainable energy information, the Center for Renewable Energy and Sustainable Technology (CREST) offers documents and databases on renewable energy, energy efficiency, and sustainable living. The site also offers related Web sites, case studies, and policy issues.

United Nations
http://www.unsystem.org
Visit this official Web site Locator for the United Nations System of Organizations to get a sense of the scope of international environmental inquiry today. Various UN organizations concern themselves with everything from maritime law to habitat protection to agriculture.

United Nations Environment Programme (UNEP)
http://www.unep.ch
Consult this home page of UNEP for links to critical topics of concern to environmentalists, including desertification, migratory species, and the impact of trade on the environment. The site will direct you to useful databases and global resource information.

www.mhcls.com/online/

World Resources Institute (WRI)
http://www.wri.org/

The World Resources Institute is committed to change for a sustainable world and believes that change in human behavior is urgently needed to halt the accelerating rate of environmental deterioration in some areas. It sponsors not only the general website above but also The Environmental Information Portal (www.earthtrends.wri.org) that provides a rich database on the interaction between human disease, pollution, and large-scale environmental, development, and demographic issues.

UNIT 2: Population, Policy, and Economy

The Hunger Project
http://www.thp.org

Browse through this nonprofit organization's site to explore the ways in which it attempts to achieve its goal: the sustainable end to global hunger through leadership at all levels of society. The Hunger Project contends that the persistence of hunger is at the heart of the major security issues that are threatening our planet.

Poverty Mapping
http://www.povertymap.net

Poverty maps can quickly provide information on the spatial distribution of poverty. This site provides maps, graphics, data, publications, news, and links that provide the public with poverty mapping from the global to the subnational level.

World Health Organization
http://www.who.int

The home page of the World Health Organization provides links to a wealth of statistical and analytical information about health and the environment in the developing world.

World Population and Demographic Data
http://geography.about.com/cs/worldpopulation/

On this site, information about world population and additional demographic data for all the countries of the world are provided.

WWW Virtual Library: Demography & Population Studies
http://demography.anu.edu.au/VirtualLibrary/

This is a definitive guide to demography and population studies. A multitude of important links to information about global poverty and hunger can be found here.

UNIT 3: Energy: Present and Future Problems

Alliance for Global Sustainability (AGS)
http://globalsustainability.org/

The AGS is a cooperative venture seeking solutions to today's urgent and complex environmental problems. Research teams from four universities study large-scale, multidisciplinary environmental problems that are faced by the world's ecosystems, economies, and societies.

Alternative Energy Institute, Inc.
http://www.altenergy.org

On this site created by a nonprofit organization, discover how the use of conventional fuels affects the environment. Also learn about research work on new forms of energy.

Energy and the Environment: Resources for a Networked World
http://zebu.uoregon.edu/energy.html

An extensive array of materials having to do with energy sources—both renewable and nonrenewable—as well as other topics of interest to students of the environment is found on this site.

Institute for Global Communication/EcoNet
http://www.igc.org/

This environmentally friendly site provides links to dozens of governmental, organizational, and commercial sites having to do with energy sources. Resources address energy efficiency, renewable generating sources, global warming, and more.

Nuclear Power Introduction
http://library.thinkquest.org/17658/pdfs/nucintro.pdf

Information regarding alternative energy forms can be accessed here. There is a brief introduction to nuclear power and a link to maps that show where nuclear power plants exist.

U.S. Department of Energy
http://www.energy.gov

Scrolling through the links provided by this Department of Energy home page will lead to information about fossil fuels and a variety of sustainable/renewable energy sources.

UNIT 4: Biosphere: Endangered Species

Endangered Species
http://www.endangeredspecie.com/

This site provides a wealth of information on endangered species anywhere in the world. Links providing data on the causes, interesting facts, law issues, case studies, and other issues on endangered species are available.

Friends of the Earth
http://www.foe.co.uk/index.html

Friends of the Earth, a nonprofit organization based in the United Kingdom, pursues a number of campaigns to protect the Earth and its living creatures. This site has links to many important environmental sites, covering such broad topics as ozone depletion, soil erosion, and biodiversity.

Natural Resources Defense Council
http://nrdc.org

The Natural Resources Defense Council (NRDC) uses law, science, and the support of more than 1 million members and activists to protect the planet's wildlife, plants, water, soils, and other resources. The site provides abundant information on global issues and political responses.

Smithsonian Institution Web Site
http://www.si.edu

Looking through this site, which will provide access to many of the enormous resources of the Smithsonian, offers a sense of the biological diversity that is threatened by humans' unsound environmental policies and practices.

World Wildlife Federation (WWF)
http://www.wwf.org

This home page of the WWF leads to an extensive array of information links about endangered species, wildlife management and preservation, and more. It provides many suggestions for how to take an active part in protecting the biosphere.

UNIT 5: Resources: Land and Water

Global Climate Change
http://www.puc.state.oh.us/consumer/gcc/index.html

The goal of this PUCO (Public Utilities Commission of Ohio) site is to serve as a clearinghouse of information related to global climate change. Its extensive links provide an explanation of the science and chronology of global climate change, acronyms, definitions, and more.

www.mhcls.com/online/

National Oceanic and Atmospheric Administration (NOAA)

http://www.noaa.gov

Through this home page of NOAA, you can find information about coastal issues, fisheries, climate, and more.

National Operational Hydrologic Remote Sensing Center (NOHRSC)

http://www.nohrsc.nws.gov

Flood images are available at this site of the NOHRSC, which works with the U.S. National Weather Service to track weather-related information.

Terrestrial Sciences

http://www.cgd.ucar.edu/tss/

The Terrestrial Sciences Section (TSS) is part of the Climate and Global Dynamics (CGD) Division at the National Center for Atmospheric Research (NCAR) in Boulder, Colorado. Scientists in the section study land-atmosphere interactions, in particular surface forcing of the atmosphere, through model development, application, and observational analyses. Here, you'll find a link to VEMAP, The Vegetation/Ecosystem Modeling and Analysis Project.

UNIT 6: The Hazards of Growth: Pollution and Climate Change

Persistent Organic Pollutants (POP)

http://www.chem.unep.ch/pops/

Visit this site to learn more about persistent organic pollutants (POPs) and the issues and concerns surrounding them.

School of Labor and Industrial Relations (SLIR): Hot Links

http://www.lir.msu.edu/hotlinks/

Michigan State University's SLIR page connects to industrial relations sites throughout the world. It has links to U.S. government statistics, newspapers and libraries, international intergovernmental organizations, and more.

Space Research Institute

http://arc.iki.rssi.ru/eng/index.htm

For a change of pace, browse through this home page of Russia's Space Research Institute for information on its Environment Monitoring Information Systems, the IKI Satellite Situation Center, and its Data Archive.

Worldwatch Institute

http://www.worldwatch.org

The Worldwatch Institute, dedicated to fostering the evolution of an environmentally sustainable society, presents this site with access to *World Watch Magazine* and *State of the World 2000*. Click on In the News and Press Releases for discussions of current problems.

We highly recommend that you review our Web site for expanded information and our other product lines. We are continually updating and adding links to our Web site in order to offer you the most usable and useful information that will support and expand the value of your Annual Editions. You can reach us at: *http://www.mhcls.com/annualeditions/*.

UNIT 1

The Global Environment: An Emerging World View

Unit Selections

1. **How Many Planets? A Survey of the Global Environment**, The Economist
2. **Five Meta-Trends Changing the World**, David Pearce Snyder
3. **Globalization's Effects on the Environment**, Jo Kwong
4. **Rescuing a Planet Under Stress**, Lester R. Brown

Key Points to Consider

- What are the connections between the attempts to develop sustainable systems and the quantity and quality of environmental data? Are there also relationships between data and the role of technology and economic systems in shaping the environmental future?

- What are some of the key "meta-trends" produced by increasing globalization of economic and other human systems? How can human societies and cultures adapt to such trends in order to prevent significant environmental disruption?

- How has the process of "globalization" altered the cultural and environmental patterns of the world? What kinds of changes brought about by an increasingly global economy have been unforeseen?

- Explain the analogy between human impact on environmental systems and an "economic bubble" in which demand so exceeds supply that the economic system collapses.

Student Website

www.mhcls.com/online

Internet References

Further information regarding these websites may be found in this book's preface or online.

Alternative Energy Institute (AEI)
http://www.altenergy.org

Earth Science Enterprise
http://www.earth.nasa.gov

IISDnet
http://www.iisd.org

National Geographic Society
http://www.nationalgeographic.com

Research and Reference (Library of Congress)
http://lcweb.loc.gov/rr/

Santa Fe Institute
http://acoma.santafe.edu

Solstice: Documents and Databases
http://solstice.crest.org/index.html

United Nations
http://www.unsystem.org

United Nations Environment Programme (UNEP)
http://www.unep.ch

World Resources Institute (WRI)
http://www.wri.org/

More than three decades after the celebration of the first Earth Day in 1970, public apprehension over the environmental future of the planet has reached levels unprecedented even during the late 1960s and early 1970s "Age of Aquarius." No longer are those concerned about the environment dismissed as "ecofreaks" and "tree-huggers." Most serious scientists have joined the rising clamor for environmental protection, as have the more traditional environmentally conscious public-interest groups. There are a number of reasons for this increased environmental awareness. Some of these reasons arise from environmental events. The dramatic hurricane season of 2005 and its links to warmer ocean waters have, for example, made it increasingly difficult to deny the effects of global warming. Atmospheric scientists are nearly unanimous in their attribution of human agencies as at least a partial cause for increasing global temperatures. But more reasons for environmental awareness arise simply from the process of globalization: the increasing unity of the world's economic, social, and information systems. Hailed by many as the salvation of the future, globalization has done little to make the world a better or safer place. Diseases once defined as "endemic" or confined to specific regions now have increased capacity for widespread dissemination and alarms have been sounded about new pandemic diseases—such as strains of flu with worldwide distribution. In addition, increased human mobility and the ease of travel has allowed human-caused disruptions to political, cultural, and economic systems to spread, and acts of terrorism now take place in locations once thought safe from such manifestations of hatred and despair. On the more positive side, the expansion of global information systems has fostered a maturation of concepts about the global nature of environmental processes.

Much of what has been learned through this increased information flow over the last two decades, particularly by American observers, has been of the environmentally ravaged world behind the old Iron Curtain—a chilling forecast of what other industrialized regions as well as the developing countries can become in the near future unless strict international environmental measures are put in place. For perhaps the first time ever, countries are beginning to recognize that environmental problems have no boundaries and that international cooperation is the only way to solve them.

The subtitle of this first unit, "An Emerging World View," is an optimistic assessment of the future: a future in which less money is spent on defense and more on environmental protection and cleanup—a new world order in which political influence might be based more on leadership in environmental and economic issues than on military might. It is probably far too early to make such optimistic predictions, to conclude that the world's nations—developed and underdeveloped—will begin to recognize that Earth's environment is a single unit. Thus far those nations have shown no tendency to recognize that humankind is a single unit and that what harms one harms all. The recent emergence of wide-scale terrorism and military action as a means of political and social policy is evidence of such a failure of recognition. Nevertheless, there is a growing international realization—aided by the information superhighway—that we are all, as environmental activists have

been saying for decades, inhabitants of Spaceship Earth and will survive or succumb together.

The articles selected for this unit have been chosen to illustrate the increasingly global perspective on environmental problems and the degree to which their solutions must be linked to political, economic, and social problems and solutions. In the lead piece of the unit, "How Many Planets?" the editors of *The Economist* attempt an analysis of what they admit is a very slippery subject by beginning with the observation that "it comes as a shock to discover how little information there is on the environment." They note the lip service paid everywhere to the concept of sustainability and acknowledge that economic growth and environmental health are not mutually inconsistent but that a great deal more work is necessary to make them compatible. They also conclude that governments, corporations, and individuals are more prepared now to think about how to use the planet than they were even 10 years ago.

Issues surrounding globalization form the subject of the next selection in the unit. In "Five Meta-Trends Changing the World," David Snyder, lifestyles editor of *The Futurist,* recognizes the "meta-trends" or multidimensional and evolutionary trends in the human-environment systems that are occurring as a result of a series of simultaneous demographic, economic, and technological trends. Snyder identifies these "meta-trends" as: (1) *Cultural modernization*, referring to the increasing "Westernization" of cultures in technological, economic, and demographic terms; (2) *Economic globalization*, meaning increasingly global competition for workers and resources and markets; (3) *Universal connectivity* or what the editor of *The Economist* has called "the death of distance," a recognition that the cell phone, the Internet, and other connective technologies have truly made the world smaller by linking people together on a virtually-instantaneous basis; (4) *Transactional transparency*, in which corporate integrity and openness will be forced to grow as watchdog groups and citizens demand a new transparency of business that will allow closer public scrutiny of business practices that impact the environment; and (5) *Social adaptation*, or responses to new medical and other technologies that will allow people to remain productive in the workplace for longer periods of time and may even produce a return to the multigenerational family of children, parents, and grandchildren that formed society's safety net prior to the onset of industrialization. Each of these trends has profound implications for the ways in which human societies around the world relate to the environmental systems they inhabit.

Some of the environmental consequences of the trend to globalization are discussed in the third article in this unit. Jo Kwong, Director of the Atlas Economic Research Foundation, notes that globalization is simply about the removal of barriers, allowing the free movement of goods, service, people, and ideas throughout the world. While this used to be viewed as a good thing, globalization has, in fact, raised many problems of both an economic and environmental nature. The author, however, tends to see the environmental issues—such as accelerated global warming through increasing economic development of the Third World and the concomitant increase in fossil fuel consumption—as producing a heightened global awareness and dialogue on global environmental issues. Globalization, Kwong notes, "can be a means to accelerate learning about the importance of market institutions to economic growth." It follows, she concludes, that environmental protection is one of the potential benefits of increasing markets and globalization.

The final selection in the opening section deals with those issues of human intellectual and political response to the global changes in society, economy, and environmental relations that are part of today's world. "Rescuing a Planet Under Stress"

Lester Brown of the Earth Policy Institute provides an overview of "mega-threats" to continued environmental, social, and economic stability. Issues such as climate change, the worldwide HIV epidemic, soil erosion and desertification, and increasing demands for food, water, and fuel are foremost, according to Brown, who equates the increasing environmental pressures to the demands of a "bubble economy" in which demands exceed supply. The way to resolve the problems inherent in this economy is to deflate demand before it bursts the bubble of supply or environmental systems. This must be done at "wartime speed", says Brown, and international agreements must be reached to stabilize population, climate, water resources, soils, and other components of the human-environment system. If this is not done, then massive environmental and economic decline will be inevitable.

All the articles in this opening section deal with environmental problems that were once confined to specific locales but—as the world has grown increasingly smaller through advanced transportation, communication, and other technologies, and as a truly global economy has developed—have now become problems on a global scale. While the potential for world-wide collapses of environmental systems is still a threat, the development of a global community with increasing awareness of the fragility of both economic and environmental systems, provides a promise for the future.

How many planets?

A survey of the global environment

The great race

Growth need not be the enemy of greenery. But much more effort is required to make the two compatible, says Vijay Vaitheeswaran

Sustainable development is a dangerously slippery concept. Who could possibly be against something that invokes such alluring images of untouched wildernesses and happy creatures? The difficulty comes in trying to reconcile the "development" with the "sustainable" bit: look more closely, and you will notice that there are no people in the picture.

That seems unlikely to stop a contingent of some of 60,000 world leaders, businessmen, activists, bureaucrats and journalists from travelling to South Africa next month for the UN-sponsored World Summit on Sustainable Development in Johannesburg. Whether the summit achieves anything remains to be seen, but at least it is asking the right questions. This survey will argue that sustainable development cuts to the heart of mankind's relationship with nature—or, as Paul Portney of Resources for the Future, an American think-tank, puts it, "the great race between development and degradation". It will also explain why there is reason for hope about the planet's future.

The best way known to help the poor today—economic growth—has to be handled with care, or it can leave a degraded or even devastated natural environment for the future. That explains why ecologists and economists have long held diametrically opposed views on development. The difficult part is to work out what we owe future generations, and how to reconcile that moral obligation with what we owe the poorest among us today.

It is worth recalling some of the arguments fielded in the run-up to the big Earth Summit in Rio de Janeiro a decade ago. A publication from UNESCO, a United Nations agency, offered the following vision of the future: "Every generation should leave water, air and soil resources as pure and unpolluted as when it came on earth. Each generation should leave undiminished all the species of animals it found existing on earth." Man, that suggests, is but a strand in the web of life, and the natural order is fixed and supreme. Put earth first, it seems to say.

Robert Solow, an economist at the Massachusetts Institute of Technology, replied at the time that this was "fundamentally the wrong way to go", arguing that the obligation to the future is "not to leave the world as we found it in detail, but rather to leave the option or the capacity to be as well off as we are." Implicit in that argument is the seemingly hard-hearted notion of "fungibility": that natural resources, whether petroleum or giant pandas, are substitutable.

Rio's fatal flaw

Champions of development and defenders of the environment have been locked in battle ever since a UN summit in Stockholm launched the sustainable-development debate three decades ago. Over the years, this debate often pitted indignant politicians and social activists from the poor world against equally indignant politicians and greens from the rich world. But by the time the Rio summit came along, it seemed they had reached a truce. With the help of a committee of grandees led by Gro Harlem Brundtland, a former Norwegian prime minister, the interested parties struck a deal in 1987: development and the environment, they declared, were inextricably linked. That compromise generated a good deal of euphoria. Green groups grew concerned over poverty, and development charities waxed lyrical about greenery. Even the World Bank joined in. Its World Development Report in 1992 gushed about "win-win" strategies, such as ending environmentally harmful subsidies, that would help both the economy and the environment.

By nearly universal agreement, those grand aspirations have fallen flat in the decade since that summit. Little headway has been made with environmental problems such as climate change and loss of biodiversity. Such progress as has been achieved has been largely due to three factors that this survey will explore in later sections: more decision-making at local level, technological innovation, and the rise of market forces in environmental matters.

The main explanation for the disappointment—and the chief lesson for those about to gather in South Africa—is that Rio overreached itself. Its participants were so anxious to reach a political consensus that they agreed to the Brundtland definition of sustainable development, which Daniel Esty of Yale University thinks has turned into "a buzz-word largely devoid of content". The biggest mistake, he reckons, is that it slides over the difficult trade-offs between environment and development in the real world. He is careful to note that there are plenty of cases where those goals are linked—but also many where they are not: "Environmental and economic policy goals are distinct, and the actions needed to achieve them are not the same."

No such thing as win-win

To insist that the two are "impossible to separate", as the Brundtland commission claimed, is nonsense. Even the World Bank now accepts that its much-trumpeted 1992 report was much too optimistic. Kristalina Georgieva, the Bank's director for the environment, echoes comments from various colleagues when she says: "I've never seen a real win-win in my life. There's always somebody, usually an elite group grabbing rents, that loses. And we've learned in the past decade that those losers fight hard to make sure that technically elegant win-win policies do not get very fat."

So would it be better to ditch the concept of sustainable development altogether? Probably not. Even people with their feet firmly planted on the ground think one aspect of it is worth salvaging: the emphasis on the future.

Nobody would accuse John Graham of jumping on green bandwagons. As an official in President George Bush's Office of Management and Budget, and previously as head of Harvard University's Centre for Risk Analysis, he has built a reputation for evidence-based policymaking. Yet he insists sustainable development is a worthwhile concept: "It's good therapy for the tunnel vision common in government ministries, as it forces integrated policymaking. In practical terms, it means that you have to take economic cost-benefit trade-offs into account in environmental laws, and keep environmental trade-offs in mind with economic development."

Jose Maria Figueres, a former president of Costa Rica, takes a similar view. "As a politician, I saw at first hand how often policies were dictated by short-term considerations such as elections or partisan pressure. Sustainability is a useful template to align short-term policies with medium- to long-term goals."

It is not only politicians who see value in saving the sensible aspects of sustainable development. Achim Steiner, head of the International Union for the Conservation of Nature, the world's biggest conservation group, puts it this way: "Let's be honest: greens and businesses do not have the same objective, but they can find common ground. We look for pragmatic ways to save species. From our own work on the ground on poverty, our members—be they bird watchers or passionate ecologists—have learned that 'sustainable use' is a better way to conserve."

Sir Robert Wilson, boss of Rio Tinto, a mining giant, agrees. He and other business leaders say it forces hard choices about the future out into the open: "I like this concept because it frames the trade-offs inherent in a business like ours. It means that single-issue activism is simply not as viable."

Kenneth Arrow and Larry Goulder, two economists at Stanford University, suggest that the old ideological enemies are converging: "Many economists now accept the idea that natural capital has to be valued, and that we need to account for ecosystem services. Many ecologists now accept that prohibiting everything in the name of protecting nature is not useful, and so are being selective." They think the debate is narrowing to the more empirical question of how far it is possible to substitute natural capital with the man-made sort, and specific forms of natural capital for one another.

The job for Johannesburg

So what can the Johannesburg summit contribute? The prospects are limited. There are no big, set-piece political treaties to be signed as there were at Rio. America's acrimonious departure from the Kyoto Protocol, a UN treaty on climate change, has left a bitter taste in many mouths. And the final pre-summit gathering, held in early June in Indonesia, broke up in disarray. Still, the gathered worthies could usefully concentrate on a handful of areas where international co-operation can help deal with environmental problems. Those include improving access for the poor to cleaner energy and to safe drinking water, two areas where concerns about human health and the environmental overlap. If rich countries want to make progress, they must agree on firm targets and offer the money needed to meet them. Only if they do so will poor countries be willing to cooperate on problems such as global warming that rich countries care about.

That seems like a modest goal, but it just might get the world thinking seriously about sustainability once again. If the Johannesburg summit helps rebuild a bit of faith in international environmental cooperation, then it will have been worthwhile. Minimising the harm that future economic growth does to the environment will require the rich world to work hand in glove with the poor world—which seems nearly unimaginable in today's atmosphere poisoned by the shortcomings of Rio and Kyoto.

To understand why this matters, recall that great race between development and degradation. Mankind has stayed comfortably ahead in that race so far, but can it go on doing so? The sheer magnitude of the economic growth that is hoped for in the coming decades (see chart 1) makes it seem inevitable that the clashes between mankind and nature will grow worse. Some are now asking whether all this economic growth is really necessary or useful in the first place, citing past advocates of the simple life.

"God forbid that India should ever take to industrialism after the manner of the West… It took Britain half the resources of the planet to achieve this prosperity. How many planets will a country like India require?", Mahatma Gandhi asked half a century ago. That question encapsulated the bundle of worries that haunts the sustainable-development debate to this day. Today, the vast majority of Gandhi's countrymen are still living the simple life—full of simple misery, malnourishment and material want. Grinding poverty, it turns out, is pretty sustainable.

If Gandhi were alive today, he might look at China next door and find that the country, once as poor as India, has been transformed beyond recognition by two decades of roaring economic growth. Vast numbers of people have been lifted out of poverty and into middle-class comfort. That could prompt him to reframe his question: how many planets will it take to satisfy China's needs if it ever achieves profligate America's affluence? One green group reckons the answer is three. The next section looks at the environmental data that might underpin such claims. It makes for alarming reading—though not for the reason that first springs to mind.

Flying blind

It comes as a shock to discover how little information there is on the environment

WHAT is the true state of the planet? It depends from which side you are peering at it. "Things are really looking up," comes the cry from one corner (usually overflowing with economists and technologists), pointing to a set of rosy statistics. "Disaster is nigh," shouts the other corner (usually full of ecologists and environmental lobbyists), holding up a rival set of troubling indicators.

According to the optimists, the 20th century marked a period of unprecedented economic growth that lifted masses of people out of abject poverty. It also brought technological innovations such as vaccines and other advances in public health that tackled many preventable diseases. The result has been a breath-taking enhancement of human welfare and longer, better lives for people everywhere on earth (see chart 2).

At this point, the pessimists interject: "Ah, but at what ecological cost?" They note that the economic growth which made all these gains possible sprang from the rapid spread of industrialisation and its resource-guzzling cousins, urbanisation, motorisation and electrification. The earth provided the necessary raw materials, ranging from coal to pulp to iron. Its ecosystems—rivers, seas, the atmosphere—also absorbed much of the noxious fallout from that process. The sheer magnitude of ecological change resulting directly from the past century's economic activity is remarkable (see table 3).

To answer that Gandhian question about how many planets it would take if everybody lived like the West, we need to know how much—or how little—damage the West's transformation from poverty to plenty has done to the planet to date. Economists point to the remarkable improvement in local air and water pollution in the rich world in recent decades. "It's Getting Better All the Time", a cheerful tract co-written by the late Julian Simon, insists that: "One of the greatest trends of the past 100 years has been the astonishing rate of progress in reducing almost every form of pollution." The conclusion seems unavoidable: "Relax! If we keep growing as usual, we'll inevitably grow greener."

The ecologically minded crowd takes a different view. "GEO3", a new report from the United Nations Environment Programme, looks back at the past few decades and sees much reason for concern. Its thoughtful boss, Klaus Töpfer (a former German environment minister), insists that his report is not "a document of doom and gloom". Yet, in summing it up, UNEP decries "the declining environmental quality of planet earth", and wags a finger at economic prosperity: "Currently, one-fifth of the world's population enjoys high, some would say excessive, levels of affluence." The conclusion seems unavoidable: "Panic! If we keep growing as usual, we'll inevitably choke the planet to death."

"People and Ecosystems", a collaboration between the World Resources Institute, the World Bank and the United Nations, tried to gauge the condition of ecosystems by examining the goods and services they produce—food, fibre, clean water, carbon storage and so on—and their capacity to continue producing them. The authors explain why ecosystems matter: half of all jobs worldwide are in agriculture, forestry and fishing, and the output from those three commodity businesses still dominates the economies of a quarter of the world's countries.

The report reached two chief conclusions after surveying the best available environmental data. First, a number of ecosystems are "fraying" under the impact of human activity. Second, ecosystems in future will be less able than in the past to deliver the goods and services human life depends upon, which points to unsustainability. But it took care to say: "It's hard, of course, to know what will be truly sustainable." The reason this collection of leading experts could not reach a firm conclusion was that, remarkably, much of the information they needed was incomplete or missing altogether: "Our knowledge of ecosystems has increased dramatically, but it simply has not kept pace with our ability to alter them."

Another group of experts, this time organised by the World Economic Forum, found itself similarly frustrated. The leader of that project, Daniel Esty of Yale, exclaims, throwing his arms in the air: "Why hasn't anyone done careful environmental measurement before? Businessmen always say, 'what matters gets measured.' Social scientists started quantitative measurement 30 years ago, and even political science turned to hard numbers 15 years ago. Yet look at environmental policy, and the data are lousy."

Gaping holes

At long last, efforts are under way to improve environmental data collection. The most ambitious of these is the Millennium Ecosystem Assessment, a joint effort among leading development agencies and environmental groups. This four-year effort is billed as an attempt to establish systematic data sets on all environmental matters across the world. But one of the researchers involved grouses that it "has very, very little new money to collect or analyse new data". It seems astonishing that governments have been making sweeping decisions on environmental policy for decades without such a baseline in the first place.

One positive sign is the growing interest of the private sector in collecting environmental data. It seems plain that leaving the task to the public sector has not worked. Information on the environment comes far lower on the bureaucratic pecking order than data on education or social affairs, which tend to be overseen by ministries with bigger budgets and more political clout. A number of countries, ranging from New Zealand to Austria, are now looking to the private sector to help collect and manage data in areas such as climate. Development banks are also considering using private contractors to monitor urban air quality, in part to get around the corruption and apathy in some city governments.

"I see a revolution in environmental data collection coming because of computing power, satellite mapping, remote sensing and other such information technologies," says Mr Esty. The arrival of hard data in this notoriously fuzzy area could cut down on environmental disputes by reducing uncertainty. One example is the long-running squabble between America's mid-western states, which rely heavily on coal, and the north-eastern states, which suffer from acid rain. Technology helped disprove claims by the mid-western states that New York's problems all resulted from home-grown pollution.

The arrival of good data would have other benefits as well, such as helping markets to work more robustly: witness America's pioneering scheme to trade emissions of sulphur dioxide, made possible by fancy equipment capable of monitoring emissions in real time. Mr Esty raises an even more intriguing possibility: "Like in the American West a hundred years ago, when barbed wire helped establish rights and prevent overgrazing, information technology can help establish 'virtual barbed wire' that secures property rights and so prevents overexploitation of the commons." He points to fishing in the waters between Australia and New Zealand, where tracking and monitoring devices have reduced over-exploitation.

Best of all, there are signs that the use of such fancy technology will not be confined to rich countries. Calestous Juma of Harvard University shares Mr Esty's excitement about the possibility of such a technology-driven revolution even in Africa: "In the past, the only environmental 'database' we had in Africa was our grandmothers. Now, with global information systems and such, the potential is enormous." Conservationists in Namibia, for example, already use satellite tracking to keep count of their elephants. Farmers in Mali receive satellite updates about impending storms on hand-wound radios. Mr Juma thinks the day is not far off when such technology, combined with ground-based monitoring, will help Africans measure trends in deforestation, soil erosion and climate change, and assess the effects on their local environment.

Make a start

That is at once a sweeping vision and a modest one. Sweeping, because it will require heavy investment in both sophisticated hardware and nuts-and-bolts information infrastructure on the ground to make sense of all these new data. As the poor world clearly cannot afford to pay for all this, the rich world must help—partly for altruistic reasons, partly with the selfish aim of discovering

in good time whether any global environmental calamities are in the making. A number of multilateral agencies now say they are willing to invest in this area as a "neglected global public good"—neglected especially by those agencies themselves. Even President Bush's administration has recently indicated that it will give environmental satellite data free to poor countries.

But that vision is also quite a modest one. Assuming that this data "revolution" does take place, all it will deliver is a reliable assessment of the health of the planet today. We will still not be able to answer the broader question of whether current trends are sustainable or not.

To do that, we need to look more closely at two very different sorts of environmental problems: global crises and local troubles. The global sort is hard to pin down, but can involve irreversible changes. The local kind is common and can have a big effect on the quality of life, but is usually reversible. Data on both are predictably inadequate. We turn first to the most elusive environmental problem of all, global warming.

Blowing hot and cold

Climate change may be slow and uncertain, but that is no excuse for inaction

WHAT would Winston Churchill have done about climate change? Imagine that Britain's visionary wartime leader had been presented with a potential time bomb capable of wreaking global havoc, although not certain to do so. Warding it off would require concerted global action and economic sacrifice on the home front. Would he have done nothing?

Not if you put it that way. After all, Churchill did not dismiss the Nazi threat for lack of conclusive evidence of Hitler's evil intentions. But the answer might be less straightforward if the following provisos had been added: evidence of this problem would remain cloudy for decades; the worst effects might not be felt for a century; but the costs of tackling the problem would start biting immediately. That, in a nutshell, is the dilemma of climate change. It is asking a great deal of politicians to take action on behalf of voters who have not even been born yet.

One reason why uncertainty over climate looks to be with us for a long time is that the oceans, which absorb carbon from the atmosphere, act as a time-delay mechanism. Their massive thermal inertia means that the climate system responds only very slowly to changes in the composition of the atmosphere. Another complication arises from the relationship between carbon dioxide (CO_2), the principal greenhouse gas (GHG), and sulphur dioxide (SO_2), a common pollutant. Efforts to reduce man-made emissions of GHGs by cutting down on fossil-fuel use will reduce emissions of both gases. The reduction in

CO_2 will cut warming, but the concurrent SO_2 cut may mask that effect by contributing to the warming.

There are so many such fuzzy factors—ranging from aerosol particles to clouds to cosmic radiation—that we are likely to see disruptions to familiar climate patterns for many years without knowing why they are happening or what to do about them. Tom Wigley, a leading climate scientist and member of the UN's Intergovernmental Panel on Climate Change (IPCC), goes further. He argues in an excellent book published by the Aspen Institute, "US Policies on Climate Change: What Next?", that whatever policy changes governments pursue, scientific uncertainties will "make it difficult to detect the effects of such changes, probably for many decades."

As evidence, he points to the negligible short- to medium-term difference in temperature resulting from an array of emissions "pathways" on which the world could choose to embark if it decided to tackle climate change (see chart 4). He plots various strategies for reducing GHGs (including the Kyoto one) that will lead in the next century to the stabilisation of atmospheric concentrations of CO_2 at 550 parts per million (ppm). That is roughly double the level which prevailed in pre-industrial times, and is often mooted by climate scientists as a reasonable target. But even by 2040, the temperature differences between the various options will still be tiny—and certainly within the magnitude of natural climatic variance. In short, in another four decades we will probably still not know if we have over- or undershot.

Ignorance is not bliss

However, that does not mean we know nothing. We do know, for a start, that the "greenhouse effect" is real: without the heat-trapping effect of water vapour, CO_2, methane and other naturally occurring GHGs, our planet would be a lifeless 30°C or so colder. Some of these GHG emissions are captured and stored by "sinks", such as the oceans, forests and agricultural land, as part of nature's carbon cycle.

We also know that since the industrial revolution began, mankind's actions have contributed significantly to that greenhouse effect. Atmospheric concentrations of GHGs have risen from around 280ppm two centuries ago to around 370ppm today, thanks chiefly to mankind's use of fossil fuels and, to a lesser degree, to deforestation and other land-use changes. Both surface temperatures and sea levels have been rising for some time.

There are good reasons to think temperatures will continue rising. The IPCC has estimated a likely range for that increase of 1.4°C–5.8°C over the next century, although the lower end of that range is more likely. Since what matters is not just the absolute temperature level but the rate of change as well, it makes sense to try to slow down the increase.

The worry is that a rapid rise in temperatures would lead to climate changes that could be devastating for many (though not all) parts of the world. Central America, most of Africa, much of south Asia and northern China could all be hit by droughts, storms and floods and otherwise made miserable. Because they are poor and have the misfortune to live near the tropics, those most likely to be affected will be least able to adapt.

The colder parts of the world may benefit from warming, but they too face perils. One is the conceivable collapse of the Atlantic "conveyor belt", a system of currents that gives much of Europe its relatively mild climate; if temperatures climb too high, say scientists, the system may undergo radical changes that damage both Europe and America. That points to the biggest fear: warming may trigger irreversible changes that transform the earth into a largely uninhabitable environment.

Given that possibility, extremely remote though it is, it is no comfort to know that any attempts to stabilise atmospheric concentrations of GHGs at a particular level will take a very long time. Because of the oceans' thermal inertia, explains Mr Wigley, even once atmospheric concentrations of GHGs are stabilised, it will take decades or centuries for the climate to follow suit. And even then the sea level will continue to rise, perhaps for millennia.

This is a vast challenge, and it is worth bearing in mind that mankind's contribution to warming is the only factor that can be controlled. So the sooner we start drawing up a long-term strategy for climate change, the better.

What should such a grand plan look like? First and foremost, it must be global. Since CO_2 lingers in the atmosphere for a century or more, any plan must also extend across several generations.

The plan must recognise, too, that climate change is nothing new: the climate has fluctuated through history, and mankind has adapted to those changes—and must continue doing so. In the rich world, some of the more obvious measures will include building bigger dykes and flood defences. But since the most vulnerable people are those in poor countries, they too have to be helped to adapt to rising seas and unpredictable storms. Infrastructure improvements will be useful, but the best investment will probably be to help the developing world get wealthier.

It is essential to be clear about the plan's long-term objective. A growing chorus of scientists now argues that we need to keep temperatures from rising by much more than 2–3°C in all. That will require the stabilisation of atmospheric concentrations of GHGs. James Edmonds of the University of Maryland points out that because of the long life of CO_2, stabilisation of CO_2 concentrations is not at all the same thing as stabilisation of CO_2 emissions. That, says Mr Edmonds, points to an unavoidable conclusion: "In the very long term, global net CO_2 emissions must eventually peak and gradually decline toward zero, regardless of whether we go for a target of 350ppm or 1,000ppm."

A low-carbon world

That is why the long-term objective for climate policy must be a transition to a low-carbon energy system. Such a transition can be very gradual and need not necessarily lead to a world powered only by bicycles and windmills, for two reasons that are often overlooked.

One involves the precise form in which the carbon in the ground is distributed. According to Michael Grubb of the Carbon Trust, a British quasi-governmental body, the long-term problem is coal. In theory, we can burn all of the conventional oil and natural gas in the ground and still meet the most ambitious goals for tackling climate change. If we do that, we must ensure that the far greater amounts of carbon trapped as coal (and unconventional resources like tar sands) never enter the atmosphere.

The snag is that poor countries are likely to continue burning cheap domestic reserves of coal for decades. That suggests the rich world should speed the development and diffusion of "low carbon" technologies using the energy content of coal without releasing its carbon into the atmosphere. This could be far off, so it still makes sense to keep a watchful eye on the soaring carbon emissions from oil and gas.

The other reason, as Mr Edmonds took care to point out, is that it is net emissions of CO_2 that need to peak and decline. That leaves scope for the continued use of fossil fuels as the main source of modern energy if only some magical way can be found to capture and dispose of the associated CO_2. Happily, scientists already have some magic in the works.

One option is the biological "sequestration" of carbon in forests and agricultural land. Another promising idea is capturing and storing CO_2—underground, as a solid or even at the bottom of the ocean. Planting "energy crops" such as switch-grass and using them in conjunction with sequestration techniques could even result in negative net CO_2 emissions, because such plants use carbon from the atmosphere. If sequestration is combined with techniques for stripping the hydrogen out of this hydrocarbon, then coal could even offer a way to sustainable hydrogen energy.

But is anyone going to pay attention to these long-term principles? After all, over the past couple of years all participants in the Kyoto debate have excelled at producing short-sighted, selfish and disingenuous arguments. And the political rift continues: the EU and Japan pushed ahead with ratification of the Kyoto treaty a month ago, whereas President Bush reaffirmed his opposition.

However, go back a decade and you will find precisely those principles enshrined in a treaty approved by the elder George Bush and since reaffirmed by his son: the UN Framework Convention on Climate Change (FCCC). This treaty was perhaps the most important outcome of the Rio summit, and it remains the basis for the international climate-policy regime, including Kyoto.

The treaty is global in nature and long-term in perspective. It commits signatories to pursuing "the stabilisation of GHG concentrations in the atmosphere at a level that would prevent dangerous interference with the climate system." Note that the agreement covers GHG concentrations, not merely emissions. In effect, this commits even gas-guzzling America to the goal of declining emissions.

Better than Kyoto

Crucially, the FCCC treaty not only lays down the ends but also specifies the means: any strategy to achieve stabilisation of GHG concentrations, it insists, "must not be disruptive of the global economy". That was the stumbling block for the Kyoto treaty, which is built upon the FCCC agreement: its targets and timetables proved unrealistic.

Any revised Kyoto treaty or follow-up accord (which must include the United States and the big developing countries) should rest on the three basic pillars. First, governments everywhere (but especially in Europe) must understand that a reduction in emissions has to start modestly. That is because the capital stock involved in the global energy system is vast and long-lived, so a dash to scrap fossil-fuel production would be hugely expensive. However, as Mr Grubb points out, that pragmatism must be flanked by policies that encourage a switch to low-carbon technologies when replacing existing plants.

Second, governments everywhere (but especially in America) must send a powerful signal that carbon is going out of fashion. The best way to do this is to levy a carbon tax. However, whether it is done through taxes, mandated restrictions on GHG emissions or market mechanisms is less important than that the signal is sent clearly, forcefully and unambiguously. This is where President Bush's mixed signals have done a lot of harm: America's industry, unlike Europe's, has little incentive to invest in low-carbon technology. The irony is that even some coal-fired utilities in America are now clamouring for CO_2 regulation so that they can invest in new plants with confidence.

The third pillar is to promote science and technology. That means encouraging basic climate and energy research, and giving incentives for spreading the results. Rich countries and aid agencies must also find ways to help the poor world adapt to climate change. This is especially important if the world starts off with small cuts in emissions, leaving deeper cuts for later. That, observes Mr Wigley, means that by mid-century "very large investments would have to have been made—and yet the 'return' on these investments would not be visible. Continued investment is going to require more faith in climate science than currently appears to be the case."

Even a visionary like Churchill might have lost heart in the face of all this uncertainty. Nevertheless, there is a glimmer of hope that today's peacetime politicians may rise to the occasion.

Miracles sometimes happen

Two decades ago, the world faced a similar dilemma: evidence of a hole in the ozone layer. Some inconclusive signs suggested that it was man-made, caused by the use of chlorofluorocarbons (CFCs). There was the distant threat of disaster, and the knowledge about a concerted global response was required. Industry was reluctant at first, yet with leadership from Britain and America the Montreal Protocol was signed in 1987. That deal has proved surprisingly successful. The manufacture of CFCs is nearly phased out, and there are already signs that the ozone layer is on the way to recovery.

This story holds several lessons for the admittedly far more complex climate problem. First, it is the rich world which has caused the problem and which must lead the way in solving it. Second, the poor world must agree to help, but is right to insist on being given time—as well as money and technology—to help it adjust. Third, industry holds the key: in the ozone-depletion story, it was only after DuPont and ICI broke ranks with the rest of the CFC manufacturers that a deal became possible. On the climate issue, BP and Shell have similarly broken ranks with Big Oil, but the American energy industry—especially the coal sector—remains hostile.

The final lesson is the most important: that the uncertainty surrounding a threat such as climate change is no excuse for inaction. New scientific evidence shows that the threat from ozone depletion had been much deadlier than was thought at the time when the world decided to act. Churchill would surely have approved.

Local difficulties

Greenery is for the poor too, particularly on their own doorstep

WHY should we care about the environment? Ask a European, and he will probably point to global warming. Ask the two little boys playing outside a newsstand in Da Shilan, a shabby neighbourhood in the heart of Beijing, and they will tell you about the city's notoriously foul air: "It's bad—like a virus!"

Given all the media coverage in the rich world, people there might believe that global scares are the chief environmental problems facing humanity today. They would be wrong. Partha Dasgupta, an economics professor at Cambridge University, thinks the current interest in global, future-oriented problems has "drawn attention away from the economic misery and ecological degradation endemic in large parts of the world today. Disaster is not something for which the poorest have to wait; it is a frequent occurrence."

Every year in developing countries, a million people die from urban air pollution and twice that number from exposure to stove smoke inside their homes. Another 3m unfortunates die prematurely every year from water-related diseases. All told, premature deaths and illnesses arising from environmental factors account for about a fifth of all diseases in poor countries, bigger than any other preventable factor, including malnutrition. The problem is so serious that Ian Johnson, the World Bank's vice-president for the environment, tells his colleagues, with a touch of irony, that he is really the bank's vice-president for health: "I say tackling the underlying environmental causes of health problems will do a lot more good than just more hospitals and drugs."

The link between environment and poverty is central to that great race for sustainability. It is a pity, then, that several powerful fallacies keep getting in the way of sensible debate. One popular myth is that trade and economic growth make poor countries' environmental problems worse. Growth, it is said, brings with it urbanisation, higher energy consumption and industrialisation—all factors that contribute to pollution and pose health risks.

In a static world, that would be true, because every new factory causes extra pollution. But in the real world, economic growth unleashes many dynamic forces that, in the longer run, more than offset that extra pollution. As chart 5 makes clear, traditional environmental risks (such as water-borne diseases) cause far more health problems in poor countries than modern environmental risks (such as industrial pollution).

Rigged rules

However, this is not to say that trade and economic growth will solve all environmental problems. Among the reasons for doubt are the "perverse" conditions under which world trade is carried on, argues Oxfam. The British charity thinks the rules of trade are "unfairly rigged against the poor", and cites in evidence the enormous subsidies lavished by rich countries on industries such as agriculture, as well as trade protection offered to manufacturing industries such as textiles. These measurements hurt the environment because they force the world's poorest countries to rely heavily on commodities—a particularly energy-intensive and ungreen sector.

Mr Dasgupta argues that this distortion of trade amounts to a massive subsidy of rich-world consumption paid by the world's poorest people. The most persuasive critique of all goes as follows: "Economic growth is not sufficient for turning environmental degradation around. If economic incentives facing producers and consumers do not change with higher incomes, pollution will continue to grow unabated with the growing scale of economic activity." Those words come not from some anti-globalist green group, but from the World Trade Organisation.

Another common view is that poor countries, being unable to afford greenery, should pollute now and clean up later. Certainly poor countries should not be made to adopt American or European environmental standards. But there is evidence to suggest that poor countries can and should try to tackle some environmental problems now, rather than wait till they have become richer.

This so-called "smart growth" strategy contradicts conventional wisdom. For many years, economists have observed that as agrarian societies industrialised, pollution increased at first, but as the societies grew wealthier it declined again. The trouble is that this applies only to some pollutants, such as sulphur dioxide, but not to others, such as carbon dioxide. Even more troublesome, those smooth curves going up, then down, turn out to be misleading. They are what you get when you plot data for poor and rich countries together at a given moment in time, but actual levels of various pollutants in any individual country plotted over time wiggle around a lot more. This suggests that the familiar bell-shaped curve reflects no immutable law, and that intelligent government policies might well help to reduce pollution levels even while countries are still relatively poor.

Developing countries are getting the message. From Mexico to the Philippines, they are now trying to curb the worst of the air and water pollution that typically accompanies industrialisation. China, for example, was persuaded by outside experts that it was losing so much potential economic output through health troubles caused by pollution (according to one World Bank study, somewhere between 3.5% and 7.7% of GDP) that tackling it was cheaper than ignoring it.

One powerful—and until recently ignored—weapon in the fight for a better environment is local people. Old-fashioned paternalists in the capitals of developing countries used to argue that poor villagers could not be relied on to look after natural resources. In fact, much academic research has shown that the poor are more often victims than perpetrators of resource depletion: it tends to be rich locals or outsiders who are responsible for the worst exploitation.

Local people usually have a better knowledge of local ecological conditions than experts in faraway capitals, as well as a direct interest in improving the quality of life in their village. A good example of this comes from the bone-dry state of Rajasthan in India, where local activism and indigenous know-how about rainwater "harvesting" provided the people with reliable water supplies—something the government had failed to do. In Bangladesh, villages with active community groups or concerned mullahs proved greener than less active neighbouring villages.

Community-based forestry initiatives from Bolivia to Nepal have shown that local people can be good custodians of nature. Several hundred million of the world's poorest people live in and around forests. Giving those villagers an incentive to preserve forests by allowing sustainable levels of harvesting, it turns out, is a far better way to save those forests than erecting tall fences around them.

To harness local energies effectively, it is particularly important to give local people secure property rights, argues Mr Dasgupta. In most parts of the developing world, control over resources at the village level is ill-defined. This often means that local elites usurp a disproportionate share of those resources, and that individuals have little incentive to maintain and upgrade forests or agricultural land. Authorities in Thailand tried to remedy this problem by distributing 5.5m land titles over a 20-year period. Agricultural output increased, access to credit improved and the value of the land shot up.

Name and shame

Another powerful tool for improving the local environment is the free flow of information. As local democracy flourishes, ordinary people are pressing for greater environmental disclosure by companies. In some countries, such as Indonesia, governments have adopted a "sunshine" policy that involves naming and shaming companies that do not meet environmental regulations. It seems to achieve results.

Bringing greenery to the grass roots is good, but on its own it will not avert perceived threats to global "public goods" such as the climate or biodiversity. Paul Portney of Resources for the Future explains: "Brazilian villagers may think very carefully and unselfishly about their future descendants, but there's no reason for them to care about and protect species or habitats that no future generation of Brazilians will care about."

That is why rich countries must do more than make pious noises about global threats to the environment. If they believe that scientific evidence suggests a credible threat, they must be willing to pay poor countries to protect such things as their tropical forests. Rather than thinking of this as charity, they should see it as payment for environmental services (say, for carbon storage) or as a form of insurance.

In the case of biodiversity, such payments could even be seen as a trade in luxury goods: rich countries would pay poor countries to look after creatures that only the rich care about. Indeed, private green groups are already buying up biodiversity "hot spots" to protect them. One such initiative, led by Conservation International and the International Union for the Conservation of Nature (IUCN), put the cost of buying and preserving 25 hot spots exceptionally rich in species diversity at less than $30 billion. Sceptics say it will cost more, as hot spots will need buffer zones of "sustainable harvesting" around them. Whatever the right figure, such creative approaches are more likely to achieve results than bullying the poor into conservation.

It is not that the poor do not have green concerns, but that those concerns are very different from those of the rich. In Beijing's Da Shilan, for instance, the air is full of soot from the many tiny coal boilers. Unlike most of the neighbouring districts, which have recently converted from coal to natural gas, this area has been considered too poor to make the transition. Yet ask Liu Shihua, a shopkeeper who has lived in the same spot for over 20 years, and he insists he would readily pay a bit more for the cleaner air that would come from using natural gas. So would his neighbours.

To discover the best reason why poor countries should not ignore pollution, ask those two little boys outside Mr

Liu's shop what colour the sky is. "Grey!" says one tyke, as if it were the most obvious thing in the world. "No, stupid, it's blue!" retorts the other. The children deserve blue skies and clean air. And now there is reason to think they will see them in their lifetime.

Working miracles

Can technology save the planet?

"NOTHING endures but change." That observation by Heraclitus often seems lost on modern environmental thinkers. Many invoke scary scenarios assuming that resources—both natural ones, like oil, and man-made ones, like knowledge—are fixed. Yet in real life man and nature are entwined in a dynamic dance of development, scarcity, degradation, innovation and substitution.

The nightmare about China turning into a resource-guzzling America raises two questions: will the world run out of resources? And even if it does not, could the growing affluence of developing nations lead to global environmental disaster?

The first fear is the easier to refute; indeed, history has done so time and again. Malthus, Ricardo and Mill all worried that scarcity of resources would snuff out growth. It did not. A few decades ago, the limits-to-growth camp raised worries that the world might soon run out of oil, and that it might not be able to feed the world's exploding population. Yet there are now more proven reserves of petroleum than three decades ago; there is more food produced than ever; and the past decade has seen history's greatest economic boom.

What made these miracles possible? Fears of oil scarcity prompted investment that led to better ways of producing oil, and to more efficient engines. In food production, technological advances have sharply reduced the amount of land required to feed a person in the past 50 years. Jesse Ausubel of Rockefeller University calculates that if in the next 60 to 70 years the world's average farmer reaches the yield of today's average (not best) American maize grower, then feeding 10 billion people will require just half of today's cropland. All farmers need to do is maintain the 2%-a-year productivity gain that has been the global norm since 1960.

"Scarcity and Growth", a book published by Resources for the Future, sums it up brilliantly: "Decades ago Vermont granite was only building and tombstone material; now it is a potential fuel, each ton of which has a usable energy content (uranium) equal to 150 tons of coal. The notion of an absolute limit to natural resource availability is untenable when the definition of resources changes drastically and unpredictably over time." Those words were written by Harold Barnett and Chandler Morse in 1963, long before the limits-to-growth bandwagon got rolling.

Giant footprint

Not so fast, argue greens. Even if we are not going to run out of resources, guzzling ever more resources could still do irreversible damage to fragile ecosystems.

WWF, an environmental group, regularly calculates mankind's "ecological footprint", which it defines as the "biologically productive land and water areas required to produce the resources consumed and assimilate the wastes generated by a given population using prevailing technology." The group reckons the planet has around 11.4 billion "biologically productive" hectares of land available to meet continuing human needs. As chart 6 overleaf shows, WWF thinks mankind has recently been using more than that. This is possible because a forest harvested at twice its regeneration rate, for example, appears in the footprint accounts at twice its area—an unsustainable practice which the group calls "ecological overshoot."

Any analysis of this sort must be viewed with scepticism. Everyone knows that environmental data are incomplete. What is more, the biggest factor by far is the land required to absorb CO_2 emissions of fossil fuels. If that problem could be managed some other way, then mankind's ecological footprint would look much more sustainable.

Even so, the WWF analysis makes an important point: if China's economy were transformed overnight into a clone of America's, an ecological nightmare could ensue. If a billion eager new consumers were suddenly to produce CO_2 emissions at American rates, they would be bound to accelerate global warming. And if the whole of the developing world were to adopt an American lifestyle tomorrow, local environmental crises such as desertification, aquifer depletion and topsoil loss could make humans miserable.

So is this cause for concern? Yes, but not for panic. The global ecological footprint is determined by three factors: population size, average consumption per person and technology. Fortunately, global population growth now appears to be moderating. Consumption per person in poor countries is rising as they become better off, but there are signs that the rich world is reducing the footprint of its consumption (as this survey's final section explains). The most powerful reason for hope—innovation—was foreshadowed by WWF's own definition. Today's "prevailing technologies" will, in time, be displaced by tomorrow's greener ones.

"The rest of the world will not live like America," insists Mr Ausubel. Of course poor people around the world covet the creature comforts that Americans enjoy, but they know full well that the economic growth needed to improve their lot will take time. Ask Wu Chengjian, an environmental official in booming Shanghai, what he thinks of the popular notion that his city might become as rich as today's Hong Kong by 2020: "Impossible—that's just not enough time." And that is Shanghai, not the impoverished countryside.

Leaps of faith

This extra time will allow poor countries to embrace new technologies that are more efficient and less environmentally damaging. That still does not guarantee a smaller ecological footprint for China in a few decades' time than for America now, but it greatly improves the chances. To see why, consider the history of "dematerialisation" and "decarbonisation" (see chart 7). Viewed across very long spans of time, productivity improvements allow economies to use ever fewer material inputs—and to emit ever fewer pollutants—per unit of economic output. Mr Ausubel concludes: "When China has today's American mobility, it will not have today's American cars," but the cleaner and more efficient cars of tomorrow.

The snag is that consumers in developing countries want to drive cars not tomorrow but today. The resulting emissions have led many to despair that technology (in the form of vehicles) is making matters worse, not better.

Can they really hope to "leapfrog" ahead to cleaner air? The evidence from Los Angeles—a pioneer in the fight against air pollution—suggests the answer is yes. "When I moved to Los Angeles in the 1960s, there was so much soot in the air that it felt like there was a man standing on your chest most of the time," says Ron Loveridge, the mayor of Riverside, a city to the east of LA that suffers the worst of the region's pollution. But, he says, "We have come an extraordinary distance in LA."

Four decades ago, the city had the worst air quality in America. The main problem was the city's infamous "smog" (an amalgam of "smoke" and "fog"). It took a while to figure out that this unhealthy ozone soup developed as a result of complex chemical reactions between nitrogen oxides and volatile organic compounds that need sunlight to trigger them off.

Arthur Winer, an atmospheric chemist at the University of California at Los Angeles, explains that tackling smog required tremendous perseverance and political will. Early regulatory efforts met stiff resistance from business interests, and began to falter when they failed to show dramatic results.

Clean-air advocates like Mr Loveridge began to despair: "We used to say that we needed a 'London fog' [a reference to an air-pollution episode in 1952 that may have killed 12,000 people in that city] here to force change." Even so, Californian officials forged ahead with an ambitious plan that combined regional regulation with stiff mandates for cleaner air. Despite uncertainties about the cause of the problem, the authorities introduced a sequence of controversial measures: unleaded and low-sulphur petrol, on-board diagnostics for cars to minimise emissions, three-way catalytic converters, vapour-recovery attachments for petrol nozzles and so on.

As a result, the city that two decades ago hardly ever met federal ozone standards has not had to issue a single alert in the past three years. Peak ozone levels are down by 50% since the 1960s. Though the population has shot up in recent years, and the vehicle-miles driven by car-crazy Angelenos have tripled, ozone levels have fallen by two-thirds. The city's air is much cleaner than it was two decades ago.

"California, in solving its air-quality problem, has solved it for the rest of the United States and the world—but it doesn't get credit for it," says Joe Norbeck of the University of California at Riverside. He is adamant that the poor world's cities can indeed leapfrog ahead by embracing some of the cleaner technologies developed specifically for the Californian market. He points to China's vehicle fleet as an example: "China's typical car has the emissions of a 1974 Ford Pinto, but the new Buicks sold there use 1990s emissions technology." The typical car sold today produces less than a tenth of the local pollution of a comparable model from the 1970s.

That suggests one lesson for poor cities such as Beijing that are keen to clean up: they can order polluters to meet high emissions standards. Indeed, from Beijing to Mexico city, regulators are now imposing rich-world rules, mandating new, cleaner technologies. In China's cities, where pollution from sooty coal fires in homes and industrial boilers had been a particular hazard, officials are keen to switch to natural-gas furnaces.

However, there are several reasons why such mandates—which worked wonders in LA—may be trickier to achieve in impoverished or politically weak cities. For a start, city officials must be willing to pay the political price of reforms that raise prices for voters. Besides, higher standards for new cars, useful though they are, cannot do the trick on their own. Often, clean technologies such as catalytic converters will require cleaner grades of petrol too. Introducing cleaner fuels, say experts, is an essential lesson from LA for poor countries. This will not come free either.

There is another reason why merely ordering cleaner new cars is inadequate: it does nothing about the vast stock of dirty old ones already on the streets. In most cities of the developing world, the oldest fifth of the vehicles on the road is likely to produce over half of the total pollution caused by all vehicles taken together. Policies that encourage a speedier turnover of the fleet therefore make more sense than "zero emissions" mandates.

Policy matters

In sum, there is hope that the poor can leapfrog at least some environmental problems, but they need more than just technology. Luisa and Mario Molina of the Massachusetts Institute of Technology, who have studied such questions closely, reckon that technology is less important than the institutional capacity, legal safeguards and financial resources to back it up: "The most important underlying factor is political will." And even a techno-optimist such as Mr Ausubel accepts that: "There is nothing automatic about technological innovation and adoption; in fact, at the micro level, it's bloody."

Clearly innovation is a powerful force, but government policy still matters. That suggests two rules for policymakers. First, don't do stupid things that inhibit innovation. Second, do sensible things that reward the development and adoption of technologies that enhance, rather than degrade, the environment.

The greatest threat to sustainability may well be the rejection of science. Consider Britain's hysterical reaction to genetically modified crops, and the European Commission's recent embrace of a woolly "precautionary principle". Precaution applied case-by-case is undoubtedly a good thing, but applying any such principle across the board could prove disastrous.

Explaining how not to stifle innovation that could help the environment is a lot easier than finding ways to encourage it. Technological change often goes hand-in-hand with greenery by saving resources, as the long history of dematerialisation shows—but not always. Sports utility vehicles, for instance, are technologically innovative, but hardly green. Yet if those SUVs were to come with hydrogen-powered fuel cells that emit little pollution, the picture would be transformed.

The best way to encourage such green innovations is to send powerful signals to the market that the environment matters. And there is no more powerful signal than price, as the next section explains.

The invisible green hand

Markets could be a potent force for greenery—if only greens could learn to love them

"MANDATE, regulate and litigate." That has been the environmentalists' rallying cry for ages. Nowhere in the green manifesto has there been much mention of the market. And, oddly, it was market-minded America that led the dirigiste trend. Three decades ago, Congress passed a sequence of laws, including the Clean Air Act, which set lofty goals and generally set rigid technological standards. Much of the world followed America's lead.

This top-down approach to greenery has long been a point of pride for groups such as the Natural Resources Defence Council (NRDC), one of America's most influential environmental outfits. And with some reason, for it has had its successes: the air and water in the developed world is undoubtedly cleaner than it was three decades ago, even though the rich world's economies have grown by leaps and bounds. This has convinced such groups stoutly to defend the green status quo.

But times may be changing. Gus Speth, now head of Yale University's environment school and formerly head of the World Resources Institute and the UNDP, as well as one of the founders of the NRDC, recently explained how he was converted to market economics: "Thirty years ago, the economists at Resources for the Future were pushing the idea of pollution taxes. We lawyers at NRDC thought they were nuts, and feared that they would derail command-and-control measures like the Clean Air Act, so we opposed them. Looking back, I'd have to say this was the single biggest failure in environmental management—not getting the prices right."

A remarkable mea culpa; but in truth, the command-and-control approach was never as successful as its advocates claimed. For example, although it has cleaned up the air and water in rich countries, it has notably failed in dealing with waste management, hazardous emissions and fisheries depletion. Also, the gains achieved have come at a needlessly high price. That is because technology mandates and bureaucratic edicts stifle innovation and ignore local realities, such as varying costs of abatement. They also fail to use cost-benefit analysis to judge trade-offs.

Command-and-control methods will also be ill-suited to the problems of the future, which are getting trickier. One reason is that the obvious issues—like dirty air and water—have been tackled already. Another is increasing technological complexity: future problems are more likely to involve subtle linkages—like those involved in ozone depletion and global warming—that will require sophisticated responses. The most important factor may be society's ever-rising expectations; as countries grow wealthier, their people start clamouring for an ever-cleaner environment. But because the cheap and simple things have been done, that is proving increasingly expensive. Hence the greens' new interest in the market.

Carrots, not just sticks

In recent years, market-based greenery has taken off in several ways. With emissions trading, officials decide on a pollution target and then allocate tradable credits to companies based on that target. Those that find it expensive to cut emissions can buy credits from those that find it cheaper, so the target is achieved at the minimum cost and disruption.

The greatest green success story of the past decade is probably America's innovative scheme to cut emissions of sulphur dioxide (SO_2). Dan Dudek of Environmental Defence, a most unusual green group, and his market-minded colleagues persuaded the elder George Bush to agree to an amendment to the sacred Clean Air Act that would introduce an emissions-trading system to achieve sharp cuts in SO_2. At the time, this was hugely controversial: America's power industry insisted the cuts were prohibitively costly, while nearly every other green group decried the measure as a sham. In the event, ED has been vindicated. America's scheme has surpassed its initial objectives, and at far lower cost than expected. So great is the interest worldwide in trading that ED is now advising groups ranging from hard-nosed oilmen at BP to bureaucrats in China and Russia.

Europe, meanwhile, is forging ahead with another sort of market-based instrument: pollution taxes. The idea is to levy charges on goods and services so that their price reflects their "externalities"—jargon for how much harm they do to the environment and human health. Sweden introduced a sulphur tax a decade ago, and found that the sulphur content of fuels dropped 50% below legal requirements.

Though "tax" still remains a dirty word in America, other parts of the world are beginning to embrace green tax reform by shifting taxes from employment to pollu-

tion. Robert Williams of Princeton University has looked at energy use (especially the terrible effects on health of particulate pollution) and concluded that such externalities are comparable in size to the direct economic costs of producing that energy.

Externalities are only half the battle in fixing market distortions. The other half involves scrapping environmentally harmful subsidies. These range from prices below market levels for electricity and water to shameless cash handouts for industries such as coal. The boffins at the OECD reckon that stripping away harmful subsidies, along with introducing taxes on carbon-based fuels and chemicals use, would result in dramatically lower emissions by 2020 than current policies would be able to achieve. If the revenues raised were then used to reduce other taxes, the cost of these virtuous policies would be less than 1% of the OECD's economic output in 2020.

Such subsidies are nothing short of perverse, in the words of Norman Myers of Oxford University. They do double damage, by distorting markets and by encouraging behaviour that harms the environment. Development banks say such subsidies add up to $700 billion a year, but Mr Myers reckons the true sum is closer to $2 trillion a year. Moreover, the numbers do not fully reflect the harm done. For example, EU countries subsidise their fishing fleets to the tune of $1 billion a year, but that has encouraged enough overfishing to drive many North Atlantic fishing grounds to near-collapse.

Fishing is an example of the "tragedy of the commons", which pops up frequently in the environmental debate. A resource such as the ocean is common to many, but an individual "free rider" can benefit from plundering that commons or dumping waste into it, knowing that the costs of his actions will probably be distributed among many neighbours. In the case of shared fishing grounds, the absence of individual ownership drives each fisherman to snatch as many fish as he can—to the detriment of all.

Of rights and wrongs

Assigning property rights can help, because providing secure rights (set at a sustainable level) aligns the interests of the individual with the wider good of preserving nature. This is what sceptical conservationists have observed in New Zealand and Iceland, where schemes for tradable quotas have helped revive fishing stocks. Similar rights-based approaches have led to revivals in stocks of African elephants in southern Africa, for example, where the authorities stress property rights and private conservation.

All this talk of property rights and markets makes many mainstream environmentalists nervous. Carl Pope, the boss of the Sierra Club, one of America's biggest green groups, does not reject market forces out of hand, but expresses deep scepticism about their scope. Pointing to the difficult problem of climate change, he asks: "Who has property rights over the commons?"

Even so, some greens have become converts. Achim Steiner of the IUCN reckons that the only way forward is rights-based conservation, allowing poor people "sustainable use" of their local environment. Paul Faeth of the World Resources Institute goes further. He says he is convinced that market forces could deliver that holy grail of environmentalism, sustainability—"but only if we get prices right."

The limits to markets

Economic liberals argue that the market itself is the greatest price-discovery mechanism known to man. Allow it to function freely and without government meddling, goes the argument, and prices are discovered and internalised automatically. Jerry Taylor of the Cato Institute, a libertarian think-tank, insists that "The world today is already sustainable—except those parts where western capitalism doesn't exist." He notes that countries that have relied on central planning, such as the Soviet Union, China and India, have invariably misallocated investment, stifled innovation and fouled their environment far more than the prosperous market economies of the world have done.

All true. Even so, markets are currently not very good at valuing environmental goods. Noble attempts are under way to help them do better. For example, the Katoomba Group, a collection of financial and energy companies that have linked up with environmental outfits, is trying to speed the development of markets for some of forestry's ignored "co-benefits" such as carbon storage and watershed management, thereby producing new revenue flows for forest owners. This approach shows promise: water consumers ranging from officials in New York City to private hydro-electric operators in Costa Rica are now paying people upstream to manage their forests and agricultural land better. Paying for greenery upstream turns out to be cheaper than cleaning up water downstream after it has been fouled.

Economists too are getting into the game of helping capitalism "get prices right." The World Bank's Ian Johnson argues that conventional economic measures such as gross domestic product are not measuring wealth creation properly because they ignore the effects of environmental degradation. He points to the positive contribution to China's GDP from the logging industry, arguing that such a calculation completely ignores the billions of dollars-worth of damage from devastating floods caused by over-logging. He advocates a more comprehensive measure the Bank is working on, dubbed "genuine GDP", that tries (imperfectly, he accepts) to measure depletion of natural resources.

That could make a dramatic difference to how the welfare of the poor is assessed. Using conventional market measures, nearly the whole of the developing world save Africa has grown wealthier in the past couple of decades. But when the degradation of nature is properly accounted for, argues Mr Dasgupta at Cambridge, the countries of Africa and south Asia are actually much worse off today than they were a few decades ago—and even China, whose economic "miracle" has been much trumpeted, comes out barely ahead.

The explanation, he reckons, lies in a particularly perverse form of market distortion: "Countries that are exporting resource-based products (often among the poorest) may be subsidising the consumption of countries that are doing the importing (often among the richest)." As evidence, he points to the common practice in poor countries of encouraging resource extraction. Whether through licenses granted at below-market rates, heavily subsidised exports or corrupt officials tolerating illegal exploitation, he reckons the result is the same: "The cruel paradox we face may well be that contemporary economic development is unsustainable in poor countries because it is sustainable in rich countries."

One does not have to agree with Mr Dasgupta's conclusion to acknowledge that markets have their limits. That should not dissuade the world from attempting to get prices right—or at least to stop getting them so wrong. For grotesque subsidies, the direction of change should be obvious. In other areas, the market itself may not provide enough information to value nature adequately. This is true of threats to essential assets, such as nature's ability to absorb and "recycle" CO_2, that have no substitute at any price. That is when governments must step in, ensuring that an informed public debate takes place.

Robert Stavins of Harvard University argues that the thorny notion of sustainable development can be reduced to two simple ideas: efficiency and intergenerational equity. The first is about making the economic pie as large as possible; he reckons that economists are well equipped to handle it, and that market-based policies can be used to achieve it. On the second (the subject of the next section), he is convinced that markets must yield to public discourse and government policy: "Markets can be efficient, but nobody ever said they're fair. The question is, what do we owe the future?"

Insuring a brighter future

How to hedge against tomorrow's environmental risks

So what do we owe the future? A precise definition for sustainable development is likely to remain elusive but, as this survey has argued, the hazy outline of a useful one is emerging from the experience of the past decade.

For a start, we cannot hope to turn back the clock and return nature to a pristine state. Nor must we freeze nature in the state it is today, for that gift to the future would impose an unacceptable burden on the poorest alive today. Besides, we cannot forecast the tastes, demands or concerns of future generations. Recall that the overwhelming pollution problem a century ago was horse manure clogging up city streets: a century hence, many of today's problems will surely seem equally irrelevant. We should therefore think of our debt to the future as including not just natural resources but also technology, institutions and especially the capacity to innovate. Robert Solow got it mostly right a decade ago: the most important thing to leave future generations, he said, is the capacity to live as well as we do today.

However, as the past decade has made clear, there is a limit to that argument. If we really care about the "sustainable" part of sustainable development, we must be much more watchful about environmental problems with critical thresholds. Most local problems are reversible and hence no cause for alarm. Not all, however: the depletion of aquifers and the loss of topsoil could trigger irreversible changes that would leave future generations worse off. And global or long-term threats, where victims are far removed in time and space, are easy to brush aside.

In areas such as biodiversity, where there is little evidence of a sustainability problem, a voluntary approach is best. Those in the rich world who wish to preserve pandas, or hunt for miracle drugs in the rainforest, should pay for their predilections. However, where there are strong scientific indications of unsustainability, we must act on behalf of the future—even at the price of today's development. That may be expensive, so it is prudent to try to minimise those risks in the first place.

A riskier world

Human ingenuity and a bit of luck have helped mankind stay a few steps ahead of the forces degrading the environment this past century, the first full one in which the planet has been exposed to industrialisation. In the century ahead, the great race between development and degradation could well become a closer call.

On one hand, the demands of development seem sure to grow at a cracking pace in the next few decades as the Chinas, Indias and Brazils of this world grow wealthy enough to start enjoying not only the necessities but also some of the luxuries of life. On the other hand, we seem to be entering a period of huge technological advances in emerging fields such as biotechnology that could greatly increase resource productivity and more than offset the effect of growth on the environment. The trouble is, nobody knows for sure.

Since uncertainty will define the coming era, it makes sense to invest in ways that reduce that risk at relatively low cost. Governments must think seriously about the future implications of today's policies. Their best bet is to encourage the three powerful forces for sustainability outlined in this survey: the empowerment of local people to manage local resources and adapt to environmental change; the encouragement of science and technology, especially innovations that reduce the ecological footprint of consumption; and the greening of markets to get prices right.

To advocate these interventions is not to call for a return to the hubris of yesteryear's central planners. These measures would merely give individuals the power to make greener choices if they care to. In practice, argues Chris Heady of the OECD, this may still not add up to sustainability "because we might still decide to be greedy, and leave less for our children."

Happily, there are signs of an emerging bottom-up push for greenery. Even such icons of western consumerism as Unilever and Procter & Gamble now sing the virtues of "sustainable consumption." Unilever has vowed that by 2005 it will be buying fish only from sustainable sources, and P&G is coming up with innovative products such as detergents that require less water, heat and packaging. It would be naive to label such actions as expressions of "corporate social responsibility": in the long run, firms will embrace greenery only if they see profit in it. And that, in turn, will depend on choices made by individuals.

Such interventions should really be thought of as a kind of insurance that tilts the odds of winning that great race just a little in humanity's favour. Indeed, even some of the world's most conservative insurance firms increasingly see things this way. As losses from weather-related disasters have risen of late (see chart 8), the industry is getting more involved in policy debates on long-term environmental issues such as climate change.

Bruno Porro, chief risk officer at Swiss Re, argues that: "The world is entering a future in which risks are more concentrated and more complex. That is why we are pressing for policies that reduce those risks through preparation, adaptation and mitigation. That will be cheaper than covering tomorrow's losses after disaster strikes."

Jeffrey Sachs of Columbia University agrees: "When you think about the scale of risk that the world faces, it is clear that we grossly underinvest in knowledge... we have enough income to live very comfortably in the developed world and to prevent dire need in the developing world. So we should have the confidence to invest in longer-term issues like the environment. Let's help insure the sustainability of this wonderful situation."

He is right. After all, we have only one planet, now and in the future. We need to think harder about how to use it wisely.

Acknowledgements

In addition to those cited in the text, the author would like to thank Robert Socolow, David Victor, Geoffrey Heal, and experts at Tsinghua University, Friends of the Earth, the European Commission, the World Business Council for Sustainable Development, the International Energy Agency, the OECD and the UN for sharing their ideas with him. A list of sources can be found on *The Economist's* website.

Five Meta-Trends Changing the World

Global, overarching forces such as modernization and widespread interconnectivity are converging to reshape our lives. But human adaptability—itself a "meta-trend"—will help keep our future from spinning out of control, assures THE FUTURIST's lifestyles editor.

By David Pearce Snyder

Last year, I received an e-mail from a long-time Australian client requesting a brief list of the "meta-trends" having the greatest impact on global human psychology. What the client wanted to know was, which global trends would most powerfully affect human consciousness and behavior around the world?

The Greek root *meta* denotes a transformational or transcendent phenomenon, not simply a big, pervasive one. A meta-trend implies multidimensional or catalytic change, as opposed to a linear or sequential change.

What follows are five meta-trends I believe are profoundly changing the world. They are evolutionary, system-wide developments arising from the simultaneous occurrence of a number of individual demographic, economic, and technological trends. Instead of each being individual free-standing global trends, they are composites of trends.

Trend 1—Cultural Modernization

Around the world over the past generation, the basic tenets of modern cultures—including equality, personal freedom, and self-fulfillment—have been eroding the domains of traditional cultures that value authority, filial obedience, and self-discipline. The children of traditional societies are growing up wearing Western clothes, eating Western food, listening to Western music, and (most importantly of all) thinking Western thoughts. Most Westerners—certainly most Americans—have been unaware of the personal intensities of this culture war because they are so far away from the "battle lines." Moreover, people in the West regard the basic institutions of modernization, including universal education, meritocracy, and civil law, as benchmarks of social progress, while the defenders of traditional cultures see them as threats to social order.

Demographers have identified several leading social indicators as key measures of the extent to which a nation's culture is modern. They cite the average level of education for men and for women, the percentage of the salaried workforce that is female, and the percentage of population that lives in urban areas. Other indicators include the percentage of the workforce that is salaried (as opposed to self-employed) and the percentage of GDP spent on institutionalized socioeconomic support services, including insurance, pensions, social security, civil law courts, worker's compensation, unemployment benefits, and welfare.

As each of these indicators rises in a society, the birthrate in that society goes down. The principal measurable consequence of cultural modernization is declining fertility. As the world's developing nations have become better educated, more urbanized, and more institutionalized during the past 20 years, their birthrates have fallen dramatically. In 1988, the United Nations forecast that the world's population would double to 12 billion by 2100. In 1992, their estimate dropped to 10 billion, and they currently expect global population to peak at 9.1 billion in 2100. After that, demographers expect the world's population will begin to slowly decline, as has already begun to happen in Europe and Japan.

Three signs that a culture is modern: its citizens' average level of education, the number of working women, and the percentage of the population that is urban. As these numbers increase, the birthrate in a society goes down, writes author David Pearce Snyder.

The effects of cultural modernization on fertility are so powerful that they are reflected clearly in local vital statistics. In India, urban birthrates are similar to those in the United States, while rural birthrates remain unmanageably high. Cultural modernization is the linchpin of human sustainability on the planet.

The forces of cultural modernization, accelerated by economic globalization and the rapidly spreading wireless telecommunications infostructure, are likely to marginalize the world's traditional cultures well before the century is over. And because the wellsprings of modernization—secular industrial economies—are so unassailably powerful, terrorism is the only means by which the defenders of traditional culture can fight to preserve their values and way of life. In the near-term future, most observers believe that ongoing cultural conflict is likely to produce at least a few further extreme acts of terrorism, security measures not withstanding. But the eventual intensity and duration of the overt, vi-

olent phases of the ongoing global culture war are largely matters of conjecture. So, too, are the expert pronouncements of the probable long-term impacts of September 11, 2001, and terrorism on American priorities and behavior.

After the 2001 attacks, social commentators speculated extensively that those events would change America. Pundits posited that we would become more motivated by things of intrinsic value—children, family, friends, nature, personal self-fulfillment—and that we would see a sharp increase in people pursuing *pro bono* causes and public-service careers. A number of media critics predicted that popular entertainment such as television, movies, and games would feature much less gratuitous violence after September 11. None of that has happened. Nor have Americans become more attentive to international news coverage. Media surveys show that the average American reads less international news now than before September 11. Event-inspired changes in behavior are generally transitory. Even if current conflicts produce further extreme acts of terrorist violence, these seem unlikely to alter the way we live or make daily decisions. Studies in Israel reveal that its citizens have become habituated to terrorist attacks. The daily routine of life remains the norm, and random acts of terrorism remain just that: random events for which no precautions or mind-set can prepare us or significantly reduce our risk.

In summary, cultural modernization will continue to assault the world's traditional cultures, provoking widespread political unrest, psychological stress, and social tension. In developed nations, where the great majority embrace the tenets of modernization and where the threats from cultural conflict are manifested in occasional random acts of violence, the ongoing confrontation between tradition and modernization seems likely to produce security mea-

sures that are inconvenient, but will do little to alter our basic personal decision making, values, or day-to-day life. Developed nations are unlikely to make any serious attempts to restrain the spread of cultural modernization or its driving force, economic globalization.

Trend 2—Economic Globalization

On paper, globalization poses the long-term potential to raise living standards and reduce the costs of goods and services for people everywhere. But the short-term marketplace consequences of free trade threaten many people and enterprises in both developed and developing nations with potentially insurmountable competition. For most people around the world, the threat from foreign competitors is regarded as much greater than the threat from foreign terrorists. Of course, risk and uncertainty in daily life is characteristically high in developing countries. In developed economies, however, where formal institutions sustain order and predictability, trade liberalization poses unfamiliar risks and uncertainties for many enterprises. It also appears to be affecting the collective psychology of both blue-collar and white-collar workers—especially males—who are increasingly unwilling to commit themselves to careers in fields that are likely to be subject to low-cost foreign competition.

Strikingly, surveys of young Americans show little sign of xenophobia in response to the millions of new immigrant workers with whom they are competing in the domestic job market. However, they feel hostile and helpless at the prospect of competing with Chinese factory workers and Indian programmers overseas. And, of course, economic history tells us that they are justifiably concerned. In those job markets that supply untariffed international industries, a "comparable global wage" for comparable types of work can be expected to emerge worldwide. This will raise workers' wages for freely traded goods and services in developing nations, while depressing wages for comparable work in mature industrial economies. To earn more than the comparable global wage, labor in developed nations will have to perform *incomparable* work, either in terms of their productivity or the superior characteristics of the goods and services that they produce. The assimilation of mature information technology throughout all production and education levels should make this possible, but developed economies have not yet begun to mass-produce a new generation of high-value-adding, middle-income jobs.

Meanwhile, in spite of the undeniable short-term economic discomfort that it causes, the trend toward continuing globalization has immense force behind it. Since World War II, imports have risen from 6% of world GDP to more than 22%, growing steadily throughout the Cold War, and even faster since 1990. The global dispersion of goods production and the uneven distribution of oil, gas, and critical minerals worldwide have combined to make international interdependence a fundamental economic reality, and corporate enterprises are building upon that reality. Delays in globalization, like the September 2003 World Trade Organization contretemps in Cancun, Mexico, will arise as remaining politically sensitive issues are resolved, including trade in farm products, professional and financial services, and the need for corporate social responsibility. While there will be enormous long-term economic benefits from globalization in both developed and developing nations, the short-term disruptions in local domestic employment will make free trade an ongoing political issue that will be manageable only so long as domestic economies continue to grow.

Trend 3—Universal Connectivity

While information technology (IT) continues to inundate us with miraculous capabilities, it has given us, so far, only one new power that appears to have had a significant impact on our collective behavior: our improved ability to communicate with each other, anywhere, anytime. Behavioral researchers have found that cell phones have blurred or changed the boundaries between work and social life and between personal and public life. Cell phones have also increased users' propensity to "micromanage their lives, to be more spontaneous, and, therefore, to be late for everything," according to Leysia Palen, computer science professor at the University of Colorado at Boulder.

Cell phones have blurred the lines between the public and the private. Nearly everyone is available anywhere, anytime—and in a decade cyberspace will be a town square, writes Snyder.

Most recently, instant messaging—via both cell phones and online computers—has begun to have an even more powerful social impact than cell phones themselves. Instant messaging initially tells you whether the person you wish to call is "present" in cyberspace—that is, whether he or she is actually online at the moment. Those who are present can be messaged immediately, in much the same way as you might look out the window and call to a friend you see in the neighbor's yard. Instant messaging gives a physical reality to cyberspace. It adds a new dimension to life: A person can now be "near," "distant," or "in cyberspace." With video instant messaging—available now, and widely available in three years—the illusion will be complete. We will have achieved what Frances Cairncross, senior editor of *The Economist*, has called "the death of distance."

Universal connectivity will be accelerated by the integration of the telephone, cell phone, and other wireless telecom media with the Internet. By 2010, all long-distance phone calls, plus a third of all local calls, will be made via the Internet, while 80% to 90% of all Internet access will be made from Web-enabled phones, PDAs, and wireless laptops. Most important of all, in less than a decade, one-third of the world's population—2 billion people—will have access to the Internet, largely via Web-enabled telephones. In a very real sense, the Internet will be the "Information Highway"—the infrastructure, or infostructure, for the computer age. The infostructure is already speeding the adoption of flexplace employment and reducing the volume of business travel, while making possible increased "distant collaboration," outsourcing, and offshoring.

Corporate integrity and openness will grow steadily under pressure from watchdog groups and ordinary citizens demanding business transparency. The leader of tomorrow must adapt to this new openness or risk business disaster.

As the first marketing medium with a truly global reach, the Internet will also be the crucible from which a global consumer culture will be forged, led by the first global youth peer culture. By 2010, we will truly be living in a global village, and cyberspace will be the town square.

Trend 4—Transactional Transparency

Long before the massive corporate malfeasance at Enron, Tyco, and WorldCom, there was a rising global movement toward greater transparency in all private and public enterprises. Originally aimed at kleptocratic regimes in Africa and the former Soviet states, the movement has now become universal, with the establishment of more stringent international accounting standards and more comprehensive rules for corporate oversight and record keeping, plus a new UN treaty on curbing public-sector corruption. Because secrecy breeds corruption and incompetence, there is a growing worldwide consensus to expose the principal transactions and decisions of *all* enterprises to public scrutiny.

But in a world where most management schools have dropped all ethics courses and business professors routinely preach that government regulation thwarts the efficiency of the marketplace, corporate and government leaders around the world are lobbying hard against transparency mandates for the private sector. Their argument: Transparency would "tie their hands," "reveal secrets to their competition," and "keep them from making a fair return for their stockholders."

Most corporate management is resolutely committed to the notion that secrecy is a necessary concomitant of leadership. But pervasive, ubiquitous computing and comprehensive electronic documentation will ultimately make all things transparent, and this may leave many leaders and decision makers feeling uncomfortably exposed, especially if they were not provided a moral compass prior to adolescence. Hill and Knowlton, an international public-relations firm, recently surveyed 257 CEOs in the United States, Europe, and Asia regarding the impact of the Sarbanes-Oxley Act's reforms on corporate accountability and governance. While more than 80% of respondents felt that the reforms would significantly improve corporate integrity, 80% said they also believed the reforms would not increase ethical behavior by corporate leaders.

While most consumer and public-interest watchdog groups are demanding even more stringent regulation of big business, some corporate reformers argue that regulations are often counterproductive and always circumventable. They believe that only 100% transparency can assure both the integrity and competency of institutional actions. In the world's law courts—and in the court of public opinion—the case for transparency will increasingly be promoted by nongovernmental organizations (NGOs) who will take advantage of the global infostructure to document and publicize environmentally and socially abusive behaviors by both private and public enterprises. The ongoing battle between institutional and socioecological imperatives will become a central theme of Web newscasts, Netpress publications, and Weblogs that have already begun to supplant traditional media networks and newspaper chains among young adults worldwide. Many of these young people will sign up with NGOs to wage undercover war on perceived corporate criminals.

In a global marketplace where corporate reputation and brand integrity will be worth billions of dollars, businesses' response to this guerrilla scrutiny will be understandably hostile. In their recently released *Study of Corporate Citizenship*, Cone/Roper, a corporate consultant on social issues, found that a majority of consumers "are willing to use their individual power to punish those companies that do not share their values." Above all, our improving comprehension of humankind's innumerable interactions with the environment will make it increasingly clear that total transparency will be crucial to the security and sustainability of a modern global economy. But there will be skullduggery, bloodshed, and heroics before total transparency finally becomes international law—15 to 20 years from now.

Trend 5—Social Adaptation

The forces of cultural modernization—education, urbanization, and institutional order—are producing social change in the developed world as well as in developing nations. During the twentieth century, it became increasingly apparent to the citizens of a growing number of modern industrial societies that neither the church nor the state was omnipotent and that their leaders were more or less ordinary people. This realization has led citizens of modern societies to assign less weight to the guidance of their institutions and their leaders and to become more self-regulating. U.S. voters increasingly describe themselves as independents, and the fastest-growing Christian congregations in America are nondenominational.

Since the dawn of recorded history, societies have adapted to their changing circumstances. Moreover, cultural modernization has freed the societies of mature industrial nations from many strictures of church and state, giving people much more freedom to be individually adaptive. And we can be reasonably certain that modern societies will be confronted with a variety of fundamental changes in circumstance during the next five, 10, or 15 years that will, in turn, provoke continuous widespread adaptive behavior, especially in America.

Reaching retirement age no longer always means playing golf and spoiling the grandchildren. Seniors in good health who enjoy working probably won't retire, slowing the prophesied workforce drain, according to author David Pearce Snyder

During the decade ahead, *infomation*—the automated collection, storage, and application of electronic data—will dramatically reduce paperwork. As outsourcing and offshoring eliminate millions of U.S. middle-income jobs, couples are likely to work two lower-pay/lower-skill jobs to replace lost income. If our employers ask us to work from home to reduce the company's office rental costs, we will do so, especially if the arrangement permits us to avoid two hours of daily commuting or to care for our offspring or an aging parent. If a wife is able to earn more money than her spouse, U.S. males are increasingly likely to become househusbands and take care of the kids. If we are in good health at age 65, and still enjoy our work, we probably won't retire, even if that's what we've been planning to do all our adult lives. If adult children must move back home after graduating from college in order to pay down their tuition debts, most families adapt accordingly.

Each such lifestyle change reflects a personal choice in response to an individual set of circumstances. And, of course, much adaptive behavior is initially undertaken as a temporary measure, to be abandoned when circumstances return to normal. During World War II, millions of women voluntarily entered the industrial workplace in the United States and the United Kingdom, for example, but returned to the domestic sector as soon as the war ended and a prosperous normalcy was restored. But the Information Revolution and the aging of mature industrial societies are scarcely temporary phenomena, suggesting that at least some recent widespread innovations in lifestyle —including delayed retirements and "sandwich households"—are precursors of long-term or even permanent changes in society.

The current propensity to delay retirement in the United States began in the mid-1980s and accelerated in the mid-1990s. Multiple surveys confirm that delayed retirement is much more a result of increased longevity and reduced morbidity than it is the result of financial necessity. A recent AARP survey, for example,

found that more than 75% of baby boomers plan to work into their 70s or 80s, regardless of their economic circumstances. If the baby boomers choose to age on the job, the widely prophesied mass exodus of retirees will not drain the workforce during the coming decade, and Social Security may be actuarially sound for the foreseeable future.

The Industrial Revolution in production technology certainly produced dramatic changes in society. Before the steam engine and electric power, 70% of us lived in rural areas; today 70% of us live in cities and suburbs. Before industrialization, most economic production was home- or family-based; today, economic production takes place in factories and offices. In preindustrial Europe and America, most households included two or three adult generations (plus children), while the great majority of households today are nuclear families with one adult generation and their children.

Current trends in the United States, however, suggest that the three great cultural consequences of industrialization—the urbanization of society, the institutionalization of work, and the atomization of the family—may all be reversing, as people adapt to their changing circumstances. The U.S. Census Bureau reports that, during the 1990s, Americans began to migrate out of cities and suburbs into exurban and rural areas for the first time in the twentieth century. Simultaneously, information work has begun to migrate out of offices and into households. Given the recent accelerated growth of telecommuting, self-employment, and contingent work, one-fourth to one-third of all gainful employment is likely to take place at home within 10 years. Meanwhile, growing numbers of baby boomers find themselves living with both their debt-burdened, underemployed adult

children and their own increasingly dependent aging parents. The recent emergence of the "sandwich household" in America resonates powerfully with the multigenerational, extended families that commonly served as society's safety nets in preindustrial times.

Leadership in Changing Times

The foregoing meta-trends are not the only watershed developments that will predictably reshape daily life in the decades ahead. An untold number of inertial realities inherent in the common human enterprise will inexorably change our collective circumstances—the options and imperatives that confront society and its institutions. Society's adaptation to these new realities will, in turn, create further changes in the institutional operating environment, among customers, competitors, and constituents. There is no reason to believe that the Information Revolution will change us any less than did the Industrial Revolution.

In times like these, the best advice comes from ancient truths that have withstood the test of time. The Greek philosopher-historian Heraclitus observed 2,500 years ago that "nothing about the future is inevitable except change." Two hundred years later, the mythic Chinese general Sun Tzu advised that "the wise leader exploits the inevitable." Their combined message is clear: "The wise leader exploits change."

David Pearce Snyder is the lifestyles editor of THE FUTURIST and principal of The Snyder Family Enterprise, a futures consultancy located at 8628 Garfield Street, Bethesda, Maryland 20817. Telephone 301-530-5807; e-mail davidpearcesnyder@earthlink.net; Web site www.the-futurist.com.

Originally published in the July/August 2004 issue of *The Futurist*, pp. 22–27. Copyright © 2004 by World Future Society, 7910 Woodmont Avenue, Suite 450, Bethesda, MD 20814. Telephone: 301/656-8274; Fax: 301/951-0394; http://www.wfs.org. Used with permission from the World Future Society.

GLOBALIZATION'S EFFECTS ON THE ENVIRONMENT

Jo Kwong

In recent years, globalization has become a remarkably polarizing issue. In particular, discussions about globalization and its environmental impacts generate ferocious debate among policy analysts, environmental activists, economists and other opinion leaders. Is globalization a solution to serious economic and social problems of the world? Or is it a profit-motivated process that leads to oppression and exploitation of the world's less fortunate?

This article examines alternative perspectives about globalization and the environment. It offers an explanation for the conflicting visions that are frequently expressed and suggests elements of an institutional framework that can align the benefits of globalization with the objective of enhanced environmental protection.

Globalization, free of the emotional rhetoric, is simply about removing barriers so goods, services, people, and ideas, can freely move from place to place. At its most rudimentary level, globalization describes a process whereby people can make their own decisions about who their trading partners are and what opportunities they wish to pursue.

While this may seem fairly innocuous, globalization certainly raises many concerns. In developed nations, some people worry about globalization's impacts on culture, traditional ways of living, and indigenous control in less developed parts of the word. They wonder, "What's to stop profit-motivated companies from developing some of the pristine environments and fragile natural resources found in the developing world?" These critics of open trade fear that residents of developing nations will be the losers in more ways than one—stripped of their land's natural resources and hopelessly in debt to exploitative developed countries. This group takes a rather paternalistic view of the problems facing the world's poor.

Others—free marketers—believe that the developed world can produce positive benefits by exporting knowledge and technology to the developing world. By avoiding mistakes made in the developed world, it is argued that developing countries can advance in manners that sidestep some of the errors that oc-

curred in others' development processes. Third-world poverty is cited as an important reason to foster greater economic growth in the developing world. To proponents of globalization, trade is seen as a way to lift the third world from poverty and enable local people to help themselves.

Moreover, there are divided views within the developing world. Some argue against so-called "eco-imperialism." "Why are others dictating whether or not we can develop our own resources? Who are these environmental activists that say billions of people in China shouldn't have cars because this will greatly accelerate global warming?" they ask. But others question, "Who are these corporations that come in and buy huge tracts of land in third-world interiors and develop large-scale forestry or oil developments, seemingly without concern about the impact on the local environment?"

In many ways, these alternative perspectives can be viewed as a "conflicts of visions" to steal a phrase from Thomas Sowell. Some people simply view the world fundamentally differently. In the globalization context, for example, one view values the protection of indigenous ways of life, even if that means living with greater poverty and fewer individual choices. Others believe economic efficiency is key – getting the most from our resources to provide the greatest amount of financial wealth and opportunity. Most likely, however, most people fall somewhere in between.

This discussion will offer an additional factor other than a "conflict of visions" that can help us understand the broad disparities in perspectives and understandings about the question, "Is globalization good for the environment?" In particular, it raises the possibility that perhaps we are not asking the right questions to address the set of concerns at hand.

In the 1990s, a number of economists sought to empirically answer the question of whether globalization helps or harms the environment. Some of the most often-cited findings are those from economists Gene Grossman and Alan Krueger. Grossman and Krueger investigated the relationship between the scale of economic activity and environmental quality for a broad set of environmental indicators. They found that environmental degradation and income have an inverted U-shaped relationship, with pollution increasing with income at low levels of income and decreasing with income at high levels of income. The turning point at which economic growth and pollution emissions switch from a positive to a negative relationship depends on the particular emissions and air quality measure tracked. For NOx, SOx and biological oxygen demand (BOD), the turning point appears to be around $5,000 per capita gross domestic product (GDP). This observation supports the view that countries can grow out of pollution problems with wealth.

These findings were followed by further studies that examined this "Environmental Kuznets Curve", as this inverted U-shaped curve was labeled, generating a new set of policy implications that supported the idea that trade can be good for the environment. If economic growth is good for the environment, policies that stimulate growth (trade liberalization, economic restructuring, and free markets) should also be good for the environment.

The most basic description of how this inverted curve can occur is to think about the types of activities that countries experience as they develop. At the most rudimentary level, people are burning cow dung and other readily available materials for heat and cooking sources. No controls are in place; the pollutants are released directly into the air. As economic activity increases and the economy reaches a point at which it can begin making investments, catalytic converters, furnaces, etc., pollution levels are reduced, and hence the inverted curve.

In "Poverty, Wealth and Waste," Barun Mitra compares patterns of waste distribution in India to those of the developed world. He addresses the myth that poor countries have lower levels of pollution:

> The painstaking efforts to recycle materials do not mean that a poor country like India is pollution-free. Indeed, the low quantity of waste generated in an economy with little capital and technological backwardness keeps the waste industry from graduating above small-scale local initiatives. And higher pollution occurs because there isn't the technology to capture highly dispersed waste such as sulfur dioxide from smokestacks or heavy metals that flow into wastewater.

A number of possible explanations for this observed relationship between pollution and income were advanced:

- As local economies grow and develop, they will inevitably change the way they use resources, creating different types of impacts upon the environment. A simple example is the pollution tradeoffs involved from our transition in transportation modes from horses to cars. Horses generated plenty of pollution in terms of manure, carcass disposal, etc. Cars, of course, generate an entirely different brand of pollution concerns. In other words, some environmental degradation along a country's development path is inevitable, especially during the take-off process of industrialization.

- Growth is associated with an increasing share of services and high-technology production, both of which tend to be more environment-friendly than production processes in earlier stages of industrialization.

- Knowledge and technology from the developed world can help ease this transition and lessen its duration, moving countries more quickly to the levels at which pollution will be decreasing. Free trade can promote a quicker diffusion of environment-friendly technologies and lead to a more efficient allocation of resources.

- The prosperity generated from economic activity will lead to more investments and higher standards of living that enable still greater investments in cleaner and newer technologies and processes. When a certain level of per capita income is reached, economic growth helps to undo the damage done in earlier years. As free trade expands, each 1 percent increase in per capita income tends to drive pollution concentrations down by 1.25 to 1.5 percent because of the movement to cleaner techniques of production.

- As individuals become richer they are willing to spend more on non-material goods, such as a cleaner environment. This point is made by Indur M. Goklany, in his description of earlier stages of development, "Society [initially] places a much higher priority on acquiring basic public health and other services such as sewage treatment, water supply, and electricity than on environmental quality, which initially worsens. But as the original priorities are met, environmental problems become higher priorities. More resources are devoted to solving those problems. Environmental degradation is arrested and then reversed."

These findings and explanations, unsurprisingly, generated an outpouring of negative response from environmental activists and anti-globalization proponents. "How can these economists be serious?" they, in effect, asked. "Do they really think it is wise to advocate policies that predictably increase pollution? Are we supposed to believe pollution will eventually decrease if we continue with the polluting activities? How absolutely ludicrous!"

Typical responses to the "growth is good" thesis include:

- Globalization will result in a "race to the bottom" as polluting companies relocate to countries with lax environmental standards.

- Trading with countries that do not have suitable environmental laws will lower environmental standards for all countries.

- Multinationals will exploit pristine environments in the developing world, reaping the resources for short-term growth, and then pulling out to repeat the process elsewhere—growth ruins the environment.

- Free trade provides a license to pollute—it is bad for the environment. Stronger environmental regulations at national and international levels are needed.

The Sierra Club summarized the widespread critiques to the Grossman and Krueger studies, drawing from research studies produced by the World Wide Fund for Nature and others. It argued that the findings were sufficiently over-generalized to dispense with any notion that they justify complacency about trade and the environment, pointing out several facts.

The empirical estimates of where "turning points" occur for different pollutants vary so widely as to cast doubt on the validity of any one set of results. For instance, where Grossman and Krueger found turning points for certain air pollutants at less than $5,000 per capita, others found turning points above $8,000 per capita.

For some air pollutants, Grossman and Krueger found that emissions levels don't follow an inverted U-curve, but following an S-curve that starts to rise again as incomes rise. For instance, they found that sulphur dioxide emissions start to rise when income increases above $14,000 per capita. The implication is that efficiency gains from improved technology at medium levels of per capita income are eventually overwhelmed by the growing size of the economy.

Since most of the world's population earns per capita incomes well below estimated turning points, global air pollution levels will continue to rise for nearly another century. By that time, emissions of some pollutants will be anywhere from two to four times higher than current levels.

Even for the limited number of pollutants that Grossman and Krueger study, they only demonstrate a correlation between changing per capita income and changing levels of environmental quality. They do not demonstrate a causal connection. The positive relationships they describe could actually be caused by noneconomic factors, such as the adoption of environmental legislation.

Both camps seem to have reasonable grounds for their views. Clearly there is a conflict of visions that is rooted in very different value systems. Can these two opposing perspectives be reconciled sufficiently to reach some type of consensus?

As noted earlier, many studies have re-examined the Environmental Kuznets curve since the publication of the Grossman and Krueger analysis in 1991, each attempting to prove or disprove the relationship between economic growth and environmental quality, or to isolate variables that may explain the observed relationships. In that same year, a fascinating monograph was published in London, called The Wealth of Nations and the Environment. Author Mikhail Bernstam set out to analyze the contention that economic growth negatively impacts the environment by examining how institutional structure impacts this relationship.

Bernstam examined and contrasted the impact of economic growth upon the environment in both capitalist and socialist countries. Interestingly, he found that the environmental Kuznets curve does in fact exist, but it does not apply to countries across the board. The Kuznets curve, he found, applies to market economies, but not to socialist ones. The difference, according to Bernstam, has its roots in the different structures of incentives and property rights of these two economic systems.

Under market economies with secure property rights and open trade, the pursuit of profits leads to the husbanding of resources. These capitalist economies use fewer resources to produce the equivalent level of output and hence do less damage to the environment. In contrast, in socialist countries, the managers of state enterprises operate under incentives that encourage them to maximize inputs, with little regard towards economic waste or damage to the environment.

More recently, a 2001 study by economists Werner Antweiler, Brian R. Copeland, and M. Scott Taylor asked, "Is Free Trade Good for the Environment?" They analyzed data on sulfur dioxide over the period 1971 to 1996, a time when trade barriers were coming down and international trade was expanding. They found that countries that opened up to trade generated faster economic growth. Although economic growth produced more pollution, the greater wealth and higher incomes also generated a demand for a cleaner environment.

To separate these effects, the Antweiler model looked at the negative environmental consequences of increases in economic activity (the scale effect), the positive environmental consequences of increases in income that lead to cleaner production methods (the technique effect), and the impact of trade-induced changes in the composition of output upon pollution concentrations (the composition effect). When the scale, technique and composition effects estimates were combined, the Antweiler et al. model yielded the conclusion that free trade is good for the environment. For example, when analyzing sulfur dioxide, the authors estimate that for each 1 percent increase in per capita income in a nation, pollution *falls* by 1 percent.

The critical explanatory factor is that wealthier countries value environmental amenities more highly and enhance their production by employing environmentally friendly technologies. However, like Bernstam, these authors specified that it is important to distinguish between communist and non-communist countries. Communist countries provided the exception to their rule about globalization's positive impacts upon the environment.

The studies, which consider the impact of institutional structures, make an important contribution to our understanding of the "economics vs. environment" debates. They suggest we consider other factors in our analysis of the effects of globalization. It is true that we often do find examples of disastrous environmental conditions, particularly when we look at socialist countries. But it is misleading to attribute the disasters to globalization. Instead, we need to examine the institutional arrangements in a particular country to see what role they play in economic development and environmental protection.

Positive Globalization

As described earlier, at its most rudimentary level, globalization simply embodies a process of free and open trade, whereby people can make decisions about who their trading partners are and what opportunities they will choose to pursue.

But the cautions of the environmental activists are worthy of consideration. Free trade, in and of itself, will not guarantee positive outcomes. We also need guiding rules that essentially create the terms for fair and civil interaction.

In *Property Rights: A Practical Guide to Freedom & Prosperity,* Terry Anderson and Laura Huggins describe the importance of institutional rules. They use the example of children playing together and inventing games. In essence, the children work together to form rules that are fair. When they cannot agree on rules, chaos typically results and their play breaks down. The same is true for civil society. Institutional rules, in the form of constitutions, common law, and so on, provide the structure for human activity.

The critical role of institutions in shaping human behavior gained international attention in 1993 when Douglass C. North received the Nobel Prize in economics. North's groundbreaking research in economic history integrated economics, sociology, statistics and history to explain the role that institutions play in economic growth.

For several decades, North looked at the question, "Why do some countries become rich, while others remain poor?" In seeking answers to this query, he came to understand that institutions establish the formal and informal sets of rules that govern the behavior of human beings in a society. His research showed that, depending on their structure and enforcement, institutional arrangements can either foster or restrain economic development.

For the past nine years, the *Index of Economic Freedom,* jointly published by the Wall Street Journal and the Heritage Foundation (Washington, DC), has provided fascinating empirical evidence of the relationship of various institutions to economic prosperity. The study analyzes and ranks the economic freedom of 161 countries according to 10 institutional factors (trade policy, property rights, regulation, and black market, for example) in an effort to trace the path to economic prosperity.

The key finding of the research, supported year after year, is that countries with the most economic freedom enjoy higher rates of long-term economic growth and prosperity than those with less economic freedom. But, more relevant to this discussion, is the finding that economic freedom, which enables people to choose who and where their trading partners are, ultimately leads to more efficient resource use.

In another comparative index, *Economic Freedom of the World 2002,* published by the Fraser Institute in conjunction with public policy institutes around the world, Nobel laureate Milton Friedman describes the importance of private property and the rule of law as a basis for economic freedom. He spells out the three key ingredients key to establishing economic freedom as follows: "First of all, and most important, the rule of law, which extends to the protection of property. Second, widespread private ownership of the means of production. Third, freedom to enter or to leave industries, freedom of competition, freedom of trade. Those are essentially the basic requirements." These same factors also provide a framework for positive environmental development.

In the 1980s, a team of economists affiliated with the Property and Environmental Research Center (PERC) in Bozeman, MT, began developing a new paradigm for environmental policy. Their model, which eventually was coined "Free Market Environmentalism" described how incentives are the key to environmental stewardship. Not surprisingly, people who face little or no consequences for environmentally destructive actions face no incentive to protect the environment. Alternatively, people who are rewarded for good stewardship are much more likely to invest in environmental protection. The key, according to economists John Baden, Richard Stroup and Terry Anderson, are the very same three elements that Milton Friedman mentioned for economic prosperity: free and open markets, clearly established property rights, and rule of law.

Free and open markets. One of the most important benefits produced by a market economy is information, conveyed in the form of prices. Prices of natural and environmental resources provide clear signals about their availability. As a resource becomes scarcer, its price increases. And of course, the reverse is also true: When a resource becomes more abundant, the price decreases.

Many people fear that the profit motive leads to the depletion or degradation of environmental resources. As counterintuitive as it may sound, the profit motive actually works to the benefit of the environment.

Businesses face incentives to carefully consider the prices of the various natural resources that they use in their production processes. If a particular resource is in short supply, its price will be higher than others that are more readily available. It makes little sense for a producer to over utilize, or "waste," a high-priced resource.

High prices also encourage the search for, and development of, appropriate substitutes or alternatives. As companies search for ways to reduce costs, they naturally tend toward utilizing lower-priced, more abundant resources. Thus, the pursuit of profits is actually a driving force to conserve resources. In essence, under free market systems, entrepreneurs compete in developing low cost, efficient means to solve contemporary resource problems.

Property rights. Clearly established property rights generate another incentive for environmental stewardship. It makes no sense for private landowners, for example, to exploit and destroy their own property. Ownership creates a long-term perspective that leads to preserving and protecting property.

Careless destruction, however, does make sense for those who are only loosely held accountable for their actions. Politicians, bureaucrats, or others, who may be short-term managers, face the incentive to maximize immediate returns, even if this means long-term environmental damage. Even managers with longer tenures realize they can simply turn to the federal government for more funds to address the problems that short-sighted decision making may have created.

Rule of Law. In many ways, the "rule of law" is the glue that holds market transactions and property rights together. Freedom to exchange is meaningless if individuals do not have secure rights to property, including the fruits of their labors. Failure of a country's legal system to provide for the security of property rights, enforcement of contracts, and the mutually agreeable settlement of disputes will undermine the operation of a market-exchange system. If individuals and businesses lack confidence that contracts will be enforced and the returns from their productive activity protected, their incentive to engage in innovative activities will be eroded.

With these elements in place, the economists' explanations prevail—globalization will enable local cultures to pick and choose the development and environmental paths that they wish to traverse. But without these institutional arrangements, the likelihood of negative consequences increases.

In countries that lack property rights and rule of law and that promote barriers to trade, an institutional structure develops that fosters destruction of the environment. For example, in Liberia, former President Charles Taylor rapidly sold off many of the nation's natural resources in order to fund his dictatorship. In the lawless structure of that country, Taylor was able to exploit the environment and his people. In a country that has clear property rights and rule of law, such corrupt options are closed off. Neither can corporations force a village, or a state, or a country to destroy its natural resources against the will of the people.

We see this illustrated in an ongoing controversy in Peru. In the 1990s, when then bankrupt Peru opened its statist economy to foreign investment, the nation drew almost $10 billion in mining capital. That sector now accounts for half of Peru's $8 billion in exports, and Peru has become one of the world's largest gold producers. Yet, the opening economy does not necessarily mean that multi nationals can run rough shod over the locals. It all depends on the institutional arrangements that are in place.

In the small town of Tambogrande, Peru, a Canadian mining company holds the rights to tap into $1 billion worth of copper and zinc beneath the town. To do so, however, requires demolishing many local homes. In a referendum held in 2002, the town residents voted to turn down the mining company's offer to build new homes in a different location. If the country's laws hold firm to the property rights of the villagers, the mining company will not be allowed to develop the copper mine without local consent. But, if the rule of law and respect of property rights are not upheld, then the foreign firm can force its will on the indigenous people.

Property rights provide a powerful incentive for people to carefully assess their options—in this case, whether the loss of their existing houses and the village is compensated for by the new homes they would be receiving. The nature of the property rights institutions indeed affects the range of outcomes. If the local government owned the rights to the housing, rather than individuals, we would expect an entirely different outcome. Local politicians likely would gain by acquiescing to the mining firm's proposal because the villagers, not the politicians, would incur the costs.

Unfortunately, in many developing countries, corruption and back door deal making, enabled by weak rule of law and property rights, proliferates. The result is that a few leaders come out ahead and the locals get short changed. Local protests are reportedly stalling at least 10 mining-investment projects in Peru that are worth $1.4 billion—and for good reason. The noted Peruvian economist, Hernando de Soto, author of the best-seller, *The Mystery of Capital*, comments that although the mines in some towns pay double the prevailing minimum wage, they do not compensate for "the loss of their sense of environmental and economic sovereignty." Consequently, the National Society of Mining, Petroleum & Energy is urging the government to adopt reforms that immediately give at least 20 percent of the royalties to on-site communities instead of sending all these funds to Lima. Manhattan Minerals, one of the companies interested in Tambogrande, thinks local communities should receive an even bigger cut, making these towns, in effect, feel more like share-holders. In other words, they need to give the locals an interest, or property right, in the operations.

In the southern Andean town of Lircay, Huanavelica, Peru's poorest state, residents are concerned that the mine will threaten adjacent agricultural lands. To show their anger, they have resorted to street demonstrations and setting fire to government installations. Their actions seem less extreme in light of previous experiences. For decades, state-owned mining created many environmental problems that residents are rightly worried about. This cultural legacy is a key factor for private mining companies as they hammer out new relationships and try to move forward.

Fortunately, their positive examples are evolving. The La Oroya copper smelter in the central Andes region was purchased by the Doe Run company—based in St. Louis. The Peruvian government gave the company 10 years to clean up the environmental mess that the government created. Doe Run has reportedly spent $40 million so far, including money for a program to reduce high blood-lead levels in area children.

Peru needs to continue to open its doors to foreign investment, or what some would call globalization, to lift its people out of poverty. It must establish institutions—rule of law, property rights and open markets that create a safe investment climate and allow corporations to prosper. Simultaneously, companies need to conduct business in a way that will benefit the local residents as well.

As another example of how incentives and disincentives can impact the environment, consider the case of India's automobile industry. Disincentives generated by the government's regulatory policies contributed to a stagnant, non-innovative industry which caused harm to the Indian economy and environment for decades.

Although Indian automobile manufacturers began producing cars in the 1930s, there was very little development and growth in that industry for over 50 years. Auto manufacturing was heavily regulated, licensed and protected. In addition, consumers faced high taxes and duties on imported automobiles and on gasoline. The upshot was that very little competition developed in India's automobile industry— autos with low fuel efficiency and high air emissions became the norm.

In recent years, however, the automobile sector has been slowly liberalizing, allowing some major multinational corporations to set up shop in India. As a result of the increased competition and relaxed barriers, more efficient and less-polluting automobiles are becoming available to Indian consumers. A free trade regime, from the outset, would have increased access to vehicles for consumers, lowered the cost of transportation, enabled the best technologies to be locally available, and improved air quality.

In other words, incentives matter. And the structure of institutions plays a key role in the nature of incentives that are in effect. Economists have raised interesting empirical questions by developing the environmental Kuznets curve, but, as the World Wide Fund for Nature (WWF) study and other suggest, there is no one curve that fits all pollutants for all places and times. Economist Bruce Yandle of Clemson University describes it this way, "There are families of relationships, and in many cases the inverted-U Environmental Kuznets Curve is the best way to approximate the link between environmental change and income growth."

Additionally, environmental activists are right in pointing out globalization's potentially negative impacts upon the environment. Income growth alone is insufficient to reduce environmental harms and may even increase these harms if a core set of institutional features are not in place.

As the Antweiler model indicates, economic growth creates the conditions for environmental protection by raising the demand for improved environmental quality and by providing the resources needed for protection. Whether environmental quality improvements materialize or not, or when, or how they develop, depends critically on government policies, social institutions, and the strength of markets. Better policies, such as removing distorting subsidies, introducing more secure property rights over resources, and using market-like mechanisms to connect the costs of pollution to prices paid for pollution-producing goods will lower peak environmental harm (flatten the underlying Environmental Kuznets Curve). These improved policies may also bring about an earlier environmental transition.

While it may seem to be an overwhelming challenge to accomplish the institutional reforms described above, the good news is that it is happening in some very unlikely parts of the world. Consider, for example, exciting changes that have recently been occurring in Rwanda, Africa.

Lawrence Reed, the founder and president of the Mackinac Center for Public Policy in Midland, Michigan, recently toured eastern Africa, home to the remaining wild mountain gorillas left in the world. Here, approximately 670 gorillas live on a string of lush, rain-forested volcanoes along the Rwanda border with Uganda and the Congo.

To Reed's surprise, native-owned and locally staffed companies conduct all gorilla safaris. Part of the fee goes to the government for salaries for national park employees and for programs that protect gorilla habitat. (These programs also are substantially supplemented by the efforts of private, non-profits that get support from around the world.) Two Rwandan entrepreneurs started the firm, Primate Safaris, three years ago. With six employees, they provide everything a gorilla safari enthusiast could hope for—a competent guide with a four-wheel drive vehicle, good meals and comfortable accommodations.

In fact, Reed's experience with Private Safaris was only the tip of the iceberg. Rwanda, he learned, is engaged in the continent's most ambitious privatization campaign. After experiencing the kind of stifling, socialist rule that consigned virtually all of Africa to grinding poverty for decades, this nation is now embracing the private sector with deliberate policy and enormous enthusiasm. Imagine Reed's surprise when, shortly after landing in Rwanda, he came across a sign at the airport outside the capital of Kigali which reads, "Privatization: A Loss? No Way." Further down the road, another sign says "Privatization fights laziness, privatization fights poverty, privatization fights smuggling, and privatization fights unemployment."

Several of the country's privatization efforts have had direct positive impacts on the environment. For example, in 1999, Shell Oil bought a portion of the assets of Petrorwanda (the bankrupt state oil company) and completely renovated 14 of the defunct firm's decrepit and environmentally hazardous gasoline stations.

An interesting development in Uganda suggests signs of similar institutional reforms. An English language, African-based band named "Afrigo" released a song entitled, "Today for Tomorrow," which celebrates the benefits of privatization. Here is a sample of the lyrics:

> Privatization, the surer route to economic emancipation/ Yeah, businessmen run businesses/government govern the nation/ You and I didn't create the situation/ Let's unite/check the economy/ a better future for our children.

Apparently, citizens of Rwanda and Uganda are embracing private property rights and other economic and political changes to better their lives and those of the next generation. Environmental protection surely will fare better in this setting than in the failed socialist systems being replaced.

Is globalization good for the environment? Viewing globalization as the destroyer or savior of the environment misses the point. The problem is not globalization per se. A lack of key institutions, namely rule of law; property rights; free and open markets is the real villain in the tale. These institutions hold people accountable for their actions, and at the same time, reward them for positive behavior. They create conditions in which market competition rewards innovation and efficiency, and in which economic development and increased wealth can fuel improved environmental quality.

Globalization—free trade and multinational investments—can advance these institutional changes, leading to enhanced social and political stability. Concerns that multinational corporations might be engaged in a "race to the environmental bottom" seem unlikely in these circumstances. To the contrary, where these institutions are in place, the result can be a "race to the top," as jurisdictions compete to improve the quality of life for their constituents.

Globalization can be a means to accelerate learning about the importance of market institutions to economic growth. Environmental protection can be one of many important benefits resulting from such a transition. Getting back to my earlier comment about "a conflict of visions," I certainly hold a contrasting view from opponents of globalization. Critics believe globalization underlies many of the problems that plague the developing world. On the other hand, I see globalization as a basic part of the solution to these problems. Greater movement of goods, services, people and ideas can lead to economic prosperity, improved environmental protection, and a host of other social benefits.

SUGGESTED FURTHER READINGS

Antweiler, Werner, Brian R. Copeland, and M. Scott Taylor. 2001. "Is Free Trade Good for the Environment?" *American Economic Review*. Vol. 91(4), pp. 877-908.

Bernstam, Mikhail. 1991. "Is Free Trade Good for the Environment." Institute of Economic Affairs, p. 7.

Bhagwati, Jagdish. 2002. *Free Trade Today*. Princeton: Princeton University Press.

Goklany, Indur M. 1998. "The Environmental Transition to Air Quality." *Regulation*. Vol. 21(4), p. 36.

Grossman, G.M. and A.B. Krueger. 1995. "Economic Growth and the Environment." *Quarterly Journal of Economics*. Vol.110(2), pp. 353-377.

Yandle, Bruce, Maya Vijayaraghavan, and Madhusudan Bhattarai. 2002. "The Environmental Kuznets Curve: A Primer." *PERC Research Study 02-1*, p. 17.

Jo Kwong, Ph.D. *is Director of Institute Relations, Atlas Economic Research Foundation in Fairfax, Virginia. This paper formed the basis for her address at Lindenwood University (St. Charles, Missouri) in the Economic Policy Lecture Series on March 29, 2004. The event was co-hosted by the Institute for Study of Economics and the Environment and the Division of Management.*

From *Society*, January/February 2005, pp. 21–28. Copyright © 2005 by Transaction Publishers. Reprinted by permission of Transaction Publishers via the Copyright Clearance Center.

Rescuing a planet under stress

Lester R. Brown

Understanding the Problem

As world population has doubled and as the global economy has expanded sevenfold over the last half-century, our claims on the Earth have become excessive. We are asking more of the Earth than it can give on an ongoing basis.

We are harvesting trees faster than they can regenerate, overgrazing rangelands and converting them into deserts, overpumping aquifers, and draining rivers dry. On our cropland, soil erosion exceeds new soil formation, slowly depriving the soil of its inherent fertility. We are taking fish from the ocean faster than they can reproduce.

We are releasing carbon dioxide into the atmosphere faster than nature can absorb it, creating a greenhouse effect. As atmospheric carbon dioxide levels rise, so does the earth's temperature. Habitat destruction and climate change are destroying plant and animal species far faster than new species can evolve, launching the first mass extinction since the one that eradicated the dinosaurs sixty-five million years ago.

Throughout history, humans have lived on the Earth's sustainable yield—the interest from its natural endowment. But now we are consuming the endowment itself. In ecology, as in economics, we can consume principal along with interest in the short run but in the long run it leads to bankruptcy.

In 2002 a team of scientists led by Mathis Wackernagel, an analyst at Redefining Progress, concluded that humanity's collective demands first surpassed the Earth's regenerative capacity around 1980. Their study, published by the U.S. National Academy of Sciences, estimated that our demands in 1999 exceeded that capacity by 20 percent. We are satisfying our excessive demands by consuming the Earth's natural assets, in effect creating a global bubble economy.

Bubble economies aren't new. U.S. investors got an up-close view of this when the bubble in high-tech stocks burst in 2000 and the NASDAQ, an indicator of the value of these stocks, declined by some 75 percent. According to the *Washington Post*, Japan had a similar experience in 1989 when the real estate bubble burst, depreciating stock and real estate assets by 60 percent. The bad-debt fallout and other effects of this collapse have left the once-dynamic Japanese economy dead in the water ever since.

The bursting of these two bubbles affected primarily people living in the United States and Japan but the global bubble economy that is based on the overconsumption of the Earth's natural capital assets will affect the entire world. When the food bubble economy, inflated by the overpumping of aquifers, bursts, it will raise food prices worldwide. The challenge for our generation is to deflate the economic bubble before it bursts.

Unfortunately, since September 11, 2001, political leaders, diplomats, and the media worldwide have been preoccupied with terrorism and, more recently, the occupation of Iraq. Terrorism is certainly a matter of concern, but if it diverts us from the environmental trends that are undermining our future until it is too late to reverse them, Osama bin Laden and his followers will have achieved their goal in a way they couldn't have imagined.

In February 2003, United Nations demographers made an announcement that was in some ways more shocking than the 9/11 attack: the worldwide rise in life expectancy has been dramatically reversed for a large segment of humanity—the seven hundred million people living in sub-Saharan Africa. The HIV epidemic has reduced life expectancy among this region's people from sixty-two to forty-seven years. The epidemic may soon claim more lives than all the wars of the twentieth century. If this teaches us anything, it is the high cost of neglecting newly emerging threats.

The HIV epidemic isn't the only emerging mega-threat. Numerous nations are feeding their growing populations by overpumping their aquifers—a measure that virtually guarantees a future drop in food production when the aquifers are depleted. In effect, these nations are creating a food bubble economy—one where food production is artificially inflated by the unsustainable use of groundwater.

Another mega-threat—climate change—isn't getting the attention it deserves from most governments, particularly that of the United States, the nation responsible for one-fourth of all carbon emissions. Washington, D.C., wants to wait until all the evidence on climate change is in, by which time it will be too late to prevent a wholesale warming of the planet. Just as governments in Africa watched HIV infection rates rise and did little about it, the United States is watching atmospheric carbon dioxide levels rise and doing little to check the increase.

Other mega-threats being neglected include eroding soils and expanding deserts, which jeopardize the livelihood and food supply of hundreds of millions of the world's people. These issues don't even appear on the radar screen of many national governments.

Thus far, most of the environmental damage has been local: the death of the Aral Sea, the burning rainforests of Indonesia, the collapse of the Canadian cod fishery, the melting of the glaciers that supply Andean cities with water, the dust bowl forming in northwestern China, and the depletion of the U.S. great plains aquifer. But as these local environmental events expand and multiply, they will progressively weaken the global economy, bringing closer the day when the economic bubble will burst.

Humanity's demands on the Earth have multiplied over the last half-century as our numbers have increased and our incomes have risen. World population grew from 2.5 billion in 1950 to 6.1 billion in 2000. The growth during those fifty years exceeded that during the four million years since our ancestors first emerged from Africa.

Incomes have risen even faster than population. According to Erik Assadourian's *Vital Signs* 2003 article, "Economic Growth Inches Up," income per person worldwide nearly tripled from 1950 to 2000. Growth in population and the rise in incomes together expanded global economic output from just under $7 trillion (in 2001 dollars) of goods and services in 1950 to $46 trillion in 2000—a gain of nearly sevenfold.

Population growth and rising incomes together have tripled world grain demand over the last half-century, pushing it from 640 million tons in 1950 to 1,855 million tons in 2000, according to the U.S. Department of Agriculture (USDA). To satisfy this swelling demand, farmers have plowed land that was highly erodible—land that was too dry or too steeply sloping to sustain cultivation. Each year billions of tons of topsoil are being blown away in dust storms or washed away in rainstorms, leaving farmers to try to feed some seventy million additional people but with less topsoil than the year before.

Demand for water also tripled as agricultural, industrial, and residential uses increased, outstripping the sustainable supply in many nations. As a result, water tables are falling and wells are going dry. Rivers are also being drained dry, to the detriment of wildlife and ecosystems.

Fossil fuel use quadrupled, setting in motion a rise in carbon emissions that is overwhelming nature's capacity to fix carbon dioxide. As a result of this carbon-fixing deficit, atmospheric carbon dioxide concentrations climbed from 316 parts per million (ppm) in 1959, when official measurement began, to 369 ppm in 2000, according to a report issued by the Scripps Institution of Oceanography at the University of California.

The sector of the economy that seems likely to unravel first is food. Eroding soils, deteriorating rangelands, collapsing fisheries, falling water tables, and rising temperatures are converging to make it more difficult to expand food production fast enough to keep up with demand. According to the USDA, in 2002 the world grain harvest of 1,807 million tons fell short of world grain consumption by 100 million tons, or 5 percent. This shortfall, the largest on record, marked the third consecutive year of grain deficits, dropping stocks to the lowest level in a generation.

Now the question is: can the world's farmers bounce back and expand production enough to fill the hundred-million-ton shortfall, provide for the more than seventy million people added each year, and rebuild stocks to a more secure level? In the past, farmers responded to short supplies and higher grain prices by planting more land and using more irrigation water and fertilizer. Now it is doubtful that farmers can fill this gap without further depleting aquifers and jeopardizing future harvests.

At the 1996 World Food Summit in Rome, Italy, hosted by the UN Food and Agriculture Organization (FAO), 185 nations plus the European community agreed to reduce hunger by half by 2015. Using 1990–1992 as a base, governments set the goal of cutting the number of people who were hungry—860 million—by roughly 20 million per year. It was an exciting and worthy goal, one that later became one of the UN Millennium Development Goals.

But in its late 2002 review of food security, the UN issued a discouraging report:

> This year we must report that progress has virtually ground to a halt. Our latest estimates, based on data from the years 1998-2000, put the number of under-nourished people in the world at 840 million.... a decrease of barely 2.5 million per year over the eight years since 1990–92.

Since 1998–2000, world grain production per person has fallen 5 percent, suggesting that the ranks of the hungry are now expanding. As noted earlier, life expectancy is plummeting in sub-Saharan Africa. If the number of hungry people worldwide is also increasing, then two key social indicators are showing widespread deterioration in the human condition.

The ecological deficits just described are converging on the farm sector, making it more difficult to sustain rapid growth in world food output. No one knows when the growth in food production will fall behind that of demand, driving up prices, but it may be much sooner than we think. The triggering events that will precipitate future food shortages are likely to be spreading water shortages interacting with crop-withering heat waves in key food-producing regions. The economic indicator most likely to signal serious trouble in the deteriorating relationship between the global economy and the Earth's ecosystem is grain prices.

Food is fast becoming a national security issue as growth in the world harvest slows and as falling water tables and rising temperatures hint at future shortages. According to the USDA more than one hundred nations import part of the wheat they consume. Some forty import rice. While some nations are only marginally dependent on imports, others couldn't survive without them. Egypt and Iran, for example, rely on imports for 40 percent of their grain supply. For Algeria, Japan, South Korea, and Taiwan, among others, it is 70 percent or more. For Israel and Yemen, over 90 percent. Just six nations—Argentina,

Australia, Canada, France, Thailand, and the United States—supply 90 percent of grain exports. The United States alone controls close to half of world grain exports, a larger share than Saudi Arabia does of oil.

Thus far the nations that import heavily are small and middle-sized ones. But now China, the world's most populous nation, is soon likely to turn to world markets in a major way. As reported by the International Monetary Fund, when the former Soviet Union unexpectedly turned to the world market in 1972 for roughly a tenth of its grain supply following a weather-reduced harvest, world wheat prices climbed from $1.90 to $4.89 a bushel. Bread prices soon rose, too.

If China depletes its grain reserves and turns to the world grain market to cover its shortfall—now forty million tons per year—it could destabilize world grain markets overnight. Turning to the world market means turning to the United States, presenting a potentially delicate geopolitical situation in which 1.3 billion Chinese consumers with a $100-billion trade surplus with the United States will be competing with U.S. consumers for U.S. grain. If this leads to rising food prices in the United States, how will the government respond? In times past, it could have restricted exports, even imposing an export embargo, as it did with soybeans to Japan in 1974. But today the United States has a stake in a politically stable China. With an economy growing at 7 to 8 percent a year, China is the engine that is powering not only the Asian economy but, to some degree, the world economy.

For China, becoming dependent on other nations for food would end its history of food self-sufficiency, leaving it vulnerable to world market uncertainties. For Americans, rising food prices would be the first indication that the world has changed fundamentally and that they are being directly affected by the growing grain deficit in China. If it seems likely that rising food prices are being driven in part by crop-withering temperature rises, pressure will mount for the United States to reduce oil and coal use.

For the world's poor—the millions living in cities on $1 per day or less and already spending 70 percent of their income on food—rising grain prices would be life threatening. A doubling of world grain prices today could impoverish more people in a shorter period of time than any event in history. With desperate people holding their governments responsible, such a price rise could also destabilize governments of low-income, grain-importing nations.

Food security has changed in other ways. Traditionally it was largely an agricultural matter. But now it is something that our entire society is responsible for. National population and energy policies may have a greater effect on food security than agricultural policies do. With most of the three billion people to be added to world population by 2050 (as estimated by the UN) being born in nations already facing water shortages, childbearing decisions may have a greater effect on food security than crop planting decisions. Achieving an acceptable balance between food and people today depends on family planners and farmers working together.

Climate change is the wild card in the food security deck. The effect of population and energy policies on food security differ from climate in one important respect: population stability can be achieved by a nation acting unilaterally. Climate stability cannot.

Instituting the Solution

Business as usual—Plan A—clearly isn't working. The stakes are high, and time isn't on our side. The good news is that there are solutions to the problems we are facing. The bad news is that if we continue to rely on timid, incremental responses our bubble economy will continue to grow until eventually it bursts. A new approach is necessary—a Plan B—an urgent reordering of priorities and a restructuring of the global economy in order to prevent that from happening.

Plan B is a massive mobilization to deflate the global economic bubble before it reaches the bursting point. Keeping the bubble from bursting will require an unprecedented degree of international cooperation to stabilize population, climate, water tables, and soils—and at wartime speed. Indeed, in both scale and urgency the effort required is comparable to the U.S. mobilization during World War II.

Our only hope now is rapid systemic change—change based on market signals that tell the ecological truth. This means restructuring the tax system by lowering income taxes and raising taxes on environmentally destructive activities, such as fossil fuel burning, to incorporate the ecological costs. Unless we can get the market to send signals that reflect reality, we will continue making faulty decisions as consumers, corporate planners, and government policymakers. Ill-informed economic decisions and the economic distortions they create can lead to economic decline.

Stabilizing the world population at 7.5 billion or so is central to avoiding economic breakdown in nations with large projected population increases that are already overconsuming their natural capital assets. According to the Population Reference Bureau, some thirty-six nations, all in Europe except Japan, have essentially stabilized their populations. The challenge now is to create the economic and social conditions and to adopt the priorities that will lead to population stability in all remaining nations. The keys here are extending primary education to all children, providing vaccinations and basic health care, and offering reproductive health care and family planning services in all nations.

Shifting from a carbon-based to a hydrogen-based energy economy to stabilize climate is now technologically possible. Advances in wind turbine design and in solar cell manufacturing, the availability of hydrogen generators, and the evolution of fuel cells provide the technologies needed to build a climate-benign hydrogen economy. Moving quickly from a carbon-based to a hydrogen-based energy economy depends on getting the price right, on incorporating the indirect costs of burning fossil fuels into the market price.

On the energy front, Iceland is the first nation to adopt a national plan to convert its carbon-based energy economy to one based on hydrogen. Denmark and Germany are leading the world into the age of wind. Japan has emerged as the world's leading manufacturer and user of solar cells. With its commercialization of a solar roofing material, it leads the world in electricity generation from solar cells and is well positioned to assist in the electrification of villages in the developing world. The Netherlands leads the industrial world in exploiting the bicycle as an alternative to the automobile. And the Canadian province of Ontario is emerging as a leader in phasing out coal. It plans to replace its five coal-fired power plants with gas-fired plants, wind farms, and efficiency gains.

Stabilizing water tables is particularly difficult because the forces triggering the fall have their own momentum, which must be reversed. Arresting the fall depends on quickly raising water productivity. In pioneering drip irrigation technology, Israel has become the world leader in the efficient use of agricultural water. This unusually labor-intensive irrigation practice, now being used to produce high-value crops in many nations, is ideally suited where water is scarce and labor is abundant.

In stabilizing soils, South Korea and the United States stand out. South Korea, with once denuded mountainsides and hills now covered with trees, has achieved a level of flood control, water storage, and hydrological stability that is a model for other nations. Beginning in the late 1980s, U.S. farmers systematically retired roughly 10 percent of the most erodible cropland, planting the bulk of it to grass, according to the USDA. In addition, they lead the world in adopting minimum-till, no-till, and other soil-conserving practices. With this combination of programs and practices, the United States has reduced soil erosion by nearly 40 percent in less than two decades.

Thus all the things we need to do to keep the bubble from bursting are now being done in at least a few nations. If these highly successful initiatives are adopted worldwide, and quickly, we can deflate the bubble before it bursts.

Yet adopting Plan B is unlikely unless the United States assumes a leadership position, much as it belatedly did in World War II. The nation responded to the aggression of Germany and Japan only after it was directly attacked at Pearl Harbor on December 7, 1941. But respond it did. After an all-out mobilization, the U.S. engagement helped turn the tide, leading the Allied Forces to victory within three and a half years.

This mobilization of resources within a matter of months demonstrates that a nation and, indeed, the world can restructure its economy quickly if it is convinced of the need to do so. Many people—although not yet the majority—are already convinced of the need for a wholesale restructuring of the economy. The issue isn't whether most people will eventually be won over but whether they will be convinced before the bubble economy collapses.

History judges political leaders by whether they respond to the great issues of their time. For today's leaders, that issue is how to deflate the world's bubble economy before it bursts. This bubble threatens the future of everyone, rich and poor alike. It challenges us to restructure the global economy, to build an eco-economy.

We now have some idea of what needs to be done and how to do it. The UN has set social goals for education, health, and the reduction of hunger and poverty in its Millennium Development Goals. My latest book, *Plan B*, offers a sketch for the re-

structuring of the energy economy to stabilize atmospheric carbon dioxide levels, a plan to stabilize population, a strategy for raising land productivity and restoring the earth's vegetation, and a plan to raise water productivity worldwide. The goals are essential and the technologies are available.

We have the wealth to achieve these goals. What we don't yet have is the leadership. And if the past is any guide to the future, that leadership can only come from the United States. By far the wealthiest society that has ever existed, the United States has the resources to lead this effort.

Yet the additional external funding needed to achieve universal primary education in the eighty-eight developing nations that require help is conservatively estimated by the World Bank at $15 billion per year. Funding for an adult literacy program based largely on volunteers is estimated at $4 billion. Providing for the most basic health care is estimated at $21 billion by the World Health Organization. The additional funding needed to provide reproductive health and family planning services to all women in developing nations is $10 billion a year.

Closing the condom gap and providing the additional nine billion condoms needed to control the spread of HIV in the developing world and Eastern Europe requires $2.2 billion—$270 million for condoms and $1.9 billion for AIDS prevention education and condom distribution. The cost per year of extending school lunch programs to the forty-four poorest nations is $6 billion per year. An additional $4 billion per year would cover the cost of assistance to preschool children and pregnant women in these nations.

In total, this comes to $62 billion. If the United States offered to cover one-third of this additional funding, the other industrial nations would almost certainly be willing to provide the remainder, and the worldwide effort to eradicate hunger, illiteracy, disease, and poverty would be under way.

The challenge isn't just to alleviate poverty, but in doing so to build an economy that is compatible with the Earth's natural systems—an eco-economy, an economy that can sustain progress. This means a fundamental restructuring of the energy economy and a substantial modification of the food economy. It also means raising the productivity of energy and shifting from fossil fuels to renewables. It means raising water productivity over the next half-century, much as we did land productivity over the last one.

It is easy to spend hundreds of billions in response to terrorist threats but the reality is that the resources needed to disrupt a modern economy are small, and a Department of Homeland Security, however heavily funded, provides only minimal protection from suicidal terrorists. The challenge isn't just to provide a high-tech military response to terrorism but to build a global society that is environmentally sustainable, socially equitable, and democratically based—one where there is hope for everyone. Such an effort would more effectively undermine the spread of terrorism than a doubling of military expenditures.

We can build an economy that doesn't destroy its natural support systems, a global community where the basic needs of all the Earth's people are satisfied, and a world that will allow us to think of ourselves as civilized. This is entirely doable. To paraphrase former President Franklin Roosevelt at another of those hinge points in history, let no one say it cannot be done.

The choice is ours—yours and mine. We can stay with business as usual and preside over a global bubble economy that keeps expanding until it bursts, leading to economic decline. Or we can adopt Plan B and be the generation that stabilizes population, eradicates poverty, and stbilizes climate. Historians will record the choice—but it is ours to make.

Lester R. Brown is president of the Earth Policy Institute. This article is adapted from his recently released book *Plan B: Rescuing a Planet Under Stress and a Civilization in Trouble*, which is available for free downloading at **www.earth-policy.org**

From *The Humanist,* Vol. 63, Issue 6, November/December 2003. Copyright 2003 by American Humanist Association. Excerpted from *Plan B: Rescuing a Planet Under Stress and a Civilization in Trouble* by Lester Brown (W.W. Norton & Co., NY, 2003). Reprinted by permission.

UNIT 2
Population, Policy, and Economy

Unit Selections

Key Points to Consider

- Why should policy makers in the more developed countries of the world become more aware of the true dimensions of the world's food problem? How can increased awareness of food scarcity and misallocation lead to solutions for both food production and environmental protection?

- What is the relationship between environmental conditions and national security? Are there non-military solutions that might be more effective in achieving global security than are military ones?

- How does "factory farming," which may increase food production, also increase problems of environmental contamination? What is the relationship between regulations developed to control factory farming and the migration of this agrobusiness system to new areas?

- What is the relationship between resource scarcity and political stability in the world's primary oil-producing region: North Africa and the Middle East? Which resource will be more important for future regional stability: oil or water?

- What is the relationship between population growth and food production? Has global population reached a level at which increasing populations are unsupportable? Or will new crops, new sources of water, and new technologies keep up with even modest population increases?

Student Website
www.mhcls.com/online

Internet References
Further information regarding these websites may be found in this book's preface or online.

The Hunger Project
http://www.thp.org
Poverty Mapping
http://www.povertymap.net
World Health Organization
http://www.who.int
World Population and Demographic Data
http://geography.about.com/cs/worldpopulation/
WWW Virtual Library: Demography & Population Studies
http://demography.anu.edu.au/VirtualLibrary/

One of the greatest setbacks on the road to the development of more stable and sensible population policies came about as a result of inaccurate population growth projections made in the late 1960s and early 1970s. The world was in for a population explosion, the experts told us back then. But shortly after the publication of the heralded works *The Population Bomb* (Paul Ehrlich, 1975) and *Limits to Growth* (D. H. Meadows et al., 1974), the growth rate of the world's population began to decline slightly. There was no cause and effect relationship at work here. The decline in growth was simply the demographic transition at work, a process in which declining population growth tends to accompany increasing levels of economic development. Since the alarming predictions did not come to pass, the world began to relax a little. However, two facts still remain: population growth in biological systems must be limited by available resources, and the availability of Earth's resources is finite.

That population growth cannot continue indefinitely is a mathematical certainty. But it is also a certainty that contemporary notions of a continually expanding economy must give way before the realities of a finite resource base. Consider the following: In developing countries, high and growing rural population densities have forced the use of increasingly marginal farmland once considered to be too steep, too dry, too wet, too sterile, or too far from markets for efficient agricultural use. Farming this land damages soil and watershed systems, creates deforestation problems, and adds relatively little to total food production. In the more developed world, farmers also have been driven—usually by market forces—to farm more marginal lands and to rely more on environmentally harmful farming methods utilizing high levels of agricultural chemicals (such as pesticides and artificial fertilizers). These chemicals create hazards for all life and rob the soil of its natural ability to renew itself. The increased demand for economic expansion has also created an increase in the use of precious groundwater reserves for irrigation purposes, depleting those reserves beyond their natural capacity to recharge and creating the potential for once-fertile farmland and grazing land to be transformed into desert. The continued demand for higher production levels also contributes to a soil erosion problem that has reached alarming proportions in all agricultural areas of the world, whether high or low on the scale of economic development. The need to increase the food supply and its consequent effects on the agricultural environment are not the only results of continued population growth. For industrialists, the larger market creates an almost irresistible temptation to accelerate production, requiring the use of more marginal resources and resulting in the destruction of more fragile ecological systems, particularly in the tropics. For consumers, the increased demand for products means increased competition for scarce resources, driving up the cost of those resources until only the wealthiest can afford what our grandfathers would have viewed as an adequate standard of living.

The articles selected for this second unit all relate, in one way or another, to the theory and reality of population growth and its relationship to public policy and economic growth. In the first selection, "Population and Consumption: What We Know, What

We Need to Know," geographer and MacArthur Fellow Robert Kates argues that the present set of environmental problems are tied to both the expanding human population in strict numerical terms and the tendency of that growing population to demand more per capita shares of the world's dwindling resources. Kates also recognizes the role of increasing levels of technology as a factor in increasing levels of environmental disturbance. Kates' conclusion is that, while economic development is a global good, increasingly developed societies must learn to curb their tastes for material things if the capacity of the environment to supply both necessary and unnecessary materials is not to be outstripped.

In the following selection—"A New Security Paradigm," Gregory D. Foster of the National Defense University, picks up on some of Kates' arguments, extending them from the economic to the political. Foster notes that while there has been a tendency to equate "national security" with military action against rogue states or terrorism, far more important in the long run may be defense measures that are not military, directed toward environmental security—an area where environmental conditions and concerns over national security converge. Such global environmental problems as climate change, Foster contends, "take us much farther along the path to ultimate causes than terrorism ever could." If nations can view pre-emptive strikes against supposed weapons of mass destruction as being in their best interests, then they should take a similar view of pre-emptive strikes against the conditions that produced an enhanced greenhouse world.

The next two articles in this section move from the theoretical to the practical and from global to regional scales in addressing issues of population, resources, and environment. In "Factory Farming in the Developing World," Danielle Nierenberg of the World Watch Institute notes that, while factory farming has allowed meat to become more common in the diet of developing countries, the farming system itself produces significant damage at both local and global scales. She suggests that part of the problem is in thinking about factory farms—or the presence of meat in the diet—as symbols of wealth. Wealth also defines class and race and the next article in the section is devoted to a discussion of "environmental justice" or the notion that race becomes an issue when dealing with environmental topics. In "Where Oil and Water Do Mix: Environmental Scarcity and Future Conflict in the Middle East and North Africa," Jason J. Morrissette and Douglas A. Borer of the University of Georgia and the Naval Postgraduate School respectively, contend that control of water resources has been a principal challenge for human societies since the emergence of agricultural civilizations. In the world's greatest arid region, stretching from the Atlantic coast of Morocco to Iran, contemporary conflicts tend to revolve around control of petroleum and natural gas resources. The reasons why "water wars" have been reduced in the arid world are rooted largely in linkages between water use and the global economy. But the economic factors that have reduced the importance of water shortages cannot extend indefinitely into the future: water is ultimately a nonrenewable resource and its use must be sustainable. The basic environmental and demographic trends of this region suggest that future conflicts will revolve not around oil but around water.

The unit's concluding two articles both treat specific environmental problems at a global scale. In "The Irony of Climate," *World Watch Institute* researcher Brian Halweil suggests the presence of irony in the fact that while natural global climate change thousands of years ago may have pushed humans into food production (agriculture) instead of hunting and gathering, the human society made possible by agriculture is now creating anthropogenic or human-induced climate change. Indeed, the nature of modern agriculture itself, driven by increasing fossil-fuel consumption and use of agricultural chemicals derived from petroleum, is a contributing factor in global warming. As local disturbances in climate alter agricultural patterns, moving food over long distances can accelerate the impact of food production upon climate—the ultimate feedback loop in terms of population and environmental carrying capacity. David Pimental and Anne Wilson, both of Cornell University, also discuss the nature of contemporary agriculture: not in the context of climate change but in that of the relationship between food production and population. They note that, while global agricultural output for centuries has been able to keep pace with growing populations, that is no longer the case. The world's population has reached the limit of arable land and fresh water and the amount of cropland per person has fallen dramatically in the last half-century. Production per person of the world's most important grain crops have been falling worldwide for the last two decades. Meeting the challenges of land and water shortages and the resulting food shortages will test both human will and human technology in the coming half-century.

All the authors of selections in this unit make it clear that the global environment is being stressed by population growth as well as environmental and economic policies that result in more environmental pressure and degradation. While it should be evident that we can no longer afford to permit the unplanned and unchecked growth of the planet's dominant species, it should also be apparent that the unchecked growth of economic systems without some kind of environmental accounting systems is just as dangerous.

Population and Consumption

What We Know, What We Need to Know

by Robert W. Kates

Thirty years ago, as Earth Day dawned, three wise men recognized three proximate causes of environmental degradation yet spent half a decade or more arguing their relative importance. In this classic environmentalist feud between Barry Commoner on one side and Paul Ehrlich and John Holdren on the other, all three recognized that growth in population, affluence, and technology were jointly responsible for environmental problems, but they strongly differed about their relative importance. Commoner asserted that technology and the economic system that produced it were primarily responsible.[1] Ehrlich and Holdren asserted the importance of all three drivers: population, affluence, and technology. But given Ehrlich's writings on population,[2] the differences were often, albeit incorrectly, described as an argument over whether population or technology was responsible for the environmental crisis.

Now, 30 years later, a general consensus among scientists posits that growth in population, affluence, and technology are jointly responsible for environmental problems. This has become enshrined in a useful, albeit overly simplified, identity known as IPAT, first published by Ehrlich and Holdren in *Environment* in 1972[3] in response to the more limited version by Commoner that had appeared earlier in *Environment* and in his famous book *The Closing Circle*.[4] In this identity, various forms of environmental or resource impacts (I) equals population (P) times affluence (A) (usually income per capita) times the impacts per unit of income as determined by technology (T) and the institutions that use it. Academic debate has now shifted from the greater or lesser importance of each of these driving forces of environmental degradation or resource depletion to debate about their interaction and the ultimate forces that drive them.

However, in the wider global realm, the debate about who or what is responsible for environmental degradation lives on. Today, many Earth Days later, international debates over such major concerns as biodiversity, climate change, or sustainable development address the population and the affluence terms of Holdrens' and Ehrlich's identity, specifically focusing on the character of consumption that affluence permits. The concern with technology is more complicated because it is now widely recognized that while technology can be a problem, it can be a

solution as well. The development and use of more environmentally benign and friendly technologies in industrialized countries have slowed the growth of many of the most pernicious forms of pollution that originally drew Commoner's attention and still dominate Earth Day concerns.

A recent report from the National Research Council captures one view of the current public debate, and it begins as follows:

For over two decades, the same frustrating exchange has been repeated countless times in international policy circles. A government official or scientist from a wealthy country would make the following argument: The world is threatened with environmental disaster because of the depletion of natural resources (or climate change or the loss of biodiversity), and it cannot continue for long to support its rapidly growing population. To preserve the environment for future generations, we need to move quickly to control global population growth, and we must concentrate the effort on the world's poorer countries, where the vast majority of population growth is occurring.

Government officials and scientists from low-income countries would typically respond:

If the world is facing environmental disaster, it is not the fault of the poor, who use few resources. The fault must lie with the world's wealthy countries, where people consume the great bulk of the world's natural resources and energy and cause the great bulk of its environmental degradation. We need to curtail overconsumption in the rich countries which use far more than their fair share, both to preserve the environment and to allow the poorest people on earth to achieve an acceptable standard of living.[5]

It would be helpful, as in all such classic disputes, to begin by laying out what is known about the relative responsibilities of both population and consumption for the environmental crisis, and what might need to be known to address them. However, there is a profound asymmetry that must fuel the frustration of the developing countries' politicians and scientists: namely, how much people know about population and how little they know about consumption. Thus, this article begins by examining these differences in knowledge and action and concludes with the alternative actions needed to go from more to enough in both population and consumption.[6]

Population

What population is and how it grows is well understood even if all the forces driving it are not. Population begins with people and their key events of birth, death, and location. At the margins, there is some debate over when life begins and ends or whether residence is temporary or permanent, but little debate in between. Thus, change in the world's population or any place is the simple arithmetic of adding births, subtracting deaths, adding immigrants, and subtracting outmigrants. While whole subfields of demography are devoted to the arcane details of these additions and subtractions, the error in estimates of population for almost all places is probably within 20 percent and for countries with modern statistical services, under 3 percent—better estimates than for any other living things and for most other environmental concerns.

Current world population is more than six billion people, growing at a rate of 1.3 percent per year. The peak annual growth rate in all history—about 2.1 percent—occurred in the early 1960s, and the peak population increase of around 87 million per year occurred in the late 1980s. About 80 percent or 4.8 billion people live in the less developed areas of the world, with 1.2 billion living in industrialized countries. Population is now projected by the United Nations (UN) to be 8.9 billion in 2050, according to its medium fertility assumption, the one usually considered most likely, or as high as 10.6 billion or as low as 7.3 billion.[7]

A general description of how birth rates and death rates are changing over time is a process called the demographic transition.[8] It was first studied in the context of Europe, where in the space of two centuries, societies went from a condition of high births and high deaths to the current situation of low births and low deaths. In such a transition, deaths decline more rapidly than births, and in that gap, population grows rapidly but eventually stabilizes as the birth decline matches or even exceeds the death decline. Although the general description of the transition is widely accepted, much is debated about its cause and details.

The world is now in the midst of a global transition that, unlike the European transition, is much more rapid. Both births and deaths have dropped faster than experts expected and history foreshadowed. It took 100 years for deaths to drop in Europe compared to the drop in 30 years in the developing world. Three is the current global average births per woman of reproductive age. This number is more than halfway between the average of five children born to each woman at the post World War II peak of population growth and the average of 2.1 births required to achieve eventual zero population growth.[9] The death transition is more advanced, with life expectancy currently at 64 years. This represents three-quarters of the transition between a life expectancy of 40 years to one of 75 years. The current rates of decline in births outpace the estimates of the demographers, the UN having reduced its latest medium ex-

pectation of global population in 2050 to 8.9 billion, a reduction of almost 10 percent from its projection in 1994.

Demographers debate the causes of this rapid birth decline. But even with such differences, it is possible to break down the projected growth of the next century and to identify policies that would reduce projected populations even further. John Bongaarts of the Population Council has decomposed the projected developing country growth into three parts and, with his colleague Judith Bruce, has envisioned policies that would encourage further and more rapid decline.[10] The first part is unwanted fertility, making available the methods and materials for contraception to the 120 million married women (and the many more unmarried women) in developing countries who in survey research say they either want fewer children or want to space them better. A basic strategy for doing so links voluntary family planning with other reproductive and child health services.

Yet in many parts of the world, the desired number of children is too high for a stabilized population. Bongaarts would reduce this desire for large families by changing the costs and benefits of childrearing so that more parents would recognize the value of smaller families while simultaneously increasing their investment in children. A basic strategy for doing so accelerates three trends that have been shown to lead to lower desired family size: the survival of children, their education, and improvement in the economic, social, and legal status for girls and women.

However, even if fertility could immediately be brought down to the replacement level of two surviving children per woman, population growth would continue for many years in most developing countries because so many more young people of reproductive age exist. So Bongaarts would slow this momentum of population growth by increasing the age of childbearing, primarily by improving secondary education opportunity for girls and by addressing such neglected issues as adolescent sexuality and reproductive behavior.

How much further could population be reduced? Bongaarts provides the outer limits. The population of the developing world (using older projections) was expected to reach 10.2 billion by 2100. In theory, Bongaarts found that meeting the unmet need for contraception could reduce this total by about 2 billion. Bringing down desired family size to replacement fertility would reduce the population a billion more, with the remaining growth—from 4.5 billion today to 7.3 billion in 2100—due to population momentum. In practice, however, a recent U.S. National Academy of Sciences report concluded that a 10 percent reduction is both realistic and attainable and could lead to a lessening in projected population numbers by 2050 of upwards of a billion fewer people.[11]

Consumption

In contrast to population, where people and their births and deaths are relatively well-defined biological events, there is no consensus as to what consumption includes. Paul Stern of the National Research Council has described the different ways physics, economics, ecology, and sociology view consumption.[12] For physicists, matter and energy cannot be consumed, so consumption is conceived as transformations of matter and energy with increased entropy. For economists, consumption is spending on consumer goods and services and thus distinguished from their production and distribution. For ecologists, consumption is obtaining energy and nutrients by eating something else, mostly green plants or other consumers of green plants. And for some sociologists, consumption is a status symbol—keeping up with the Joneses—when individuals and households use their incomes to increase their social status through certain kinds of purchases. These differences are summarized in the box below.

In 1977, the councils of the Royal Society of London and the U.S. National Academy of Sciences issued a joint statement on consumption, having previously done so on population. They chose a variant of the physicist's definition:

> *Consumption is the human transformation of materials and energy. Consumption is of concern to the extent that it makes the transformed materials or energy less available for future use, or negatively impacts biophysical systems in such a way as to threaten human health, welfare, or other things people value.*[13]

On the one hand, this society/academy view is more holistic and fundamental than the other definitions; on the other hand, it is more focused, turning attention to the environmentally damaging. This article uses it as a working definition with one modification, the addition of information to energy and matter, thus completing the triad of the biophysical and ecological basics that support life.

In contrast to population, only limited data and concepts on the transformation of energy, materials, and information exist.[14] There is relatively good global knowledge of energy transformations due in part to the common units of conversion between different technologies. Between 1950 and today, global energy production and use increased more than fourfold.[15] For material transformations, there are no aggregate data in common units on a global basis, only for some specific classes of materials including materials for energy production, construction, industrial minerals and metals, agricultural crops, and water.[16] Calculations of material use by volume, mass, or value lead to different trends.

Trend data for per capita use of physical structure materials (construction and industrial minerals, metals, and forestry products) in the United States are relatively complete. They show an inverted S shaped (logistic) growth pattern: modest doubling between 1900 and the depression of the 1930s (from two to four metric tons), followed by a steep quintupling with economic recovery until the early 1970s (from two to eleven tons), followed by a leveling off since then with fluctuations related to economic downturns (see Figure 1).[17] An aggregate analysis of all current material production and consumption in the United States averages more than 60 kilos per person per day (excluding water). Most of this material flow is split between energy and related products (38 percent) and minerals for construction (37 percent), with the remainder as industrial minerals (5 percent), metals (2 percent), products of fields (12 percent), and forest (5 percent).[18]

A massive effort is under way to catalog biological (genetic) information and to sequence the genomes of microbes, worms, plants, mice, and people. In contrast to the molecular detail, the number and diversity of organisms is unknown, but a conservative estimate places the number of species on the order of 10 million, of which only one-tenth have been described.[19] Although there is much interest and many anecdotes, neither concepts nor data are available on most cultural information. For example, the number of languages in the world continues to decline while the number of messages expands exponentially.

What Is Consumption?

Physicist: "What happens when you transform matter/energy"

Ecologist: "What big fish do to little fish"

Economist: "What consumers do with their money"

Sociologist: "What you do to keep up with the Joneses"

Trends and projections in agriculture, energy, and economy can serve as surrogates for more detailed data on energy and material transformation.[20] From 1950 to the early 1990s, world population more than doubled (2.2 times), food as measured by grain production almost tripled (2.7 times), energy more than quadrupled (4.4 times), and the economy quintupled (5.1 times). This 43-year record is similar to a current 55-year projection (1995–2050) that assumes the continuation of current trends or, as some would note, "business as usual." In this 55-year projection, growth in half again of population (1.6 times) finds almost a doubling of agriculture (1.8 times), more than twice as much energy used (2.4 times), and a quadrupling of the economy (4.3 times).[21]

Thus, both history and future scenarios predict growth rates of consumption well beyond population. An attractive similarity exists between a demographic transition that moves over time from high births and high deaths to low births and low deaths with an energy, materials, and information transition. In this transition, societies will use increasing amounts of energy and materials as consumption increases, but over time the energy and materials input per unit of consumption decrease and information substitutes for more material and energy inputs.

Some encouraging signs surface for such a transition in both energy and materials, and these have been variously labeled as decarbonization and dematerialization.[22] For more than a century, the amount of carbon per unit of energy produced has been decreasing. Over a shorter period, the amount of energy used to produce a unit of production has also steadily declined. There is also evidence for dematerialization, using fewer materials for a unit of production, but only for industrialized countries and for some specific materials. Overall, improvements in technology

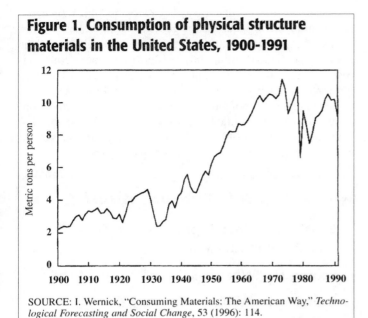

Figure 1. Consumption of physical structure materials in the United States, 1900-1991

SOURCE: I. Wernick, "Consuming Materials: The American Way," *Technological Forecasting and Social Change*, 53 (1996): 114.

and substitution of information for energy and materials will continue to increase energy efficiency (including decarbonization) and dematerialization per unit of product or service. Thus, over time, less energy and materials will be needed to make specific things. At the same time, the demand for products and services continues to increase, and the overall consumption of energy and most materials more than offsets these efficiency and productivity gains.

What to Do about Consumption

While quantitative analysis of consumption is just beginning, three questions suggest a direction for reducing environmentally damaging and resource-depleting consumption. The first asks: *When is more too much for the life-support systems of the natural world and the social infrastructure of human society?* Not all the projected growth in consumption may be resource-depleting— "less available for future use"—or environmentally damaging in a way that "negatively impacts biophysical systems to threaten human health, welfare, or other things people value."[23] Yet almost any human-induced transformations turn out to be either or both resource-depleting or damaging to some valued environmental component. For example, a few years ago, a series of eight energy controversies in Maine were related to coal, nuclear, natural gas, hydroelectric, biomass, and wind generating sources, as well as to various energy policies. In all the controversies, competing sides, often more than two, emphasized environmental benefits to support their choice and attributed environmental damage to the other alternatives.

Despite this complexity, it is possible to rank energy sources by the varied and multiple risks they pose and, for those concerned, to choose which risks they wish to minimize and which they are more willing to accept. There is now almost 30 years of experience with the theory and methods of risk assessment and 10 years of experience with the identification and setting of en-

vironmental priorities. While there is still no readily accepted methodology for separating resource-depleting or environmentally damaging consumption from general consumption or for identifying harmful transformations from those that are benign, one can separate consumption into more or less damaging and depleting classes and *shift* consumption to the less harmful class. It is possible to *substitute* less damaging and depleting energy and materials for more damaging ones. There is growing experience with encouraging substitution and its difficulties: renewables for nonrenewables, toxics with fewer toxics, ozone-depleting chemicals for more benign substitutes, natural gas for coal, and so forth.

The second question, *Can we do more with less?*, addresses the supply side of consumption. Beyond substitution, shrinking the energy and material transformations required per unit of consumption is probably the most effective current means for reducing environmentally damaging consumption. In the 1997 book, *Stuff: The Secret Lives of Everyday Things*, John Ryan and Alan Durning of Northwest Environment Watch trace the complex origins, materials, production, and transport of such everyday things as coffee, newspapers, cars, and computers and highlight the complexity of reengineering such products and reorganizing their production and distribution.[24]

Yet there is growing experience with the three Rs of consumption shrinkage: reduce, recycle, reuse. These have now been strengthened by a growing science, technology, and practice of industrial ecology that seeks to learn from nature's ecology to reuse everything. These efforts will only increase the existing favorable trends in the efficiency of energy and material usage. Such a potential led the Intergovernmental Panel on Climate Change to conclude that it was possible, using current best practice technology, to reduce energy use by 30 percent in the short run and 50–60 percent in the long run.[25] Perhaps most important in the long run, but possibly least studied, is the potential for and value of substituting information for energy and materials. Energy and materials per unit of consumption are going down, in part because more and more consumption consists of information.

The third question addresses the demand side of consumption—*When is more enough?*[26] Is it possible to reduce consumption by more satisfaction with what people already have, by *satiation*, no more needing more because there is enough, and by *sublimation*, having more satisfaction with less to achieve some greater good? This is the least explored area of consumption and the most difficult. There are, of course, many signs of *satiation* for some goods. For example, people in the industrialized world no longer buy additional refrigerators (except in newly formed households) but only replace them. Moreover, the quality of refrigerators has so improved that a 20-year or more life span is commonplace. The financial pages include frequent stories of the plight of this industry or corporation whose markets are saturated and whose products no longer show the annual growth equated with profits and progress. Such enterprises are frequently viewed as failures of marketing or entrepreneurship rather than successes in meeting human needs sufficiently and efficiently. Is it possible to reverse such views, to create a standard of satiation, a satisfaction in a need well met?

Can people have more satisfaction with what they already have by using it more intensely and having the time to do so? Economist Juliet Schor tells of some overworked Americans who would willingly exchange time for money, time to spend with family and using what they already have, but who are constrained by an uncooperative employment structure.[27] Proposed U.S. legislation would permit the trading of overtime for such compensatory time off, a step in this direction. *Sublimation*, according to the dictionary, is the diversion of energy from an immediate goal to a higher social, moral, or aesthetic purpose. Can people be more satisfied with less satisfaction derived from the diversion of immediate consumption for the satisfaction of a smaller ecological footprint?[28] An emergent research field grapples with how to encourage consumer behavior that will lead to change in environmentally damaging consumption.[29]

A small but growing "simplicity" movement tries to fashion new images of "living the good life."[30] Such movements may never much reduce the burdens of consumption, but they facilitate by example and experiment other less-demanding alternatives. Peter Menzel's remarkable photo essay of the material goods of some 30 households from around the world is powerful testimony to the great variety and inequality of possessions amidst the existence of alternative life styles.[31] Can a standard of "more is enough" be linked to an ethic of "enough for all"? One of the great discoveries of childhood is that eating lunch does not feed the starving children of some far-off place. But increasingly, in sharing the global commons, people flirt with mechanisms that hint at such—a rationing system for the remaining chlorofluorocarbons, trading systems for reducing emissions, rewards for preserving species, or allowances for using available resources.

A recent compilation of essays, *Consuming Desires: Consumption, Culture, and the Pursuit of Happiness*,[32] explores many of these essential issues. These elegant essays by 14 well-known writers and academics ask the fundamental question of why more never seems to be enough and why satiation and sublimation are so difficult in a culture of consumption. Indeed, how is the culture of consumption different for mainstream America, women, inner-city children, South Asian immigrants, or newly industrializing countries?

Why We Know and Don't Know

In an imagined dialog between rich and poor countries, with each side listening carefully to the other, they might ask themselves just what they actually know about population and consumption. Struck with the asymmetry described above, they might then ask: "Why do we know so much more about population than consumption?"

The answer would be that population is simpler, easier to study, and a consensus exists about terms, trends, even policies. Consumption is harder, with no consensus as to what it is, and with few studies except in the fields of marketing and advertising. But the consensus that exists about population comes from substantial research and study, much of it funded by governments and groups in rich countries, whose asymmetric concern readily identifies the troubling fertility behavior of others and only reluctantly considers their own consumption behavior. So while consumption is harder, it is surely studied less (see Table 1).

The asymmetry of concern is not very flattering to people in developing countries. Anglo-Saxon tradition has a long history of dominant thought holding the poor responsible for their con-

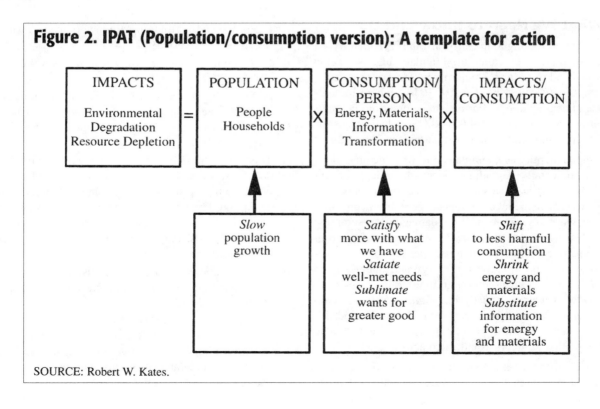

Figure 2. IPAT (Population/consumption version): A template for action

IMPACTS		POPULATION		CONSUMPTION/ PERSON		IMPACTS/ CONSUMPTION
Environmental Degradation Resource Depletion	=	People Households	X	Energy, Materials, Information Transformation	X	

Slow population growth

Satisfy more with what we have *Satiate* well-met needs *Sublimate* wants for greater good

Shift to less harmful consumption *Shrink* energy and materials *Substitute* information for energy and materials

SOURCE: Robert W. Kates.

dition—they have too many children—and an even longer tradition of urban civilization feeling besieged by the barbarians at their gates. But whatever the origins of the asymmetry, its persistence does no one a service. Indeed, the stylized debate of population versus consumption reflects neither popular understanding nor scientific insight. Yet lurking somewhere beneath the surface concerns lies a deeper fear.

Table 1. A comparison of population and consumption

Population	Consumption
Simpler, easier to study	More complex
Well-funded research	Unfunded, except marketing
Consensus terms, trends	Uncertain terms, trends
Consensus policies	Threatening policies

SOURCE: Robert W. Kates.

Consumption is more threatening, and despite the North–South rhetoric, it is threatening to all. In both rich and poor countries alike, making and selling things to each other, including unnecessary things, is the essence of the economic system. No longer challenged by socialism, global capitalism seems inherently based on growth—growth of both consumers and their consumption. To study consumption in this light is to risk concluding that a transition to sustainability might require profound changes in the making and selling of things and in the opportunities that this provides. To draw such conclusions, in the absence of convincing alternative visions, is fearful and to be avoided.

What We Need to Know and Do

In conclusion, returning to the 30-year-old IPAT identity—a variant of which might be called the Population/Consumption (PC) version—and restating that identity in terms of population and consumption, it would be: $I = P*C/P*I/C$, where I equals environmental degradation and/or resource depletion; P equals the number of people or households; and C equals the transformation of energy, materials, and information (see Figure 2).

With such an identity as a template, and with the goal of reducing environmentally degrading and resource-depleting influences, there are at least seven major directions for research and policy. To reduce the level of impacts per unit of consumption, it is necessary to separate out more damaging consumption and shift to less harmful forms, *shrink* the amounts of environmentally damaging energy and materials per unit of consumption, and *substitute* information for energy and materials. To reduce consumption per person or household, it is necessary to *satisfy* more with what is already had, *satiate* well-met consumption needs, and *sublimate* wants for a greater good. Finally, it is possible to *slow* population growth and then to *stabilize* population numbers as indicated above.

However, as with all versions of the IPAT identity, population and consumption in the PC version are only proximate

driving forces, and the ultimate forces that drive consumption, the consuming desires, are poorly understood, as are many of the major interventions needed to reduce these proximate driving forces. People know most about slowing population growth, more about shrinking and substituting environmentally damaging consumption, much about shifting to less damaging consumption, and least about satisfaction, satiation, and sublimation. Thus the determinants of consumption and its alternative patterns have been identified as a key understudied topic for an emerging sustainability science by the recent U.S. National Academy of Science study.[33]

But people and society do not need to know more in order to act. They can readily begin to separate out the most serious problems of consumption, shrink its energy and material throughputs, substitute information for energy and materials, create a standard for satiation, sublimate the possession of things for that of the global commons, as well as slow and stabilize population. To go from more to enough is more than enough to do for 30 more Earth Days.

Robert W. Kates is an independent scholar in Trenton, Maine; a geographer; university professor emeritus at Brown University; and an executive editor of *Environment*. The research for "Population and Consumption: What We Know, What We Need to Know" was undertaken as a contribution to the recent National Academies/National Research Council report, *Our Common Journey: A Transition Toward Sustainability*. The author retains the copyright to this article. Kates can be reached at RR1, Box 169B, Trenton, ME 04605.

NOTES

1. B. Commoner, M. Corr, and P. Stamler, "The Causes of Pollution," *Environment*, April 1971, 2–19.
2. P. Ehrlich, *The Population Bomb* (New York: Ballantine, 1966).
3. P. Ehrlich and J. Holdren, "Review of The Closing Circle," *Environment*, April 1972, 24–39.
4. B. Commoner, *The Closing Circle* (New York: Knopf, 1971).
5. P. Stern, T. Dietz, V. Ruttan, R. H. Socolow, and J. L. Sweeney, eds., *Environmentally Significant Consumption: Research Direction* (Washington, D.C.: National Academy Press, 1997), 1.
6. This article draws in part upon a presentation for the 1997 De Lange-Woodlands Conference, an expanded version of which will appear as: R. W. Kates, "Population and Consumption: From More to Enough," in *In Sustainable Development: The Challenge of Transition*, J. Schmandt and C. H. Wards, eds. (Cambridge, U.K.: Cambridge University Press, forthcoming), 79–99.
7. United Nations, Population Division, *World Population Prospects: The 1998 Revision* (New York: United Nations, 1999).
8. K. Davis, "Population and Resources: Fact and Interpretation," K. Davis and M. S. Bernstam, eds., in *Resources, Environment and Population: Present Knowledge, Future Options*, supplement to *Population and Development Review*, 1990: 1–21.

9. Population Reference Bureau, *1997 World Population Data Sheet of the Population Reference Bureau* (Washington, D.C.: Population Reference Bureau, 1997).

10. J. Bongaarts, "Population Policy Options in the Developing World," *Science*, 263: (1994), 771–776; and J. Bongaarts and J. Bruce, "What Can Be Done to Address Population Growth?" (unpublished background paper for The Rockefeller Foundation, 1997).

11. National Research Council, Board on Sustainable Development, *Our Common Journey: A Transition Toward Sustainability* (Washington, D.C.: National Academy Press, 1999).

12. See Stern, et al., note 5 above.

13. Royal Society of London and the U.S. National Academy of Sciences, "Towards Sustainable Consumption," reprinted in *Population and Development Review*, 1977, 23 (3): 683–686.

14. For the available data and concepts, I have drawn heavily from J. H. Ausubel and H. D. Langford, eds., *Technological Trajectories and the Human Environment.* (Washington, D.C.: National Academy Press, 1997).

15. L. R. Brown, H. Kane, and D. M. Roodman, *Vital Signs 1994: The Trends That Are Shaping Our Future* (New York: W. W. Norton and Co., 1994).

16. World Resources Institute, United Nations Environment Programme, United Nations Development Programme, World Bank, *World Resources*, 1996–97 (New York: Oxford University Press, 1996); and A. Gruebler, *Technology and Global Change* (Cambridge, Mass.: Cambridge University Press, 1998).

17. I. Wernick, "Consuming Materials: The American Way," *Technological Forecasting and Social Change*, 53 (1996): 111–122.

18. I. Wernick and J. H. Ausubel, "National Materials Flow and the Environment," *Annual Review of Energy and Environment*, 20 (1995): 463–492.

19. S. Pimm, G. Russell, J. Gittelman, and T. Brooks, "The Future of Biodiversity," *Science*, 269 (1995): 347–350.

20. Historic data from L. R. Brown, H. Kane, and D. M. Roodman, note 15 above.

21. One of several projections from P. Raskin, G. Gallopin, P. Gutman, A. Hammond, and R. Swart, *Bending the Curve: Toward Global Sustainability*, a report of the Global Scenario Group, Polestar Series, report no. 8 (Boston: Stockholm Environmental Institute, 1995).

22. N. Nakicénovíc, "Freeing Energy from Carbon," in *Technological Trajectories and the Human Environment*, eds., J. H. Ausubel and H. D. Langford. (Washington, D.C.: National Academy Press, 1997); I. Wernick, R. Herman, S. Govind, and J. H. Ausubel, "Materialization and Dematerialization: Measures and Trends," in J. H. Ausubel and H. D.

Langford, eds., *Technological Trajectories and the Human Environment* (Washington, D.C.: National Academy Press, 1997), 135–156; and see A. Gruebler, note 16 above.

23. Royal Society of London and the U.S. National Academy of Science, note 13 above.

24. J. Ryan and A. Durning, *Stuff: The Secret Lives of Everyday Things* (Seattle, Wash.: Northwest Environment Watch, 1997).

25. R. T. Watson, M. C. Zinyowera, and R. H. Moss, eds., *Climate Change 1995: Impacts, Adaptations, and Mitigation of Climate Change—Scientific-Technical Analyses* (Cambridge, U.K.: Cambridge University Press, 1996).

26. A sampling of similar queries includes: A. Durning, *How Much Is Enough?* (New York: W. W. Norton and Co., 1992); Center for a New American Dream, *Enough!: A Quarterly Report on Consumption, Quality of Life and the Environment* (Burlington, Vt.: The Center for a New American Dream, 1997); and N. Myers, "Consumption in Relation to Population, Environment, and Development," *The Environmentalist*, 17 (1997): 33–44.

27. J. Schor, *The Overworked American* (New York: Basic Books, 1991).

28. A. Durning, *How Much Is Enough?: The Consumer Society and the Future of the Earth* (New York: W. W. Norton and Co., 1992); Center for a New American Dream, note 26 above; and M. Wackernagel and W. Ress, *Our Ecological Footprint: Reducing Human Impact on the Earth* (Philadelphia. Pa.: New Society Publishers, 1996).

29. W. Jager, M. van Asselt, J. Rotmans, C. Vlek, and P. Costerman Boodt, *Consumer Behavior: A Modeling Perspective in the Contest of Integrated Assessment of Global Change*, RIVM report no. 461502017 (Bilthoven, the Netherlands: National Institute for Public Health and the Environment, 1997); and P. Vellinga, S. de Bryn, R. Heintz, and P. Molder, eds., *Industrial Transformation: An Inventory of Research*. IHDP-IT no. 8 (Amsterdam, the Netherlands: Institute for Environmental Studies, 1997).

30. H. Nearing and S. Nearing. *The Good Life: Helen and Scott Nearing's Sixty Years of Self-Sufficient Living* (New York: Schocken, 1990); and D. Elgin, *Voluntary Simplicity: Toward a Way of Life That Is Outwardly Simple, Inwardly Rich* (New York: William Morrow, 1993).

31. P. Menzel, *Material World: A Global Family Portrait* (San Francisco: Sierra Club Books, 1994).

32. R. Rosenblatt, ed., *Consuming Desires: Consumption, Culture, and the Pursuit of Happiness* (Washington, D.C.: Island Press, 1999).

33. National Research Council, Board on Sustainable Development, *Our Common Journey: A Transition Toward Sustainability* (Washington, D.C.: National Academy Press, 1999).

A New Security Paradigm

It's easy to equate "national security" or "global security" with military defense against rogue states and terrorism, but a leading U.S. military expert says that view is far too narrow—and could lead to catastrophe if not changed.

Gregory D. Foster

Whatever else the year 2004 might be noted for by future historians—the U.S. political wars, the genocide in Darfur, the strategic debacle in Iraq—it may well turn out to have been a seminal year for the field of environmental security—the intellectual, operational, and policy space where environmental conditions and security concerns converge.

So too, one hopes, might it have been the year when U.S. policymakers and the American public began to awaken, however belatedly, to the need for an entirely new approach to security and for the fundamental strategic transformation necessary to achieve such security.

The highlight of the year in this regard was, for two essentially countervailing reasons, the award of the Nobel Peace Prize to Kenyan environmental activist Wangari Maathai. On one hand, by broadening the definitional bounds of peace, the award gave new legitimacy to those who would embrace unconventional conceptions of security, especially involving the environment. "This is the first time environment sets the agenda for the Nobel Peace Prize, and we have added a new dimension to peace," said committee chairman Ole Danbolt Mjoes in announcing the award. "Peace on earth depends on our ability to secure our living environment."

On the other hand, no less noteworthy were critics of the award, whose expressions of disparagement typified and reaffirmed the stultifying hold of traditionalist thinking on the conduct of international relations. Espen Barth Eide, former Norwegian deputy foreign minister, observed: "The one thing the Nobel Committee does is define the topic of this epoch in the field of peace and security. If they widen it too much, they risk undermining the core function of the Peace Prize; you end up saying everything that is good is peace." Traditionalists everywhere, including most of the U.S. policy establishment, no doubt took succor from such self-righteous indignation and re-

solved to perpetuate the received truths of the past that have made real peace so elusive and illusory to date.

Beyond the Peace Prize, two other events ten months apart served as defining bookends for what could turn out to have been the undeclared Year of Environmental Security. The first was an attention-grabbing article, "The Pentagon's Weather Nightmare," that appeared in the February 9, 2004 issue of *Fortune* magazine. Describing a report two futurists—Peter Schwartz and Doug Randall of Global Business Network—had recently prepared for the Defense Department on the national security implications of abrupt climate change, the article generated a flurry of intense but short-lived excitement and speculation on whether, why, and to what extent the Pentagon was finally taking climate change seriously.

The second bookend event came at the end of the year with the issuance of the final report of the internationally distinguished, 16-member High-Level Panel on Threats, Challenges and Change that UN Secretary-General Kofi Annan had appointed in November 2003 to examine the major threats and challenges the world faces in the broad field of peace and security.

These two particular events, potentially significant enough in their own right, should be viewed in the larger context of several other magnifying events that occurred over the course of the year.

For starters, Sir David King, chief science adviser to British prime minister Tony Blair, raised eyebrows and hackles with a controversial article in the January 9, 2004 issue of *Science* magazine. King cited climate change as "the most severe problem that we are facing today—more serious even than the threat of terrorism," and accused the U.S. government of "failing to take up the challenge of global warming." In a subsequent speech to the American Association for the Advancement of Science, he added: "Climate change is real. Millions will increasingly be exposed to hunger, drought, flooding and debilitating diseases such as malaria. Inaction due to questions over the science [a thinly veiled reference to Bush administration foot-dragging] is no longer defensible."

In March, former UN chief weapons inspector Hans Blix added further fuel to the fire in a BBC television interview with David Frost: "I think we still overestimate the danger of terror. There are other things that are of equal, if not greater, magnitude, like the environmental global risks." This statement reinforced an equally pointed one Blix had made a year earlier: "To me the question of the environment is more ominous than that of peace and war.... I'm more worried about global warming than I am of any major military conflict."

In May, the blockbuster 20th Century Fox disaster movie *The Day After Tomorrow,* portraying the cataclysmic global consequences of accelerated climate change, was released to theaters nationwide (with European release scheduled for October). Some, such as Sir David King and former vice president Al Gore, promoted or endorsed the movie, clearly not because of its admittedly unrealistic compression of time and exaggeration of catastrophic effects, but because of its potential for awakening and sensitizing the public to the plausibility and seriousness of abrupt climate change. Others fiercely criticized the movie for trivializing such a vital issue. Anti-doomsayer Gregg Easterbrook, senior editor of *The New Republic,* assailed the "cheapo, third-rate disaster movie" for its "imbecile-caliber" science: "By presenting global warming in a laughably unrealistic way, the movie will only succeed in making audiences think that climate change is a big joke, when in fact the real science case for greenhouse-gas reform gets stronger all the time."

In a major September address in London, Tony Blair, faced with continuing criticism from his opposition, called climate change "the world's greatest environmental challenge ... a challenge so far-reaching in its impact and irreversible in its destructive power, that it alters radically human existence." "Apart from a diminishing handful of skeptics," he said, "there is a virtual worldwide scientific consensus on the scope of the problem."

Then in October, the United Nations Environment Programme's Division of Early Warning and Assessment issued a thought-provoking new report, *Understanding Environment, Conflict, and Cooperation,* that resulted from the deliberations of participants in a new initiative to leverage environmental activities, policies, and actions for promoting international conflict prevention, peace, and cooperation. The subject matter of the report is not new, but the question it implies is: whether new life can be breathed into what was, throughout most of the 1990s, a lively debate over whether and how the environment and security are related and interact. Since the Kyoto negotiations of 1997, that debate has been largely moribund, to the detriment of both U.S. policy and strategic discourse more generally.

Revivifying Environmental Security

Even if the events recounted above had not occurred this past year, the findings and recommendations of the UN High-Level

Threat Panel and the introduction into the public imagination of abrupt climate change as a matter of prospective national security concern would stand as forceful stimuli for policy practitioners, scholars, and the general public to accord environmental security more serious and immediate attention.

This article goes to press before the actual release of the High-Level Panel's final report; but publicly available preliminary work by the United Nations Foundation's United Nations & Global Security Initiative, in cooperation with the. Environmental Change & Security Project of the Woodrow Wilson Center, prefigures how the Panel's thinking is likely to be guided on environmental matters. This introductory passage from a discussion summary presented to the Panel is indicative of that thrust:

> *Environmental changes can threaten global, national, and human security. Environmental issues include land degradation, climate change, water quality and quantity, and the management and distribution of natural-resource assets (such as oil, forests, and minerals). These factors can contribute directly to conflict, or can be linked to conflict, by exacerbating other causes such as poverty, migration, small arms, and infectious diseases. For example, experts predict that climate change will trigger enormous physical and social changes like water shortages, natural disasters, decreased agricultural productivity, increased rates and scope of infectious diseases, and shifts in human migration; these changes could significantly impact international security by leading to competition for natural resources, destabilizing weak states, and increasing humanitarian crises. However, managing environmental issues and natural resources can also build confidence and contribute to peace through cooperation across lines of tension.*

Add to this Secretary-General Annan's own words in announcing the High-Level Panel to the UN General Assembly in September 2003, and it seems clear that the Panel will endorse the environment-security linkage and acknowledge that environmental degradation, resource scarcity, and climate change are threats or challenges that face the world and demand collective response:

> *All of us know there are new threats that must be faced—or, perhaps, old threats in new and dangerous combinations: new forms of terrorism, and the proliferation of weapons of mass destruction. But, while some consider these threats as self-evidently the main challenge to world peace and security, others feel more immediately menaced by small arms employed in civil conflict, or by so-called "soft threats" such as the persistence of extreme poverty, the disparity of income between and within societies, and the spread of infectious diseases, or climate change and environmental degradation.*

The February 2004 *Fortune* article was a dispassionate but revealing summary of a Pentagon-commissioned study that, though unclassified, ordinarily wouldn't have received much—if any—public exposure. Substantively, the article did two things. First, judging from the volume and intensity of follow-on commentary it generated, it clearly raised expectations—positive and negative—about the content and ramifications of the Pentagon report. Was the military actually interested in climate change? Why? Enough to do something about it? To what end and with what effect (especially on the military's principal mission)?

Second, the article—and the report it reported—upped the ante in the continuing debate over climate change. In addressing *abrupt* climate change, it accentuated an emerging thesis that gives urgency to what otherwise is considered (by some, perhaps many) to be a long-term, gradual phenomenon that, if real, can be passed off, without present political or economic regret, for future generations to deal with. And in tying abrupt climate change to national security, the article and report give added—ultimate—importance to the subject. National security is, after all, the public-policy holy of holies—the iconic totem that takes precedence over all else. National security is about endangerment and survival, the thinking goes. So if something can be shown to have national security implications (however defined), then perhaps it too is about such things; perhaps it too, therefore, warrants serious attention and the commitment of resources.

For people familiar with the U.S. military's normal modes of communication, the release of the Schwartz-Randall report to *Fortune* was unusual enough to cause speculation about whether the man who commissioned it, Andrew W. Marshall—the Pentagon's director of net assessment for the past 30 years—may have been signaling concerns that went well beyond the report's scientific message: first, that the institution he works for is intractably parochial and resistant to change; second, that the Pentagon is particularly inbred and close-minded about matters as esoteric and ideologically encumbered as the environment; third, that since imaginative futurists had prepared the report, it could more easily be dismissed as speculative fantasy by bureaucratic pragmatists who prefer to think they are grounded in reality; fourth, that however long he (Marshall) may have served in the Pentagon, he has little clout in influencing the military to actually take action based on his office's analytical products; fifth, that going public therefore offers more hope for forcing internal Pentagon awakening (if not change) in response to external pressure from arguably less parochial outside parties such as Congress and the media; and sixth, that perhaps the most potent force for movement on this particular front is the business community, which has the most to both gain and lose from climate change—especially when the political regime in power opts for dogmatic inaction in deference to the cosmic invisible hand of the marketplace.

The importance of this episode, as well as its relevance for the future, lies in both the message and the method of the Schwartz-Randall report itself. The implicit message is that even worse than climate change is the not unrealistic possibility of *abrupt* climate change. For those who had not heard of it, the article made clear that abrupt climate change is not just global warming speeded up, but a wholly different kind of event triggered by the baseline climate change we already know. In brief,

the global warming now taking place could conceivably lead to a halting of the ocean currents that now keep Europe temperate—global *warming* thus ironically leading to regionally much colder conditions and widespread accelerations of the catastrophic effects already commonly associated with "normal" climate change: floods, droughts, windstorms, wave events, wildfires, disease epidemics, species loss, famine, and more. The explicit message is that the concatenation of such effects could then lead to additional, national security consequences—most notably military confrontations between states over access to scarce food, water, and energy supplies, or what the authors describe as a "world of warring states."

Paradoxically, portraying what is relevant to national security as essentially that which invites or involves military force is perhaps necessary to grab the attention of purported experts on the subject, but it thereby also betrays the shallowness and narrowness of the canonical security paradigm most of us have unthinkingly bought into. This state of affairs is reinforced by the methodology of the Schwartz-Randall report, which seeks not to predict whether, when, or how abrupt climate change and its attendant effects would occur, but merely to present a plausible scenario of what might happen if and when it does. In the authors' words, "The duration of this event could be decades, centuries, or millennia and it could begin this year or many years in the future." Despite this caveat, the theme of abrupt climate change as a national security concern may be sufficiently eye-opening and provocative that, in conjunction with the other motivating forces of the year just past, it could take public consciousness of environmental security to a new level.

Rethinking Security, Reassessing the Threat

However many people there may now be who recognize that environmental conditions precipitate or contribute to other conditions—violent conflict, civil unrest, instability, regime or state failure—regularly associated with security as usually defined, they are vastly outnumbered by those who either openly oppose the environment-security linkage or ignore it as irrelevant or inconsequential.

These oppositionists come from two different but overlapping camps: ideological conservatives and the mainstream traditionalists who dominate the national security community. This distinction is crucial because the latter—the technocratic mandarins inside and outside government—have the final exegetical say about what security is and what therefore is allowed to be a legitimate part of the security dialogue.

Oppositionists treat the environment as a purely *ideological* issue, and climate change as the most ideological of all—accordingly as dismissible as feminism, the homosexual agenda, or any other reflection of "political correctness." This despite the fact that, in a purer ontological sense, the environment is an inherently strategic matter, and climate change the most strategic of all. The environment is everywhere. It respects no borders, physical or otherwise. In its reach, its effects, and its consequences, it is truly global. And, fully understood, it brings into question all of our prevailing notions of sovereignty, territorial integrity, and even aggression and intervention. Nonetheless, just as to a hammer everything looks like a nail, to an ideologue everything looks ideological—to be accepted or rejected on the basis not of reason but of internalized dogma.

One of the major issues that has most divided those who debate the environment-security relationship is how broadly or narrowly to define security. Oppositionists invariably take the narrow road—basically equating security with defense, just as they similarly equate power with force. To them, security has axiomatic meaning that derives from its historical roots. Ironically, on this particular count the oppositionists are abetted by a shadow contingent of like-minded liberal environmentalists, who believe that linking the environment to security is dangerous because it will inevitably militarize the former and relinquish vital resources needed for environmental, protection to an already bloated, profligate military establishment. In their fear of militarizing the environment, they risk getting into bed with betes noir who are fully committed to militarizing our entire strategic posture.

The counterpoise to this narrow construction of security begins with the recognition that security is, at root, a psychological and sociological phenomenon that starts—and ends—with the individual. To be secure is, literally, to be free—from harm and danger, threat and intimidation, doubt and fear, need and want. In the hierarchy of human needs, security is one of the most basic impulses—exceeded in its primacy only by the even more basic physiological needs for food, water, shelter, and the like, each of which is dependent on environmental well-being. Such primal needs translate into the natural rights that all human beings deserve to enjoy and that governments, as we have learned from America's founders, are instituted to secure.

Individual or human security, then, is the necessary precondition for national security, not merely its residual by-product. Accordingly, *assured security* stands as the primary overarching strategic aim a democratic society such as ours must seek to attain. In this supernal sense, security is something much more robust than defense. It encompasses the totality of conditions enumerated in America's security credo, the Preamble to the Constitution—not just the common defense, but no less importantly, national unity (a more perfect union), justice, domestic tranquility, the general welfare, and liberty. Only where all of these conditions exist in adequate measure is there true security. Where even one—liberty, say, or the general welfare—is sacrificed or compromised for another—say the common defense—the result is some degree of insecurity. Thus, in the final analysis, everything is related to security; everything is related to *national* security.

However broadly or narrowly security is defined, whatever endangers it or places it at risk is a threat; and whatever constitutes or qualifies as a threat is crucial because, in the idealized protocol of traditional national security planning, threats are the ostensible starting point for determining the requirements that produce capabilities and programs for countering these threats. (In reality, of course, capabilities and programs acquire their own bureaucratic life and thus are more likely to determine than to derive from threats.)

Oppositionists generally accept as legitimate threats only those parties or phenomena that, beyond their perceived potential for harm, are considered capable of or the product of malevolent intent. *Intentionality* is the key legitimizing element. Terrorism fills this bill, just as state-based adversaries traditionally have. Weapons of mass destruction seem to qualify because, though inert entities in themselves, in human hands they can be ominous instruments of harm. Climate change and assorted forms of environmental degradation, though, typically don't pass muster as credible threats, no matter how much death and destruction they can wreak. Instead, they are implicitly written off as pure acts of nature, assuming metaphysical proportions that place them beyond human control and therefore outside the bounds of either preventive or retributive concern.

Such blinkered threat assessment is entirely characteristic of the policy establishment. To cite just a few notable examples:

- The 2002 White House national security strategy, in 34 pages of text, mentions the word environment in only one short paragraph about U.S. trade negotiations.
- In his February 2004 "Worldwide Threat Briefing" to Congress, Director of Central Intelligence George Tenet devoted five pages of testimony each to terrorism, Iraq, and proliferation, three paragraphs to global narcotics, a paragraph each to population trends, infectious disease, and humanitarian food insecurity, but nothing at all to environmental matters.
- The much bally-hooed, future-oriented Hart-Rudman Commission, whose members extolled their own prescience for adumbrating 9/11-type terrorist attacks on the United States, gave only the most cursory treatment to 21st-century environmental challenges in its initial September 1999 report. Arguing innocuously that pollution can be—and implicitly will be—counteracted by economic growth and the spread of remediation technologies, the commission essentially dismissed the subject with this (dare we say, ideological) statement: "There is fierce disagreement over several major environmental issues. Many are certain that global warming will produce major social traumas within 25 years, but the scientific evidence does not yet support such a conclusion. Nor is it clear that recent weather patterns result from anthropogenic activity as opposed to natural fluctuations."
- Somewhat in contrast, the National Intelligence Council's *Global Trends 2015* report, issued in December 2000 (before the following year's 9/11 attacks), identified natural resources and the environment as one of the most important "drivers and trends that will shape the world of 2015." Focusing principally on food, water, and energy security developments, the experts who collaborated on the report acknowledged the persistence and growth of global environmental problems in the years ahead, a growing consensus on the need to deal with such problems, and the prospect that "global warming will challenge the international community."

Typifying the thinking of policymakers and other members of the national security mandarinate, such assessments also seem more representative than not of general public sentiment. A particularly revealing indication of this is the most recent Chicago Council on Foreign Relations study of U.S. public opinion on international issues, *Global Views 2004*. Asked to identify the most critical threats to U.S. vital interests, the public ranks global warming a distant seventh (37% of respondents), behind the likes of international terrorism (75%), chemical and biological weapons (66%), unfriendly countries becoming nuclear powers (64%), immigration into the United States (58%), and other developments. Another recent (February 2004) poll by Gallup found that environmental concerns don't even make the public's top-eleven list of possible threats to U.S. vital interests—international terrorism and the spread of weapons of mass destruction far outpacing all other prospective threats.

That environmental matters should be of such little overall public concern is a reflection of how limited and unstrategic our thinking about security actually is. Perhaps if we were to pay more attention to the documented effects of particular conditions and events, rather than to the nebulous, abstract notion of intentionality implicitly embedded in our prevailing standards of threat-worthiness, we could see the world differently—and more accurately.

Look, for example, at comparative fatalities from the highly credible threat of terrorism and the highly dubious threat of natural disasters. Since 1968, there have been 19,114 incidents of terrorism world wide, resulting in a total of 23,961 deaths and 62,502 associated injuries. However disturbing these figures may be, they pale in comparison to those resulting from natural disasters.

The average *annual* death toll over the past century due to drought, famine, floods, windstorms, temperature extremes, wave surges, and wildfires has been 243,577. Thus, even if we ignore earthquakes, volcanic eruptions, and disease epidemics, and don't count injuries or other harmful effects (such as homelessness), three times as many people die each year on average in natural disasters that could be linked to—and exacerbated by—climate change as have been killed and injured together in 37 years of terrorist incidents. And lest the use of a century-long average seem skewed, consider that just since 1990, there have been more than 207,000 fatalities from the foregoing types of disasters in South Asia alone, more than 23,000 in Central America and Mexico, and tens of thousands more in other parts of the world.

These figures are startling in their empirical exactitude, more so if one accepts estimates that average annual economic losses to such disasters were on the order of $660 billion in the 1990s. They lead us to consider a final argument that ideological conservatives invoke to discredit environmental and climate concerns—the need for sounder, more defensible science—and the associated argument national security mandarins use to deny or ignore the environment-security linkage—the lack of unequivocal evidence that environmental conditions actually cause diminished security in the form of violence.

Both arguments are excuses for denial and inaction; and both are suffused with hypocrisy. Those who demand conclusive *proof* that environmental conditions *cause* violence set a disingenuously unattainable legitimizing standard that permits them to perpetuate their own established preference for dealing with visible, immediate, politically remunerative symptoms. Terrorism is a cardinal example of this—singularly symptomatic,

never causative, except at some advanced, derivative level, where violence produces further violence.

Those who call for science as the only proper basis for public policy—at least climate policy—pretend to be motivated by a rigorous quest for objective (nonpolitical, non-ideological) truth. Yet they shamelessly accept or reject truth claims, labeling them "scientific" or not, based on whether those claims support or contradict their pre-established ideological beliefs. President Bush, for example, has repeatedly stated that climate policy must be based on better science (that is not yet available). But when asked about embryonic stem-cell research in this past year's second presidential debate, he stated that science is important, but it must be balanced by ethics. So, when the issue is stem-cell research—or perhaps abortion or homosexuality or capital punishment—ethics can take precedence over non-cooperative science; but when the issue is climate change or the environment more generally, not so. Maybe the earth really is flat.

Of more immediate relevance to this discussion is the practice common to many who call for better climate science. Paradoxically, they are perfectly content to unquestioningly accept and espouse demonstrably unscientific *assertions* from the military—especially concerning the degradation of military readiness that allegedly results from the so-called "encroachment" of environmental restrictions (e.g., species protection) on military installations. This despite the fact that the General Accounting Office has strongly criticized the military for failing to document whether and how much encroachment has actually degraded readiness.

Senator James Inhofe (R-OK) and the Senate Republican Policy Committee both exemplify this particular hypocrisy. Inhofe has said that "catastrophic global warming is a hoax"—"alarmism not based on objective science" even as he has said that "readiness problems ... are caused by an ever-growing maze of environmental procedures and regulations in which we are losing the ability to prepare our patriot children, our war fighters, for war." Similarly the Senate Republican Policy Committee claims that "what scientists do agree on [with regard to climate change] is not policy-relevant, and on policy-relevant issues, there is little scientific agreement," while also asserting: "Among the most burdensome [examples of encroachment] are environmental laws and lawsuits that hinder or even ban military training and testing—thereby impairing readiness.... The evidence of detrimental impact is ample."

Searching for a Strategic Response

What the foregoing contradictions suggest, among other things, is that the prevailing paradigm of security, according primacy as it does to the military and the use of force, long ago hijacked us intellectually and continues to hold us hostage; and, moreover, that in the absence of countervailing strategic thought of any consequence, ideology inevitably rushes in to fill the intellectual void, as it has in the case of environmental security, thereby forcing out rationality and blinding us to the future. The only remedy for this state of affairs, the only hope that the environment, climate change in particular, and, for that matter, other unconventional threats and challenges might be taken seriously as matters of serious security concern, is for fundamental strategic transformation to take place.

It seems insultingly obvious that strategic threats demand strategic response. But let us grasp the magnitude of this statement, for in the media age in which we live, there is virtually nothing—however obscure, however remote that is without almost instantaneous strategic consequence: Let us further understand why being strategic is therefore so intrinsically important. First, it is a moral obligation of government—to take the long view, to grasp the big picture, to anticipate and prevent, to appreciate the hidden, residual consequences of action or inaction, to recognize and capitalize on the interrelatedness of all things otherwise seemingly discrete and unrelated.

Second, being strategic inoculates us against crisis. Where crisis occurs, be it a terrorist incident or a natural disaster, strategic thinking has failed—with the unwanted result that decisionmaking must be artificially compressed and forced, and resources diverted from their intended purposes. Thus does *crisis prevention* stand alongside assured security as an overarching strategic aim of democratic society.

Third, being strategic provides the intellectual basis for both the strategic leadership expected of a superpower and the enduring, broad-based consensus necessary to galvanize a diverse, pluralistic society in common cause in the face of uncertainty, complexity, and ambiguity.

Four strategic imperatives should guide our future. The first let us call *targeted causation management*—focusing our thinking and our actions on identifying and eradicating the underlying causes of insecurity, thereby curing the disease rather than treating the symptoms. Environmental degradation and climate change take us much farther along the path to ultimate causes than terrorism ever could, especially if we acknowledge that the social, political, economic, and military conditions we prefer to deal with and attribute violence to may mask disaffection and unrest more deeply attributable to an environmentally degraded quality of life.

A second strategic imperative, *institutionalized anticipatory response,* calls for institutionalizing—giving permanence and legitimacy to the capacity and inclination for preventive action. This would enhance the prospects that conditions and events can be dealt with when they are manageable, before they mutate out of control and demand forceful response. Examples could range from a Manhattan Project-like effort to develop alternative energy sources and technologies, to greater inter-jurisdictional intelligence sharing, to massive disaster-resistant infrastructure development in the developing world.

A third strategic imperative is *appropriate situational tailoring*—dealing with conditions and events on their own geographic, cultural, and political terms rather than, as we are wont to do, inviting failure by imposing our preferred capabilities and approaches on the situations at hand. In a purely institutional sense, such tailoring might take the form, for example, of new multilateral collective security regimes in each region of the world, with major environmental preparedness and enforcement arms.

The fourth strategic imperative is *comprehensive operational integration*—achieving fuller organizational, doctrinal, procedural, and technological integration across military-non-military, governmental-non-governmental, and national-international lines. In a conceptual policy sense, this might assume the form of an overarching strategic architecture for unifying the activities of five organizational and cultural pillars—sustainable development, sustainable energy, sustainable business, sustainable consumption, and sustainable security. In a purely structural sense, the recognition that reorganization may be required to give birth and life to needed rethinking might produce such measures as the addition of a new Cabinet-level secretary of energy and environmental affairs to formal National Security Council membership, the creation of a UN under secretary-general for environmental affairs (or environmental security), or the expansion of the United Nations Environment Programme into an organization with operational capabilities and enforcement authority. In any event, all such measures would have to be underwritten by a firm commitment to more thoroughgoing transparency and multilateralism.

Finally, let us turn to the military. On the one hand, military action represents the least strategic option available for addressing environmental security (or virtually anything else for that matter). At least this is true so long as the military continues to be configured and oriented as it is and always has been—that is, for warfighting. On the other hand, the military is so central to our governing conception of security that true strategic transformation can take place only if it includes, or perhaps is preceded by, far-reaching military transformation—making real what until now has been only tiresome rhetoric from the Pentagon.

If the military has shown itself serious to date about environmental matters—even to the extent of crediting itself with being an excellent steward of nature—it is entirely a reflection of a distinctly engineering and management orientation dedicated principally to installation cleanup and remediation. Environmental security—the stuff of operations and intelligence, rather than of engineering and logistics—has been largely alien to the military ethos and identity. One need only consider the military's efforts under President Clinton to seek and gain selected exemptions from the Kyoto Protocol, or its tireless (if not entirely successful) attempts under President Bush to seek exemption from an array of environmental laws alleged to degrade readiness.

Two overriding considerations must guide military transformation. The first is the realization that what we ought to want is a military that is not just militarily effective—an instrument of force that serves the state—but that is *strategically effective*—an instrument of power that serves the larger aims of society and even humanity. The second overriding consideration is the concomitant realization that the military must be, and be seen to be, not a warfighting machine so much as a self-contained, self-sufficient enterprise that is capable of being projected over long distances for sustained periods of time to effectively manage all stages of a full range of complex emergencies.

Such considerations, taken to heart, ideally would produce a completely revamped military organized, manned, equipped, and trained primarily for nation-building, peacekeeping, humanitarian assistance, and disaster response, and only residually for warfighting. Such a military not only would possess the requisite capabilities for fulfilling the strategic imperatives enumerated above; it also would project the all-important imagery of a force truly committed to the pursuit of peace rather than to the enduringly illogical proposition that peace can be purchased by practicing war.

If we are to think and act strategically, which we must, we do well to recall the declaration from the Gayanashagowa, the Great Law of Peace of the Six Nations Iroquois Confederacy: "In our every deliberation we must consider the impact of our decisions on the next seven generations." And in applying this strategic precept to the matter at hand, which we must, we do no less well to take up the challenge issued recently by former Soviet President Mikhail Gorbachev. Interviewed some months ago, he was asked what he thought of the American doctrine of preemption. To which he responded:

Those who talk about leadership of the world all the time ought to exercise it. Rather than develop strategic doctrines of military preemption—as we've seen in Iraq, where no weapons of mass destruction have yet been found—let's act where the intelligence is clear: on climate change and other issues such as water, where today 2 billion people in the world don't have access to clean water. Let's talk instead about preempting global warming and the looming water crisis.

Indeed. Words for the self-proclaimed world's only superpower to act on.

Gregory D. Foster is a professor at the Industrial College of the Armed Forces, National Defense University, Washington, D.C., where he previously has served as George C. Marshall Professor and J. Carlton Ward Distinguished Professor and Director of Research. The views presented here are strictly his own.

Factory Farming in the Developing World

In some critical respects, this is not progress at all.

by Danielle Nierenberg

Walking through Bobby Inocencio's farm in the hills of Rizal province in the Philippines is like taking a step back to a simpler time. Hundreds of chickens (a cross between native Filipino chickens and a French breed) roam around freely in large, fenced pens. They peck at various indigenous plants, they eat bugs, and they fertilize the soil, just as domesticated chickens have for ages.

The scene may be old, but Inocencio's farm is anything but simple. What he has recreated is a complex and successful system of raising chickens that benefits small producers, the environment, and even the chickens. Once a "factory farmer," Inocencio used to raise white chickens for Pure Foods, one of the biggest companies in the Philippines.

Thousands of birds were housed in long, enclosed metal sheds that covered his property. Along with the breed stock and feeds he had to import, Inocencio also found himself dealing with a lot of imported diseases and was forced to buy expensive antibiotics to keep the chickens alive long enough to take them to market. Another trick of the trade Inocencio learned was the use of growth promotants that decrease the time it takes for chickens to mature.

Eventually he noticed that fewer and fewer of his neighbors were raising chickens, which threatened the community's food security by reducing the locally available supply of chickens and eggs. As the community dissolved and farms (and farming methods) that had been around for generations went virtually extinct, Inocencio became convinced that there had to be a different way to raise chickens and still compete in a rapidly globalizing marketplace. "The business of the white chicken," he says, "is controlled by the big guys." Not only do small farmers have to compete with the three big companies that control white chickens in the Philippines, but they must also contend

with pressure from the World Trade Organization (WTO) to open up trade. In the last two decades the Filipino poultry production system has transitioned from mainly backyard farms to a huge industry. In the 1980s the country produced 50 million birds annually. Today that figure has increased some ten-fold. The large poultry producers have benefited from this population explosion, but average farmers have not. So Inocencio decided to go forward by going back and reviving village-level poultry enterprises that supported traditional family farms and rural communities.

Inocencio's farm and others like it show that the Philippines can support indigenous livestock production and stand up to the threat of the factory farming methods now spreading around the world. Since 1997, his Teresa Farms has been raising free range chickens and teaching other farmers how to do the same. He says that the way he used to raise chickens, by concentrating so many of them in a small space, is dangerous. Diseases such as avian flu, leukosis J (avian leukemia), and Newcastle disease are spread from white chickens to the Filipino native chicken populations, in some cases infecting eggs before the chicks are even born. "The white chicken," says Inocencio, "is weak, making the system weak. And if these chickens are weak, why should we be raising them? Limiting their genetic base and using breeds that are not adapted to conditions in the Philippines is like setting up the potential for a potato blight on a global scale." Now Teresa Farms chickens are no longer kept in long, enclosed sheds, but roam freely in large tree-covered areas of his farm that he encloses with recycled fishing nets.

Inocencio's chickens also don't do drugs. Antibiotics, he says, are not only expensive but encourage disease. He found the answer to the problem of preventing diseases in chickens literally in his own back yard. His chickens eat spices and native plants with antibacterial and other medicinal properties. Chili, for instance, is mixed in grain to treat respiratory problems, stimulate appetite during heat stress, de-worm the birds, and to treat Newcastle disease. Native plants growing on the farm, including *ipil-ipil* and *damong maria*, are also used as low-cost alternatives to antibiotics and other drugs.

There was a time when most farms in the Philippines, the United States, and everywhere else functioned much like Bobby Inocencio's. But today the factory model of raising animals in intensive conditions is spreading around the globe.

A New Jungle

Meat once occupied a very different dietary place in most of the world. Beef, pork, and chicken were considered luxuries, and were eaten on special occasions or to enhance the flavor of other foods. But as agriculture became more mechanized, so did animal production. In the United States, livestock raised in the West was herded or transported east to slaughterhouses and packing mills. Upton Sinclair's *The Jungle*, written almost a century ago when the United States lacked many food-safety and labor regulations, described the appalling conditions of slaughterhouses in Chicago in the early 20th century and was a shocking expose of meat production and the conditions inflicted on both animals and humans by the industry. Workers were treated much like animals themselves, forced to labor long hours for very little pay under dangerous conditions, and with no job security.

If *The Jungle* were written today, however, it might not be set in the American Midwest. Today, developing nations like the Philippines are becoming the centers of large-scale livestock production and processing to feed the world's growing appetite for cheap meat and other animal products. But the problems Sinclair pointed to a century ago, including hazardous working conditions, unsanitary processing methods, and environmental contamination, still exist. Many have become even worse. And as environmental regulations in the European Union and the United States become stronger, large agribusinesses are moving their animal production operations to nations with less stringent enforcement of environmental laws.

These intensive and environmentally destructive production methods are spreading all over the globe, to Mexico, India, the former Soviet Union, and most rapidly throughout Asia. Wherever they crop up, they create a web of related food safety, animal welfare, and environmental problems. Philip Lymbery, campaign director of the World Society for the Protection of Animals, describes the growth of industrial animal production this way: Imagine traditional livestock production as a beach and factory farms as a tide. In the United States, the tide has completely covered the beach, swallowing up small farms and concentrating production in the hands of a few large companies. In Taiwan, it is almost as high. In the Philippines, however, the tide is just hitting the beach. The industrial, factory-farm methods of raising and slaughtering animals—methods that were conceived and developed in the United States and Western

Europe—have not yet swept over the Philippines, but they are coming fast.

An Appetite for Destruction

Global meat production has increased more than fivefold since 1950, and factory farming is the fastest growing method of animal production worldwide. Feedlots are responsible for 43 percent of the world's beef, and more than half of the world's pork and poultry are raised in factory farms. Industrialized countries dominate production, but developing countries are rapidly expanding and intensifying their production systems. According to the United Nations Food and Agriculture Organization (FAO), Asia (including the Philippines) has the fastest developing livestock sector. On the islands that make up the Philippines, 500 million chickens and 20 million hogs are slaughtered each year.

Despite the fact that many health-conscious people in developed nations are choosing to eat less meat, worldwide meat consumption continues to rise. Consumption is growing fastest in the developing countries. Two-thirds of the gains in meat consumption in 2002 were in the developing world, where urbanization, rising incomes, and globalized trade are changing diets and fueling appetites for meat and animal products. Because eating meat has been perceived as a measure of economic and social development, the Philippines and other poor nations are eager to climb up the animal-protein ladder. People in the Philippines still eat relatively little meat, but their consumption is growing. As recently as 1995, the average Filipino ate 21 kilograms of meat per year. Since then, average consumption has soared to almost 30 kilograms per year, although that is still less than half the amount in Western countries, where per-capita consumption is 80 kilograms per year.

This push to increase both production and consumption in the Philippines and other developing nations is coming from a number of different directions. Since the end of World War II, agricultural development has been considered a part of the foreign aid and assistance given to developing nations. The United States and international development agencies have been leaders in promoting the use of pesticides, artificial fertilizers, and other chemicals to boost agricultural production in these countries, often at the expense of the environment. American corporations like Purina Mills and Tyson Foods are also opening up feed mills and farms so they can expand business in the Philippines.

But Filipinos are also part of the push to industrialize agriculture. "This is not an idea only coming from the West," says Dr. Abe Agulto, president of the Philippine Society for the Protection of Animals, "but also coming from us." Meat equals wealth in much of the world and many Filipino businesspeople have taken up largescale livestock production to supply the growing demand for meat. But small farmers don't get much financial support in the Philippines. It's not farms like Bobby Inocencio's that are likely to get government assistance, but the big production facilities that can crank out thousands of eggs, chicks, or piglets a year.

The world's growing appetite for meat is not without its consequences, however. One of the first indications that meat production can be hazardous arises long before animals ever reach the slaughterhouse. Mountains of smelly and toxic manure are created by the billions of animals raised for human consumption in the world each year. In the United States, people in North Carolina know all too well the effects of this liquid and solid waste. Hog production there has increased faster than anywhere else in the nation, from 2 million hogs per year in 1987 to 10 million hogs per year today. Those hogs produce more than 19 million tons of manure each year and most of it gets stored in lagoons, or large uncovered containment pits. Many of those lagoons flooded and burst when Hurricane Floyd swept through the region in 1999. Hundreds of acres of land and miles of waterway were flooded with excrement, resulting in massive fish kills and millions of dollars in cleanup costs. The lagoons' contents are also known to leak out and seep into groundwater.

Some of the same effects can now be seen in the Philippines. Not far from Teresa Farms sits another, very different, farm that produces the most frequently eaten meat product in the world. Foremost Farms is the largest piggery, or pig farm, in all of Asia. An estimated 100,000 pigs are produced there every year.

High walls surround Foremost and prevent people in the community from getting in or seeing what goes on inside. What they do get a whiff of is the waste. Not only do the neighbors smell the manure created by the 20,000 hogs kept at Foremost or the 10,000 hogs kept at nearby Holly Farms, but their water supply has also been polluted by it. In fact, they've named the river where many of them bathe and get drinking water the River Stink. Apart from the stench, some residents have complained of skin rashes, infections, and other health problems from the water. And instead of keeping the water clean and installing effective waste treatment, the firms are just digging deeper drinking wells and giving residents free access to them. Many in the community are reluctant to complain about the smell because they fear losing their water supply. Even the mayor of Bulacan, the nearby village, has said "we give these farms leeway as much as possible because they provide so much economically."

It would be easy to assume that some exploitive foreign corporation owns Foremost, but in fact the owner is Lucia Tan, a Filipino. Tan is not your average Filipino, however, but the richest man in the Philippines. In addition to Foremost Farms, he owns San Miguel beer and Philippine Airlines. Tan might be increasing his personal wealth, but his farm and others like it are gradually destroying traditional farming methods and threatening indigenous livestock breeds in the Philippines. As a result, many small farmers can no longer afford to produce hogs for sale or for their own consumption, which forces them to become consumers of Tan's pork. Most of the nation's 11 million hogs are still kept in back yards, but because of farms like Foremost, factory farming is growing. Almost one-quarter of the breeding herd is now factory farmed. More than 1 million pigs are raised in factory farms every year in Bulacan alone.

Chicken farms in the Philippines are also becoming more intensive. The history of intensive poultry production in the Philippines is not long. Forty years ago, the nation's entire population was fed on native eggs and chickens produced by family farmers. Now, most of those farmers are out of business. They have lost not only their farms, but livestock diversity and a way of life as well.

The loss of this way of life to the industrialized farm-to-abattoir system has made the process more callous at every stage. Adopting factory farming methods works to diminish farmers' concern for the welfare of their livestock. Chickens often can't walk properly because they have been pumped full of growth-promoting antibiotics to gain weight as quickly as possible. Pigs are confined to gestation crates where they can't turn around. Cattle are crowded together in feedlots that are seas of manure.

Most of the chickens in the country are from imported breed stock and the native Filipino chicken has practically disappeared because of viral diseases spread by foreign breeds. Almost all of the hens farmed commercially for their eggs are confined in wire battery cages that cram three or four hens together, giving each bird an area less than the size of this page to stand on.

Unlike laying hens, chickens raised for meat in the Philippines are not housed in cages. But they're not pecking around in back yards, either. Over 90 percent of the meat chickens raised in the Philippines live in long sheds that house thousands of birds. At this time, most Filipino producers allow fowl to have natural ventilation and lighting and some roaming room, but they are under pressure to adopt more "modern" factory-farm standards to increase production.

The problems of a system that produces a lot of animals in crowded and unsanitary conditions can also be seen off the farm. The *baranguay* (neighborhood) of Tondo in Manila is best known for the infamous "Smoky Mountain" garbage dump that collapsed on scavengers in 2000, killing at least 200 people. But another hazard also sits in the heart of Tondo. Surrounded by tin houses, stores, and bars, the largest government-owned slaughtering facility in the country processes more than 3,000 swine, cattle, and *caraboa* (water buffalo) per day, all brought from farms just outside the city limits. The slaughterhouse does have a waste treatment system where the blood and other waste is supposed to be treated before it is released into the city's sewer system and nearby Manila Bay. Unfortunately, that's not what's going on. Instead, what can't be cut up and sold for human consumption is dumped into the sewer.

Some 60 men are employed at the plant. They stun, bludgeon, and slaughter animals by hand and at a breakneck pace. They wear little protective gear as they slide around on floors slippery with blood, which makes it hard to stun animals on the first try, or sometimes even the second, or to butcher meat without injuring themselves.

The effects of producing meat this way also show up in rising cases of food-borne illness, emerging animal diseases that can spread to humans, and in an increasingly overweight Filipino population that doesn't remember where meat comes from.

There are few data on the incidence of food-borne illness in the Philippines or most other developing nations, and even fewer about how much of it might be related to eating unsafe meat. What food safety experts do know is that food-borne illness is one of the most widespread health problems worldwide.

And it could be an astounding 300–350 times more frequent than reported, according to the World Health Organization. Developing nations bear the greatest burden because of the presence of a wide range of parasites, toxins, and biological hazards and the lack of surveillance, prevention, and treatment measures—all of which ensnarl the poor in a chronic cycle of infection. According to the FAO, the trend toward increased commercialization and intensification of livestock production is leading to a variety of food safety problems. Crowded, unsanitary conditions and poor waste treatment in factory farms exacerbate the rapid movement of animal diseases and food-borne infections. *E. coli* 0157:H7, for instance, is spread from animals to humans when people eat food contaminated by manure. Animals raised in intensive conditions often arrive at slaughterhouses covered in feces, thus increasing the chance of contamination during slaughtering and processing.

Cecilia Ambos is one of the meat inspectors at the Tondo slaughterhouse. Cecilia or another inspector is required to be on site at all times, but she says she rarely has to go to the killing floor. Inspections of carcasses only occur, she said, if one of the workers alerts the inspector. That doesn't happen very often, and not because the animals are all perfectly healthy. Consider that the men employed at the plant are paid about $5 per day, which is less than half of the cost of living—and are working as fast as they can to slaughter a thousand animals per shift. It's unlikely that they have the time or the knowledge to notice problems with the meat.

Since the 1960s, farm-animal health in the United States has depended not on humane farming practices but on the use of antibiotics. Many of the same drugs used to treat human illnesses are also used in animal production, thus reducing the arsenal of drugs available to fight food-borne illnesses and other health problems. Because antibiotics are given to livestock to prevent disease from spreading in crowded conditions and to increase growth, antibiotic resistance has become a global threat. In the Philippines, chicken, egg, and hog producers use antibiotics not because their birds or hogs are sick, but because drug companies and agricultural extension agents have convinced them that these antibiotics will ensure the health of their birds or pigs and increase their weight.

Livestock raised intensively can also spread diseases to humans. Outbreaks of avian flu in Hong Kong during the past five years have led to massive culls of thousands of chickens. When the disease jumped the species barrier for the first time in 1997, six of the eighteen people infected died. Avian flu spread to people living in Hong Kong again this February, killing two. Dr. Gary Smith, of the University of Pennsylvania School of Veterinary Medicine, also warns that "it is not high densities [of animals] that matter, but the increased potential for transmission between firms that we should be concerned about. The nature of the farming nowadays is such that there is much more movement of animals between farms than there used to be, and much more transport of associated materials between farms taking place rapidly. The problem is that the livestock industry is operating on a global, national, and county level." The foot-and-mouth disease epidemic in the United Kingdom is a perfect example of how just a few cows can spread a disease across an entire nation.

Modern Methods, Modern Policies?

The expansion of factory farming methods in the Philippines i raising the probability that it will become another fast food na tion. Factory farms are supplying much of the pork and chicke preferred by fast food restaurants there. American-style fas food was unknown in the Philippines until the 1970s, when Jol libee, the Filipino version of McDonald's, opened its doors Now, thanks to fast food giants like McDonald's, Kentucky Fried Chicken, Burger King, and others, the traditional diet o rice, vegetables, and a little meat or fish is changing—and so are rates of heart disease, diabetes, and stroke, which have riser to numbers similar to those in the United States and othe western nations.

The Filipino government doesn't see factory farming as a threat. To the contrary, many officials hope it will be a solution to their country's economic woes, and they're making it easie for large farms to dominate livestock production. For instance the Department of Agriculture appears to have turned a blinc eye when many farms have violated environmental and anima welfare regulations. The government has also encouraged bis farms to expand by giving them loans. But as the farms ge bigger and produce more, domestic prices for chicken and por fail, forcing more farmers to scale up their production methods And because the Philippines (and many other nations) are pre vented by the Global Agreement on Tariffs and Trade and the WTO from imposing tariffs on imported products, the Philip pines is forced to allow cheap, factory-farmed American por and poultry into the country. These products are then sold a lower prices than domestic meat.

Rafael Mariano, a leader in the Peasant Movement for the Philippines (KMP), has not turned away from the problems caused by factory farming in the Philippines. He and the 800,000 farmers he works with believe that "factory farming is not acceptable, we have our own farming." But farmers, he says, are told by big agribusiness companies that their method are old fashioned, and that to compete in the global market they must forget what they have learned from generations o farming. Rafael and KMP are working to promote traditiona methods of livestock production that benefit small farmers anc increase local food security. This means doing what farmer used to do: raising both crops and animals. In mixed crop-livestock farms, animals and crops are parts of a self-sustaining system. Some farmers in the Philippines raise hogs, chickens, ti lapia, and rice on the same farm. The manure from the hogs anc chickens is used to fertilize the algae in ponds needed for bot tilapia and rice to grow. These farms produce little waste, pro vide a variety of food for the farm, and give farmers social se curity when prices for poultry, pork, and rice go down.

The Philippines is not the only country at risk from the spread of factory farms. Argentina, Brazil, Canada, China India, Mexico, Pakistan, South Africa, Taiwan, and Thailanc are all seeing growth in industrial animal production. As regu-

The Return of the Native... Chicken

Bobby Inocencio believes that the happier his chickens are, the healthier they will be, both on the farm and at the table. Native Filipino chickens are a tough sell commercially, he says, because they typically weigh in at only one kilogram apiece. But Inocencio's chickens are part native and part SASSO (a French breed), and grow to two kilos in just 63 days in a free-range system. They are also better adapted to the climate of the Philippines, unlike white chickens that are more vulnerable to heat. As a result, Inocencio's chickens not only are nutritious, but taste good. Raising white chickens, he says, forced small farmers to become "consumers of a chicken that doesn't taste like anything." Further, his chickens don't contain any antibiotics and are just 5 percent fat, compared to 35 percent in the white chicken. Because they are not raised in the very high densities of factory farms, these chickens actually enrich the environment with their manure. They also provide a reliable source of income for local farmers and give Filipinos a taste of how things used to be.

lations controlling air and water pollution from such farms are strengthened in one country, companies simply pack up and move to countries with more lenient rules. Western European nations now have among the strongest environmental regulations in the world; farmers can only apply manure during certain times of the year and they must follow strict controls on how much ammonia is released from their farms. As a result, a number of companies in the Netherlands and Germany are moving their factory farms—but to the United States, not to developing countries. According to a recent report in the *Dayton Daily News*, cheap land and less restrictive environmental regulations in Ohio are luring European livestock producers to the Midwest. There, dairies with fewer than 700 cows are not required to obtain permits, which would regulate how they control manure. But 700 cows can produce a lot of manure. In 2001, five Dutch-owned dairies were cited by the Ohio Environmental Protection Agency for manure spills. "Until there are international regulations controlling waste from factory farms," says William Weida, director of the Global Reaction Center for the Environment/Spira Factory Farm project, "it is impossible to prevent farms from moving to places with less regulation." Mauricio Rosales of FAO's Livestock, Environment, and Development Project also stresses the need for siting farms

where they will benefit both people and the environment. "Zoning," he says, "is necessary to produce livestock in the most economically viable places, but with the least impact." For instance, when livestock live in urban or peri-urban areas, the potential for nutrient imbalances is high. In rural areas manure can be a valuable resource because it contains nitrogen and phosphorous, which fertilize the soil. In cities, however, manure is a toxic, polluting nuisance.

The triumph of factory farming is not inevitable. In 2001, the World Bank released a new livestock strategy which, in a surprising reversal of its previous commitment to funding of large-scale livestock projects in developing nations, said that as the livestock sector grows "there is a significant danger that the poor are being crowded out, the environment eroded, and global food safety and security threatened." It promised to use a "people-centered approach" to livestock development projects that will reduce poverty, protect environmental sustainability, ensure food security and welfare, and promote animal welfare. This turnaround happened not because of pressure from environmental or animal welfare activists, but because the large-scale, intensive animal production methods the Bank once advocated are simply too costly. Past policies drove out smallholders because economies of scale for large units do not internalize the environmental costs of producing meat. The Bank's new strategy includes integrating livestock–environment interactions in to environmental impact assessments, correcting regulatory distortions that favor large producers, and promoting and developing markets for organic products. These measures are steps in the right direction, but more needs to be done by lending agencies, governments, non-governmental organizations, and individual consumers. Changing the meat economy will require a rethinking of our relationship with livestock and the price we're willing to pay for safe, sustainable, humanely-raised food.

Meat is more than a dietary element, it's a symbol of wealth and prosperity. Reversing the factory farm tide will require thinking about farming systems as more than a source of economic wealth. Preserving prosperous family farms and their landscapes and raising healthy, humanely treated animals, should also be viewed as a form of affluence.

Further Reading:
World Society for the Protection of Animals,
www.wspa.org.uk

Danielle Nierenberg *is a Staff Researcher at the Worldwatch Institute.*

Where Oil and Water Do Mix: Environmental Scarcity and Future Conflict in the Middle East and North Africa

JASON J. MORRISSETTE and DOUGLAS A. BORER

"Many of the wars of the 20th century were about oil, but wars of the 21st century will be over water."

—Isamil Serageldin
World Bank Vice President

In the eyes of a future observer, what will characterize the political landscape of the Middle East and North Africa? Will the future mirror the past or, as suggested by the quote above, are significant changes on the horizon? In the past, struggles over territory, ideology, colonialism, nationalism, religion, and oil have defined the region. While it is clear that many of those sources of conflict remain salient today, future war in the Middle East and North Africa also will be increasingly influenced by economic and demographic trends that do not bode well for the region. By 2025, world population is projected to reach eight billion.[1] As a global figure, this number is troubling enough; however, over 90 percent of the projected growth will take place in developing countries in which the vast majority of the population is dependent on local renewable resources. For instance, World Bank estimates place the present annual growth rate in the Middle East and North Africa at 1.9 percent versus a worldwide average of 1.4 percent.[2] In most of these countries, these precious renewable resources are controlled by small segments of the domestic political elite, leaving less and less to the majority of the population. As a result, if present population and economic trends continue, we project that many future conflicts throughout the region will be directly linked to what academic researchers term "environmental scarcity"[3]—the scarcity of renewable resources such as arable land, forests, and fresh water.

The purpose of this article is twofold. In the first section, we conceptualize how environmental scarcity is linked to domestic political unrest and the subsequent crisis of domestic political legitimacy that may ultimately result in conflict. We review the academic literature which suggests that competition over water is the key environmental variable that will play an increasing role in future domestic challenges to governments throughout the region. We then describe how these crises of domestic political legitimacy may result in both intrastate and interstate conflict. Even though the Middle East can generally be characterized as an arid climate, two great river systems, the Nile and the Tigris/Euphrates, serve to anchor the major population centers in the region. Conflict over the water of the Nile may someday come to pass between Egypt, Sudan, and Ethiopia; while Turkey, Syria, and Iraq all are located along the Tigris/Euphrates watershed and compete for its resources. Further conflict over water may embroil Israel, Syria, and the Palestinians.

Despite many existing predictions of war over water, we investigate the intriguing question: How have governments in the Middle East thus far avoided conflict over dwindling water supplies? In the second section of the article, we discuss the concept of "virtual water" and use this concept to illustrate the important linkages between water usage and the global economy, showing how existing tangible water shortages have been ameliorated by a combination of economic factors, which may or may not be sustainable into the future.

Environmental Scarcity and Conflict: An Overview

Mostafa Dolatyar and Tim Gray identify water resources as "the principal challenge for humanity from the early days of civilization."[4] The 1998 United Nations Development Report estimates that almost a third of the 4.4 billion people currently living in the developing world have no access to clean water. The report goes on to note that the world's population tripled in the 20th century, resulting in a corresponding sixfold increase in the use of water resources. Moreover, infrastructure problems related to water supply abound in much of the developing world; the United Nations estimates that between 30 and 50 percent of the water presently diverted for irrigation purposes is lost through leaking pipes alone. In turn, roughly 20 countries in the developing world presently suffer from water stress (defined as having less than 1,000 cubic meters of available freshwater per capita), and 25 more are expected to join that list by 2050.[5] In response to these trends, the United Nations resolved in 2002 to reduce by half the proportion of people in the developing world who are unable to reach—or afford—safe drinking water.

In turn, numerous scholars in recent years have conceptualized water in security terms as a key strategic resource in many regions of the world. Thomas Naff maintains that water scarcity holds significant potential for conflict in large part because it is fundamentally essential to life. Naff identifies six basic characteristics that distinguish water as a vital and potentially contentious resource. (1) Water is necessary for sustaining life and has no substitute for human or animal use. (2) Both in terms of domestic and international policy, water issues are typically addressed by policymakers in a piecemeal fashion rather than comprehensively. (3) Since countries typically feel compelled by security concerns to control the ground on or under which water flows, by its nature, water is also a terrain security issue. (4) Water issues are frequently perceived as zero-sum, as actors compete for the same limited water resources. (5) As a result of the competition for these limited resources, water presents a constant potential for conflict. (6) International law concerning water resources remains relatively "rudimentary" and "ineffectual."[6] As these factors suggest, water is a particularly volatile strategic issue, especially when it is in severe shortage.

Arguing that environmental concerns have gained prominence in the post-Cold War era, Alwyn R. Rouyer establishes a basic paradigm of contemporary environmental conflict. Rouyer argues that "rapid population growth, particularly in the developing world, is putting severe stress on the earth's physical environment and thus creating a growing scarcity of renewable resources, including water, which in turn is precipitating violent civil and international conflict that will escalate in severity as scarcity increases."[7] Rouyer goes on to assert that this potential conflict over scarce resources will likely be most disruptive in states with rapidly expanding populations in which policymakers lack the political and economic capability to minimize environmental damage.

"Almost a third of the 4.4 billion people currently living in the developing world have no access to clean water."

Security concerns linked fundamentally to environmental scarcity are far from a contrivance of the post-Cold War era, however. Ulrich Küffner asserts that conflicts over water "have occurred between many countries in all climatic regions, but between countries in arid regions they appear to be unavoidable. Claims over water have led to serious tensions, to threats and counter threats, to hostilities, border clashes, and invasions."[8] Moreover, as Miriam Lowi notes, "Well before the emergence of the na-

tion-state, the arbitrary political division of a unitary river basin ... led to problems regarding the interests of the states and/or communities located within the basin and the manner in which conflicting interests should be resolved."[9] Lowi fundamentally frames the issue of water scarcity in terms of a dilemma of collective action and failed cooperation—the archetypal "Tragedy of the Common"—in which communal resources are abused by the greediness of individuals. In many regions of the world, the international agreements and coordinating institutions necessary to lower the likelihood of conflict over water are either inadequate or altogether nonexistent.[10]

Thomas Homer-Dixon argues that the environmental resource scarcity that potentially results in conflict, including water scarcity, fundamentally derives from one of three sources. The first, *supply-induced scarcity*, is caused when a resource is either degraded (for example, when cropland becomes unproductive due to overuse) or depleted (for example, when cropland is converted into suburban housing). Throughout most of the Middle East and North Africa countries, both environmental and resource degradation and depletion are of relevant concern. For instance, many of these countries face significant decreases in the agricultural productivity of their arable soil as a result of ongoing trends of desertification, soil erosion, and pollution. This problem is coupled with the continued loss of croplands to urbanization, as rural dwellers move to cities in search of employment and opportunity. The second source of environmental scarcity, *demand-induced scarcity*, is caused by either an increase in per-capita consumption or by simple population growth. If the supply remains constant, and demand increases by existing users consuming more, or more users each consuming the same amount, eventually scarcity will result as demand overtakes supply. The third type of environmental scarcity is known as *structural scarcity*, a phenomenon that results when resource supplies are unequally distributed. In this case the "haves" in any given society generally control and consume an inordinate amount of the existing supply, which results in the more numerous "have-nots" experiencing the scarcity.[11]

These three sources of scarcity routinely overlap and interact in two common patterns: "resource capture" and ecological marginalization. Resource capture occurs when both demand-induced and supply-induced scarcities interact to produce structural scarcity. In this pattern, powerful groups within society foresee future shortages and act to ensure the protection of their vested interests by using their control of state structures to capture control of a valuable resource. An example of this pattern occurred in Mauritania (one of Algeria's neighbors) in the 1970s and 1980s when the countries bordering the Senegal River built a series of dams to boost agricultural production. As a result of the new dams, the value of land adjacent to the river rapidly increased—an economic development that motivated Mauritanian Moors to abandon their traditional vocation as cattle grazers located in the arid land in the north and, instead, to migrate south onto lands next to the river. However, black Mauritanians already occupied the land on the river's edge. As a result, the Moorish political elite that controlled the Mauritanian government rewrote the legislation on citizenship and land rights to effectively block black Mauritanians from land ownership. By declaring blacks as non-citizens, the Islamic Moors managed to capture the land through nominally legal (structural) means. As a result, high levels of violence later arose between Mauritania and Senegal, where hundreds of thousands of the black Mauritanians had become refugees after being driven from their land.[12]

The second pattern, ecological marginalization, occurs when demand-induced and structural scarcities interact in a way that results in supply-induced scarcity. An example of this pattern comes from the Philippines, a country whose agricultural lands traditionally have been controlled by a small group of dominant landowners who, prior to the election of former President Estrada, have controlled Filipino politics since colonial times. Population growth in the 1960s and 1970s forced many poor peasants to settle in the marginal soils of the upland interior. This more mountainous land could not sustain the lowland slash-and-burn farming practices that they brought with them. As a result, the Philippines suffered serious ecological damage in the form of water pollution, soil erosion, landslides, and changes in the hydrological cycle that led to further hardship for the peasantry as the land's capacity shrunk. As a result of their economic marginalization, many upland peasants became increasingly susceptible to the revolutionary rhetoric promoted by the communist-led New People's Army, or they supported the "People Power" movement that ousted US-backed Ferdinand Marcos from power in 1986.[13]

Thus, as shown in the Philippines, social pressures created by environmental scarcity can have a direct influence on the ruling legitimacy of the state, and may cause state power to crumble. Indeed, reductions in agricultural and economic production can produce objective socio-economic hardship; however, deprivation does not necessarily produce grievances against the government that result in serious domestic unrest or rebellion. One can look at the relative stability in famine-stricken North Korea as a poignant example of a polity whose citizens have suffered widespread physical deprivation under policies of the existing regime, but who are unwilling or unable to risk their lives to challenge the state.

This phenomenon is partly explained by conflict theorists who argue that individuals and groups have feelings of "relative deprivation" when they perceive a gap between what they believe they deserve and what in reality they actually have achieved.[14] In other words, can a government meet the expectations of the masses enough to avoid conflict? For example, in North Korea—a regime that tightly controls the information that its people receive—many people understand that they are suffering,

but they may not know precisely how much they are suffering relative to others, such as their brethren in the South. The North Korean government indoctrinates its people to expect little other than hardship, which in turn it blames on outside enemies of the state. Thus, the people of North Korea have very low expectations, which their government has been able to meet. More important, then, is the question of whom do the people perceive as being responsible for their plight? If the answer is the people's own government—whether as a result of supply-induced, demand-induced, or structural resource scarcity—then social discord and rebellion are more likely to result in intrastate conflict, as citizens challenge the ruling legitimacy of the state itself. If the answer is someone else's government, then interstate conflict may result.

On numerous occasions, history has shown that governments whose people are suffering can remain in power for long periods of time by pointing to external sources for the people's hardship.[15] As noted above regarding political legitimacy, perception is politically more important than any standard of objective truth.[16] When faced with a crisis of legitimacy derived from environmental resource scarcity, any political regime essentially has a choice of two options in dealing with the situation. The regime may choose temporarily not to respond to looming challenges to its authority because water-induced stress may in fact pass when sufficient heavy rainfall occurs. However, most regimes in the Middle East and North Africa have sought more proactive ways to ensure their survival. Indeed, a people might forgive its government for one drought, but if governmental action is not taken, a subsequent drought-induced crisis of legitimacy could result in significant social upheaval by an unforgiving public. Furthermore, if the government itself is perceived to be the direct source of the scarcity—through structural arrangements, resource capture, or other means—these trends of social unrest are likely to be exacerbated. Thus, in order to survive, most states have developed policies to increase their water supplies and to address issues of environmental scarcity. The problem with doing so throughout most of the Middle East and North Africa, however, is that increasing supply in one state often creates environmental scarcity problems in another. If Turkey builds dams, Iraq and Syria are vulnerable; if Ethiopia or the Sudan builds dams, Egypt feels threatened. Thus far, interstate water problems leading to war have been avoided due to the economic interplay between oil wealth and the importation of "virtual water," which will be discussed at greater length below.

As noted above, resource scarcity issues centered on water are particularly prominent in the Middle East and North Africa. Ewan Anderson notes that resource geopolitics in the Middle East "has long been dominated by one liquid—oil. However, another liquid, water, is now recognized as the fundamental political weapon in the region."[17] Ecologically speaking, water scarcity in the Middle East and North Africa results from four primary causes: fundamentally dry climatic conditions, drought, desiccation (the degradation of land due to the drying up of the soil), and water stress (the low availability of water resulting from a growing population).[18] These resource scarcity problems are exacerbated in the Middle East by such factors as poor water quality and inadequate—and, at times, purposefully discriminatory—resource planning. As a result of these ecological and political trends, Nurit Kliot states, "water, not oil, threatens the renewal of military conflicts and social and economic disruptions" in the Middle East.[19] In the case of the Arab-Israeli conflict, Alwyn Rouyer suggests that "water has become inseparable from land, ideology, and religious prophecy."[20] Martin Sherman echoes these sentiments in the following passage, describing specifically the Arab-Israeli conflict:

> In recent years, particularly since the late 1980s, water has become increasingly dominant as a bone of contention between the two sides. More than one Arab leader, including those considered to be among the most moderate, such as King Hussein of Jordan and former UN Secretary General, Boutros Boutros-Ghali of Egypt, have warned explicitly that water is the issue most likely to become the cause of a future Israeli-Arab war.[21]

"Water is a particularly volatile strategic issue, especially when it is in severe shortage."

While Jochen Renger contends that a conflict waged explicitly over water may not lie on the immediate horizon, he notes that "it is likely that water might be used as leverage during a conflict."[22] As a result of such geopolitical trends, managing these water resources in the Middle East and North Africa—and, in turn, managing the conflict over these resources—should be considered a primary concern of both scholars and policymakers.

Keeping the Peace: The Importance of Virtual Water

The warning signals that war over water may replace war over oil and other traditional sources of conflict are very real in recent history. Yet, for more than 25 years, despite increasing demand, water has not been the primary cause of war in the Middle East and North Africa. The scenarios outlined in the section above have yet to fully address the fundamental questions of why and how governments in the region have thus far avoided major interstate conflict over water. In order to understand the likelihood of war, we must address the foundation of the past peace, testing whether or not this foundation remains strong for the foreseeable future. How have the governments of the region been able to avoid the apparently inevitable consequences of conflict that derive from the interlinked problems of water deficits, population

growth, and weak economic performance? In this section of the article, we turn our attention to the important linkages between water usage and the global economy, showing how existing water shortages have been ameliorated by a combination of economic factors.

To understand the politics of water in the Middle East and North Africa, one must first look at the region's most fungible resource: oil. For much of the post-World War II era, the growing need for oil to fuel economic growth has served as the dominant motivating factor in US security policy in the Middle East. Conventional wisdom in the United States holds that US dependency on Middle Eastern oil is a strategic weakness. Indeed, the specter of a regional hegemonic power that controls the oil and that is also hostile to the United States strikes fear into the hearts of policymakers in Washington. Thus, for roughly the past 50 years, the United States has sought to prop up "moderate" (meaning pro-US) regimes while denying hegemony to "radical" (meaning anti-US) regimes.[23] However, we contend that both policymakers and the public at large in the United States generally misunderstand the politics of oil as they relate to water in the Middle East.

In absolute terms, problems arising from US vulnerability to foreign oil are basically true—it would be better to be free of dependency on oil from any foreign source than to be dependent. However, the other side of the equation is often forgotten: oil-producing states are dependent on the United States and other major oil importers for their economic livelihood. More bluntly put, oil-exporting states are dependent on the influx of dollars, euros, and yen to purchase goods, services, and commodities that they lack. Thus, oil-producing countries in the Middle East and North Africa, few of whom have managed to successful diversify their economies beyond the petroleum sector, exist in an interdependent world economy. The world depends on their oil, and they depend on the world's goods and services—including that most valuable life-sustaining resource, water.

On the surface, this perhaps seems to be a contentious claim. Outgoing oil tankers do not return with freshwater used to grow crops, and Middle East countries do not rely on the importation of bottled water for their daily consumption needs. However, according to hydrologists, each individual needs approximately 100 cubic meters of water each year for personal needs, and an additional 1,000 cubic meters are required to grow the food that person consumes. Thus, every person alive requires approximately 1,100 cubic meters of water every year. In 1970, the water needs of most Middle Eastern and North African countries could be met from sources within the region. During the colonial and early post-colonial eras, regional governments and their engineers had effectively managed supply to deliver new water to meet the requirements of the growing urban populations, industrial requirements, agricultural needs, and other demand-induced factors. What is clear is that in the past 30 years, the status of the region's water resources has significantly

Country	Total Water Resources per Capita (cubic meters) in 2000	Percent of Population with Access to Adequate, Improved Water Source, 2000	GDP per Capita, 2000 Estimate
Algeria	477	94%	$5,500
Egypt	930	95%	$3,600
Iran	2,040	95%	$6,300
Iraq	1,544	85%	$2,500
Israel	180	99%	$18,900
Jordan	148	96%	$3,500
Kuwait	0	100%	$15,000
Lebanon	1,124	100%	$5,000
Libya	148	72%	$8,900
Morocco	1,062	82%	$3,500
Oman	426	39%	$7,700
Qatar	–	100%	$20,300
Saudi Arabia	119	95%	$10,500
Sudan	5,312	–	$1,000
Syria	2,845	80%	$3,100
Tunisia	434	–	$6,500
Turkey	3,162	83%	$6,800
Yemen	241	69%	$820

Sources: World Bank Development Indicators, Country-at-a-Glance Tables, Freshwater Resources, and *CIA World Factbook*, at http://www.worldbank.org and http://www.cia.gov.

Figure 1. Water Resources and Economics in the Middle East and North Africa.

worsened as populations have increased (an example of demand-induced scarcity). Since the mid-1970s, most countries have been able to supply daily consumption and industrial needs; however, as indicated in Figure 1, the approximate 1,000 cubic meters of water per capita that is required for self-sufficient agricultural production represents a seemingly impossible challenge for some Middle Eastern and North African economies.

Simply put, many countries of the region cannot presently meet the irrigation requirements needed to feed their own growing populations.[24] Furthermore, for those countries that have sufficient resources to meet this need in aggregate (such as Syria), resource capture and structural distribution problems keep water out of the hands of many citizens. If this situation has been deteriorating for nearly three decades, the question remains: Why has there been no war over water? The answer, according to Tony Allen, lies in an extremely important hidden source of water, which he describes as "virtual water."[25] Virtual water is the water contained in the food that the region imports—from the United States, Australia, Argentina, New Zealand, the countries of the European Union, and other major food-exporting countries. If each person of

the world consumes food that requires 1,000 cubic meters of water to grow, plus 100 additional cubic meters for drinking, hygiene, and industrial production, it is still possible that any country that cannot supply the water to produce food may have sufficient water to meet its needs—if it has the economic capacity to buy, or the political capacity to beg, the remaining virtual water in the form of imported food.

According to Allen, more water flows into the countries of the Middle East and North Africa as virtual water each year than flows down the Nile for Egypt's agriculture. Virtual water obtained in the food available on the global market has enabled the governments of the region's countries to augment their inadequate and declining water resources. For instance, despite its meager freshwater resources of 180 cubic meters per capita, Israel—otherwise self-sufficient in terms of food production—manages its problems of water scarcity in part by importing large supplies of grain each year. As noted in Figure 1, this pattern is replicated by eight other countries in the region that have less than 1,100 cubic meters of water per person. Thus, the global cereal grain commodity markets have proven to be a very accessible and effective system for importing virtual water needs. In the Middle East and North Africa, politicians and resource managers have thus far found this option a better choice than resorting to war over water with their neighbors. As a result, the strategic imperative for maintaining peace has been met through access to virtual water in the form of food imports from the global market.[26]

The global trade in food commodities has been increasingly accessible, even to poor economies, for the past 50 years. During the Cold War, food that could not be purchased was often provided in the form of grants by either the United States or the Soviet Union, and in times of famine, international relief efforts in various parts of the globe have fed the starving. Over time, competition by the generators of the global grain surplus—the United States, Australia, Argentina, and the European Community—brought down the global price of grain. As a result, the past quarter-century, the period during which water conflicts in the Middle East and North Africa have been most insistently predicted, was also a period of global commodity markets awash with surplus grain. This situation allowed the region's states to replace domestic water supply shortages with subsidized virtual water in the form of purchases from the global commodities market. For example, during the 1980s, grain was being traded at about $100 (US) a ton, despite costing about $200 a ton to produce.[27] Thus, US and European taxpayers were largely responsible for funding the cost of virtual water (in the form of significant agricultural subsidies they paid their own farmers) which significantly benefited the countries of the Middle East and North Africa.

For the most part we concur with Allen's evaluation that countries have not gone to war primarily over water, and that they have not done so because they have been able to purchase virtual water on the international market. However, the key question for the future is, Will this situation continue? If the answer is yes, and grain will remain affordable to the countries of the region, then it is relatively safe to conclude that conflict derived from environmental scarcity (in the form of water deficits) will not be a significant problem in the foreseeable future. However, if the answer is no, and grain will not be as affordable as it has been in the past, then future conflict scenarios based on environmental scarcity must be seriously considered.

Global Economic Restructuring: The World Trade Organization's Impact on Subsidies

Regrettably, a trend toward the answer "no" appears to be gaining some momentum due to ongoing structural changes in the global economy. The year 1995 witnessed a dramatic change in the world grain market, when wheat prices rose rapidly, eventually reaching $250 a ton by the spring of 1996. With the laws of supply and demand kicking in, this increased price resulted in greater production; by 1998, world wheat prices had fallen back to $140 a ton,

but had risen again to over $270 by June 2001.[28] These rapid wheat price fluctuations reemphasize the strategic importance and volatility of virtual water. If the global price of food staples remains affordable, many countries in the Middle East and North Africa may struggle to meet the demand-induced scarcity resulting from their growing populations, but they most likely will succeed. However, if basic food staple prices rise significantly in the coming decades and the existing economic growth patterns that have characterized the region's economies over the past 30 years remain constant, an outbreak of war is more likely.

It is clear that recent structural changes in the world economy do not favor the continuation of affordable food prices for the region's countries in the future. As noted above, wheat that costs $200 a ton to produce has often been sold for $100 a ton on world markets. This situation is possible only when the supplier is compensated for the lost $100 per ton in the form of a subsidy. Historically, these subsidies have been paid by the governments of major cereal grain-producing countries, primarily the United States and members of the European Union. Indeed, for the last 100 years, farm subsidies have been a bedrock public policy throughout the food-exporting countries of the first world. However, with the steady embrace of global free-trade economics and the establishment of the World Trade Organization (WTO), agricultural subsidies have come under pressure in most major grain-producing countries. According to a recent US Department of Agriculture (USDA) study, "The elimination of agriculture trade and domestic policy distortions could raise world agriculture prices about 12 percent.[29]

> *"Many countries of the region cannot presently meet the irrigation requirements needed to feed their own growing populations."*

Thus, as the WTO gains systematic credibility over the coming decades, its free-trade policies will further erode the practice of farm price supports, and it is highly unlikely that the aggregate farm subsidies of the past will continue at historic levels in the future. Under the new WTO regime, global food production will be increasingly based on the real cost of production plus whatever profit is required to keep farmers in business. Therefore, as global food prices rise in the future, and American and European governments are restricted by the new global trading regime from subsidizing their farmers, the price of virtual water in the Middle East and North Africa and throughout the food-importing world will also rise. According to the USDA report mentioned above, both developed and developing countries will gain from WTO liberalization. Developed countries that are major food exporters will gain immediately from the projected $31 billion in increased global food prices, of which they will share $28.5 billion ($13.3 billion to the United States), with $2.6 billion going to food exporters in the developing world. However, the report also claims food-importing countries will gain because global food price increases will spur more efficient production in their own economies, thus enabling them a "potential benefit" of $21 billion.[30] Even if accepted at face value, it is clear that such benefits will occur mostly in those developing countries with an abundance of water resources. Indeed, developing countries that produce fruits, vegetables, and other high-value crops for export to first-world markets may indeed benefit from the reduction of farm subsidies, which today undercut their competitive advantage. But when it comes to basic foodstuffs—wheat, corn, and rice—the cereal grains that sustain life for most people, the developing world cannot compete with the highly efficient mechanized corporate farms of the first world.

In future research, basic intelligence is needed on two fronts. First, we must obtain a clearer understanding of the capacity of global commodity markets to meet future virtual water needs in the form of food. Second, we must identify which Middle East and North Africa governments will most likely have the economic capacity to meet their virtual water needs though food purchases—or, perhaps more important, which ones will not. In short, is there food available in the global market, and can countries afford to buy it? Countries that cannot afford virtual water may choose instead to pursue war as a means of achieving their national interest goals. Clearly the strongest countries, or those least susceptible to intrastate or interstate conflict arising from environmental scarcity, are those that have significant water resources or the economic capacity to purchase virtual water. However, it is also clear that the relative condition of peace that has existed in the Middle East and North Africa has been maintained historically through deeply buried linkages between American and European taxpayers, their massive farm subsidies programs, and world food prices. In the future, it appears that these hidden links may be radically altered if not broken by the World Trade Organization, and, as a result, the likelihood of conflict will increase.

Conclusion: Why War Will Come

Having moved away from the conventional understanding of water strictly as a zero-sum environmental resource by reconceptualizing it in more fungible economic terms, we nevertheless believe two incompatible social trends will collide to make war in the Middle East and North Africa virtually inevitable in the future. The first trend is economic globalization. As capitalism becomes ever more embraced as the global economic philosophy, and the world increasingly embraces free-trade economics, economic growth is both required and is inevitable. The WTO will facilitate this aggregate global growth, which, on the plus side, will undoubtedly increase the ba-

sic standard of living for the average world citizen. However, the global economy will be required to meet the needs of an estimated eight billion citizens in the year 2025. Achieving growth will demand an ever-greater share of the world's existing natural resources, including water. Thus, if present regional economic and demographic trends continue, resource shortfalls will occur, with water being the most highly stressed resource in the Middle East and North Africa.

Globalization is both a cause and a consequence of the rapid spread of information technology. Thus, in the globalized world, the figurative distance between cultures, philosophies of rule, and, perhaps more important, a basic understanding of what is possible in life, becomes much shorter. Personal computers, the internet, cellular phones, fax machines, and satellite television are all working in partnership to rewire the psychological infrastructure of the citizens of the Middle East and North Africa, and the world at large. As a result, by making visible what is possible in the outside world, this cognitive liberation will bring heightened material expectations of a better life, both economically and politically. Consequently, citizens will demand more from their governments. This emerging reality will collide head-on with the second trend—political authoritarianism—that characterizes most Middle East and North Africa governments.

Throughout the region there are few governments that allow for public expression of dissent. Although Turkey, Algeria, Tunisia, and Egypt are democracies in name, these states have exhibited a propensity to revert to authoritarian tactics when deemed necessary to limit political activity among their respective populaces.[31] Likewise, while Israel is institutionally a democracy, ethnic minorities are all but excluded from the democratic process. The remainder of the Middle East and North Africa states can be described only as authoritarian regimes. In retrospect, the most fundamental common denominator of all authoritarian regimes throughout history is their fierce resistance to change. Change is seen as a threat to the regime because most authoritarian regimes base their right to rule in some form of infallibility: the infallibility of the sultan, the king, or the ruling party and its ideology. Any admission that change is needed strikes at the foundation of this inflexible infallibility. Historically, most change has occurred in the Middle East and North Africa during times of intrastate unrest and interstate war. In the coming decades, globalization will bring change that will be resisted by governments of the region. As a result, to the distant observer the future will resemble the past: periods of wholesale peace will be a rare occurrence, intense competition and low-intensity conflict will be the norm, and major wars will occur at sporadic intervals.

The wild card in this equation may be post-2004 Iraq. Operation Iraqi Freedom and the ouster of Saddam Hussein have altered the strategic political landscape. If a sustainable democracy indeed emerges in

Iraq, the country may turn away from future conflicts with its neighbors. Potential conflict between Turkey and Iraq over water may now be averted due to the fact that both countries may choose nonviolent solutions to their disputes. If President Bush's vision of a democratic Middle East comes to fruition, war may be averted. After all, there is a rich body of scholarly research regarding the "democratic peace" that suggests liberal democracies are significantly less likely to resort to war to resolve interstate disputes, and post-Saddam Iraq could serve as a key litmus test for the future of democratic reform in the region. However, it is also highly unlikely that regime change will come quickly to the moderate authoritarian states of the region that are also US allies. Decisionmakers in Washington may be able to dictate the political future of Iraq, but even America's mighty arsenal of political, economic, and military power cannot alter the basic demographic and environmental trends in the region.

NOTES

1. Alex Marshall, ed., *The State of World Population 1997* (New York: United Nations Population Fund, 1997), p. 70.
2. "The World Bank: Middle East and North Africa Data Profile," *The World Bank Group Country Data* (2000), http://www.worldbank.org/data/countrydata/countrydata.html.
3. The leading scholar in this area is Thomas Homer-Dixon. For example, see his recent book (coedited with Jessica Blitt), *Ecoviolence: Links Among Environment, Population, and Security* (New York: Rowman & Littlefield, 1998), which focuses on Chiapas, Gaza, South Africa, Pakistan, and Rwanda.
4. Mostafa Dolatyar and Tim S. Gray, *Water Politics in the Middle East: A Context for Conflict or Co-operation?* (New York: St. Martin's Press, 2000), p. 6.
5. *Human Development Report: Consumption for Human Development* (New York: United Nations Development Programme, Oxford Univ. Press, 1998), p. 55; "Water Woes Around the World," MSNBC, 9 September 2002, http://www.msnbc.com/news/802693.asp.
6. Thomas Naff, "Conflict and Water Use in the Middle East," in *Water in the Arab World: Perspectives and Prognoses,* ed. Peter Rogers and Peter Lydon (Cambridge, Mass.: Harvard Univ. Press, 1994), p. 273.
7. Alwyn R. Rouyer, *Turning Water into Politics: The Water Issue in the Palestinian-Israeli Conflict* (New York: St. Martin's Press, 2000), p. 7.
8. Ulrich Küffner, "Contested Waters: Dividing or Sharing?" in *Water in the Middle East: Potential for Conflicts and Prospects for Cooperation,* ed. Waltina Scheumann and Manuel Schiffler (New York: Springer, 1998), p. 71.
9. Miriam R. Lowi, *Water and Power: The Politics of a Scarce Resource in the Jordan River Basin* (Cambridge, Eng.: Cambridge Univ. Press, 1993), p. 1.
10. Ibid., pp. 2ff.
11. Thomas Homer-Dixon and Jessica Blitt, "Introduction: A Theoretical Overview," in *Ecoviolence: Links Among Environment, Population, and Scarcity,* ed. Thomas Homer-Dixon and Jessica Blitt (New York: Rowman & Littlefield, 1998), p. 6.
12. Thomas Homer-Dixon and Valerie Percival, "The Case of Senegal-Mauritania," in *Environmental Scarcity and Violent Conflict: Briefing Book* (Washington: American Association for the Advancement of Science and the University of Toronto, 1996), pp. 35-38.
13. Douglas Borer witnessed this agricultural problem while visiting rural areas on the Bataan peninsula in late 1985 and early 1986. The members of the New People's Army which he met were uninterested in Marxism, but they were very interested in ridding themselves of the Marcos regime. See Thomas Homer-Dixon and Valerie Percival, "The Case of the Philippines," ibid., p. 49.
14. Ted Gurr, *Why Men Rebel* (Princeton, N.J.: Princeton Univ. Press, 1970).
15. One need only look 90 miles southward from the Florida coast to find proof of this reality in Castro's Cuba.
16. Thus, Saddam Hussein was able to remain in power in Iraq until 2003 due to two essential factors. First, as noted in a recent article by James Quinlivan, Saddam had created "groups with special loyalties to the regime and the creation of parallel military organizations and multiple internal security agencies," that made Iraq essentially a "coup-proof" regime. (See James T. Quinlivan, "Coup-Proofing: Its Practice and Consequences in the Middle East," *International Security,* 24 [Fall 1999], 131-65.) Second, Saddam had convinced a significant portion of his people that the United States (and Britain) were responsible for their suffering. Thus, as long as these perceptions held and Saddam was able to command loyalty of the inner regime, his ouster from power by domestic sources remained unlikely.
17. Ewan W. Anderson, "Water: The Next Strategic Resource," in *The Politics of Scarcity: Water in the Middle East,"* ed. Joyce R. Starr and Daniel C. Stoll (Boulder, Colo.: Westview Press, 1988), p. 1.
18. Hussein A. Amery and Aaron T. Wolf, "Water, Geography, and Peace in the Middle East," in *Water in the Middle East: A Geography of Peace,* ed. Hussein A. Amery and Aaron T. Wolf (Austin: Univ. of Texas Press, 2000).
19. Nubit Kliot, *Water Resources and Conflict in the Middle East* (London and New York: Routledge, 1994), p. v, as quoted in Dolatyar and Gray, p. 9.
20. Rouyer, p. 9.
21. Martin Sherman, *The Politics of Water in the Middle East: An Israeli Perspective on the Hydro-Political Aspects of the Conflict* (New York: St. Martin's Press, 1999), p. xi.
22. Jochen Renger, "The Middle East Peace Process: Obstacles to Cooperation Over Shared Waters," in *Water in the Middle East: Potential for Conflict and Prospects for Cooperation,* ed. Waltina Scheumann and Manuel Schiffler (New York: Springer, 1998), p. 50.
23. Thus, even though the Saudi government is much more Islamized in religious terms than that of the Iraqis or Syrians, as long as the Saudi government is pro-US and serves US interests in supplying cheap oil, it receives the benevolent "moderate" label, while the more secularized Iraqis and Syrians have been labeled with the prerogative labels "radical" or "rogue-states."
24. Tony Allan, "Watersheds and Problemsheds: Explaining the Absence of Armed Conflict over Water in the Middle East," in *MERNIA: Middle East Review of International Affairs Journal,* 2 (March 1998), http:// biu.ac.il/SOC/besa/meria/journal/1998/issue1/jv2n1a7.html.
25. Ibid.
26. Ibid.

27. Ibid.
28. Prices from 26 June 2001 quoted at `http://www.usafutures.com/commodityprices.htm`.
29. "Agricultural Policy Reform in the WTO—The Road Ahead," in *ERS Agricultural Economics Report*, No. 802, ed. Mary E. Burfisher (Washington: US Department of Agriculture, May 2001), p. iii.
30. Ibid., p. 6.
31. For instance, as of 2004, Freedom House (`http://www.freedomhouse.org/`) classifies Algeria, Egypt, and Tunisia as "not free," and Turkey as only "partly free."

Jason J. Morrissette is a doctoral candidate and instructor of record in the School of Public and International Affairs at the University of Georgia. He is currently writing his dissertation on the political economy of water scarcity and conflict.

Douglas A. Borer (Ph.D., Boston University, 1993) is an Associate Professor in the Department of Defense Analysis at the Naval Postgraduate School. He recently served as Visiting Professor of Political Science at the US Army War College. Previously he was Director of International Studies at Virginia Tech, and he has taught overseas in Fiji and Australia. Dr. Borer is a former Fulbright Scholar at the University of Kebangsaan Malaysia, and has published widely in the areas of security, strategy, and foreign policy.

From *Parameters*, Winter 2004–2005, pp. 86–101. Copyright © 2005 by U.S. Army War College. Reprinted by permission of the authors.

The Irony of Climate

Archaeologists suspect that a shift in the planet's climate thousands of years ago gave birth to agriculture. Now climate change could spell the end of farming as we know it.

BRIAN HALWEIL

High in the Peruvian Andes, a new disease has invaded the potato fields in the town of Chacllabamba. Warmer and wetter weather associated with global climate change has allowed late blight—the same fungus that caused the Irish potato famine—to creep 4,000 meters up the mountainside for the first time since humans started growing potatoes here thousands of years ago. In 2003, Chacllabamba farmers saw their crop of native potatoes almost totally destroyed. Breeders are rushing to develop tubers resistant to the "new" disease that retain the taste, texture, and quality preferred by Andean populations.

Meanwhile, old-timers in Holmes County, Kansas, have been struggling to tell which way the wind is blowing, so to speak. On the one hand, the summers and winters are both warmer, which means less snow and less snowmelt in the spring and less water stored in the fields. On the other hand, there's more rain, but it's falling in the early spring, rather than during the summer growing season. So the crops might be parched when they need water most. According to state climatologists, it's too early to say exactly how these changes will play out—if farmers will be able to push their corn and wheat fields onto formerly barren land or if the higher temperatures will help once again to turn the grain fields of Kansas into a dust bowl. Whatever happens, it's going to surprise the current generation of farmers.

Asian farmers, too, are facing their own climate-related problems. In the unirrigated rice paddies and wheat fields of Asia, the annual monsoon can make or break millions of lives. Yet the reliability of the monsoon is increasingly in doubt. For instance, El Niño events (the cyclical warming of surface waters in the eastern Pacific Ocean) often correspond with weaker monsoons, and El Niños will likely increase with global warming. During the El Niño-induced drought in 1997, Indonesian rice farmers pumped water from swamps close to their fields, but food losses were still high: 55 percent for dryland maize and 41 percent for wetland maize, 34 percent for wetland rice, and 19 percent for cassava. The 1997 drought was followed by a particularly wet winter that delayed planting for two

months in many areas and triggered heavy locust and rat infestations. According to Bambang Irawan of the Indonesian Center for Agricultural Socio-Economic Research and Development, in Bogor, this succession of poor harvests forced many families to eat less rice and turn to the less nutritious alternative of dried cassava. Some farmers sold off their jewelry and livestock, worked off the farm, or borrowed money to purchase rice, Irawan says. The prospects are for more of the same: "If we get a substantial global warming, there is no doubt in my mind that there will be serious changes to the monsoon," says David Rhind, a senior climate researcher with NASA's Goddard Institute for Space Studies.

Archaeologists believe that the shift to a warmer, wetter, and more stable climate at the end of the last ice age was key for humanity's successful foray into food production. Yet, from the American breadbasket to the North China Plain to the fields of southern Africa, farmers and climate scientists arc finding that generations-old patterns of rainfall and temperature are shifting. Farming may be the human endeavor most dependent on a stable climate—and the industry that will struggle most to cope with more erratic weather, severe storms, and shifts in growing season lengths. While some optimists are predicting longer growing seasons and more abundant harvests as the climate warms, farmers are mostly reaping surprises.

Toward the Unknown (Climate) Region

For two decades, Hartwell Allen, a researcher with the University of Florida in Gainesville and the U.S. Department of Agriculture, has been growing rice, soybeans, and peanuts in plastic, greenhouse-like growth chambers that allow him to play God. He can control—"rather precisely"—the temperature, humidity, and levels of atmospheric carbon. "We grow the plants under a daily maximum/minimum cyclic temperature that would mimic the real world cycle," Allen says. His lab has tried regimes of 28 degrees C day/18 degrees C night, 32/22, 36/26, 40/30, and 44/34. "We ran one experiment to 48/38, and got very few surviving plants," he says. Allen found that while a doubling of carbon dioxide and a slightly increased temperature stimulate seeds to germinate and the plants to grow larger and lusher, the higher temperatures are deadly when the plant starts producing pollen. Every stage of the process—pollen transfer, the growth of the tube that links the pollen to the seed, the viability of the pollen itself—is highly sensitive. "It's all or nothing, if pollination isn't successful," Allen notes. At temperatures above 36 degrees C during pollination, peanut yields dropped about six percent per degree of temperature increase. Allen is particularly concerned about the implications for places like India and West Africa, where peanuts are a dietary staple and temperatures during the growing season are already well above 32 degrees C: "In these regions the crops are mostly rainfed. If global warming also leads to drought in these areas, yields could be even lower."

As plant scientists refine their understanding of climate change and the subtle ways in which plants respond, they are beginning to think that the most serious threats to agriculture will not be the most dramatic: the lethal heatwave or severe drought or endless deluge. Instead, for plants that humans have bred to thrive in specific climatic conditions, it is those subtle shifts in temperatures and rainfall during key periods in the crops' lifecycles that will be most disruptive. Even today, crop losses associated with background climate variability are significantly higher than those caused by disasters such as hurricanes or flooding.

John Sheehy at the International Rice Research Institute in Manila has found that damage to the world's major grain crops begins when temperatures climb above 30 degrees C during flowering. At about 40 degrees C, yields are reduced to zero. "In rice, wheat, and maize, grain yields are likely to decline by 10 percent for every 1 degree C increase over 30 degrees. We are already at or close to this threshold," Sheehy says, noting regular heat damage in Cambodia, India, and his own center in the Philippines, where the average temperature is now 2.5 degrees C higher than 50 years ago. In particular, higher night-time temperatures forced the plants to work harder at respiration and thus sapped their energy, leaving less for producing grain. Sheehy estimates that grain yields in the tropics might fall as much as 30 percent over the next 50 years, during a period when the region's already malnourished population is projected to increase by 44 percent. (Sheehy and his colleagues think a potential solution is breeding rice and other crops to flower early in the morning or at night so that the sensitive temperature process misses the hottest part of the day. But, he says, "we haven't been successful in getting any real funds for the work.") The world's major plants can cope with temperature shifts to some extent, but since the dawn of agriculture farmers have selected plants that thrive in stable conditions.

Climatologists consulting their computer climate models see anything but stability, however. As greenhouse gases trap more of the sun's heat in the Earth's atmosphere, there is also more energy in the climate system, which means more extreme swings—dry to wet, hot to cold. (This is the reason that there can still be severe winters on a warming planet, or that March 2004 was the third-warmest month on record after one of the coldest winters ever.) Among those projected impacts that climatologists have already observed in most regions: higher maximum temperatures and more hot days, higher minimum temperatures and fewer cold days, more variable and extreme rainfall events, and increased summer drying and associated risk of drought in continental interiors. All of these conditions will likely accelerate into the next century.

Cynthia Rosenzweig, a senior research scholar with the Goddard Institute for Space Studies at Columbia University, argues that although the climate models will always be improving, there are certain changes we can already predict with a level of confidence. First, most studies indicate "intensification of the hydrological cycle," which essentially means more droughts and floods, and more variable and extreme rainfall. Second, Rosenzweig says, "basically every study has shown that there will be increased incidence of crop pests." Longer growing seasons mean more generations of pests during the summer, while shorter and warmer winters mean that fewer adults, larvae, and eggs will die off.

Third, most climatologists agree that climate change will hit farmers in the developing world hardest. This is partly a result of geography. Farmers in the tropics already find themselves near the temperature limits for most major crops, so any warming is likely to push their crops over the top. "All increases in temperature, however small, will lead to decreases in production," says Robert Watson, chief scientist at the World Bank and former chairman of the Intergovernmental Panel on Climate Change. "Studies have consistently shown that agricultural regions in the developing world are more vulnerable, even before we consider the ability to cope," because of poverty, more limited irrigation technology, and lack of weather tracking systems. "Look at the coping strategies, and then it's a real double whammy," Rosenzweig says. In sub-Saharan Africa—ground zero of global hunger, where the number of starving people has doubled in the last 20 years—the current situation will undoubtedly be exacerbated by the climate crisis. (And by the 2080s, Watson says, projections indicate that even temperate latitudes will begin to approach the upper limit of the productive temperature range.)

Coping With Change

"Scientists may indeed need decades to be sure that climate change is taking place," says Patrick Luganda, chairman of the Network of Climate Journalists in the Greater Horn of Africa. "But, on the ground, farmers have no choice but to deal with the daily reality as best they can." Luganda says that several years ago local farming communities in Uganda could determine the onset of rains and their cessation with a fair amount of accuracy. "These days there is no guarantee that the long rains will start, or stop, at the usual time," Luganda says. The Ateso people in north-central Uganda report the disappearance of asisinit, a swamp grass favored for thatch houses because of its beauty and durability. The grass is increasingly rare because farmers have started to plant rice and millet in swampy areas in response to more frequent droughts. (Rice farmers in Indonesia coping with droughts have done the same.) Farmers have also begun to sow a wider diversity of crops and to stagger their plantings to hedge against abrupt climate shifts. Luganda adds that repeated crop failures have pushed many farmers into the urban centers: the final coping mechanism.

The many variables associated with climate change make coping difficult, but hardly futile. In some cases, farmers may need to install sprinklers to help them survive more droughts. In other cases, plant breeders will need to look for crop varieties that can withstand a greater range of temperatures. The good news is that many of the same changes that will help farmers cope with climate change will also make communities more self-sufficient and reduce dependence on the long-distance food chain.

Planting a wider range of crops, for instance, is perhaps farmers' best hedge against more erratic weather. In parts of Africa, planting trees alongside crops—a system called agroforestry that might include shade coffee and cacao, or leguminous trees with corn—might be part of the answer. "There is good reason to believe that these systems will be more resilient than a maize monoculture," says Lou Verchot, the lead scientist on climate change at the International Centre for Research in Agroforestry in Nairobi. The trees send their roots considerably deeper than the crops, allowing them to survive a drought that might damage the grain crop. The tree roots will also pump water into the upper soil layers where crops can tap it. Trees improve the soil as well: their roots create spaces for water flow and their leaves decompose into compost. In other words, a farmer who has trees won't lose everything. Farmers in central Kenya are using a mix of coffee, macadamia nuts, and cereals that results in as many as three marketable crops in a good year. "Of course, in any one year, the monoculture will yield more money," Verchot admits, "but farmers need to work on many years." These diverse crop mixes are all the more relevant since rising temperatures will eliminate much of the traditional coffee- and tea-growing areas in the Caribbean, Latin America, and Africa. In Uganda, where coffee and tea account for nearly 100 percent of agricultural exports, an average temperature rise of 2 degrees C would dramatically reduce the harvest, as all but the highest altitude areas become too hot to grow coffee.

In essence, farms will best resist a wide range of shocks by making themselves more diverse and less dependent on outside inputs. A farmer growing a single variety of wheat is more likely to lose the whole crop when the temperature shifts dramatically than a farmer growing several wheat varieties, or better yet, several varieties of plants besides wheat. The additional crops help form a sort of ecological bulwark against blows from climate change. "It will be important to devise more resilient agricultural production systems that can absorb and survive more variability," argues Fred Kirschenmann, director of the Leopold Center for Sustainable Agriculture at Iowa State University. At his own family farm in North Dakota, Kirschenmann has struggled with two years of abnormal weather that nearly eliminated one crop and devastated another. Diversified farms will cope better with drought, increased pests, and a range

of other climate-related jolts. And they will tend to be less reliant on fertilizers and pesticides, and the fossil fuel inputs they require. Climate change might also be the best argument for preserving local crop varieties around the world, so that plant breeders can draw from as wide a palette as possible when trying to develop plants that can cope with more frequent drought or new pests.

Farms with trees planted strategically between crops will not only better withstand torrential downpours and parching droughts, they will also "lock up" more carbon. Lou Verchot says that the improved fallows used in Africa can lock up 10-20 times the carbon of nearby cereal monocultures, and 30 percent of the carbon in an intact forest. And building up a soil's stock of organic matter—the dark, spongy stuff in soils that stores carbon and gives them their rich smell—not only increases the amount of water the soil can hold (good for weathering droughts), but also helps bind more nutrients (good for crop growth).

Best of all, for farmers at least, systems that store more carbon are often considerably more profitable, and they might become even more so if farmers get paid to store carbon under the Kyoto Protocol. There is a plan, for instance, to pay farmers in Chiapas, Mexico, to shift from farming that involves regular forest clearing to agroforestry. The International Automobile Federation is funding the project as part of its commitment to reducing carbon emissions from sponsored sports car races. Not only that, "increased costs for fossil fuels will accelerate demand for renewable energies," says Mark Muller of the Institute for Agriculture and Trade Policy in Minneapolis, Minnesota, who believes that farmers will find new markets for biomass fuels like switchgrass that can be grown on the farm, as well as additional royalties from installing wind turbines on their farms.

However, "carbon farming is a temporary solution," according to Marty Bender of the Land Institute's Sunshine Farm in Salina, Kansas. He points to a recent paper in *Science* showing that even if America's soils were returned to their pre-plow carbon content—a theoretical maximum for how much carbon they could lock up—this would be equal to only two decades of American carbon emissions. "That is how little time we will be buying," Bender says, "despite the fact that it may take a hundred years of aggressive, national carbon farming and forestry to restore this lost carbon." (Cynthia Rosenzweig also notes that the potential to lock up carbon is limited, and that a warmer planet will reduce the amount of carbon that soils can hold: as land heats up, invigorated soil microbes respire more carbon dioxide.)

"We really should be focusing on energy efficiency and energy conservation to reduce the carbon emissions by our national economy," Bender concludes. That's why Sunshine Farm, which Bender directs, has been farming without fossil fuels, fertilizers, or pesticides in order to reduce its contribution to climate change and to find an inherently local solution to a global problem. As the name implies, Sunshine Farm runs essentially on sunlight. Homegrown sunflower seeds and soybeans become biodiesel that fuels tractors and trucks. The farm raises nearly three-fourths of the feed—oats, grain sorghum, and alfalfa—for its draft horses, beef cattle, and poultry. Ma-

The atmosphere, which sets the stage for climate dynamics, is a global commons. It has no gatekeeper—everyone can access it—and it has a limited capacity to absorb emissions before climate stability is undermined. Societies have developed a variety of gate-keeping solutions to help manage commons resources. One, privatization, would be difficult to apply to the atmosphere, although carbon-trading schemes might reduce emissions if caps are set low enough. Governments might also act as gatekeepers by means of treaties, such as the Kyoto Protocol, that limit emissions.

nure and legumes in the crop rotation substitute for energy-gobbling nitrogen fertilizers. A 4.5-kilowatt photovoltaic array powers the workshop tools, electric fencing, water pumps, and chick brooding pens. The farm has eliminated an amount of energy equivalent to that used to make and transport 90 percent of its supplies. (Including the energy required to make the farm's machinery lowers the figure to 50 percent, still a huge gain over the standard American farm.)

But these energy savings are only part of this distinctly local solution to an undeniably global problem, Bender says. "If local food systems could eliminate the need for half of the energy used for food processing and distribution, then that would save 30 percent of the fossil energy used in the U.S. food system," Bender reasons. "Considering that local foods will require some energy use, let's round the net savings down to 25 percent. In comparison, on-farm direct and indirect energy consumption constitutes 20 percent of energy use in the U.S. food system. Hence, local food systems could potentially save more energy than is used on American farms."

In other words, as climate tremors disrupt the vast intercontinental web of food production and rearrange the world's major breadbaskets, depending on food from distant suppliers will be more expensive and more precarious. It will be cheaper and easier to cope with local weather shifts, and with more limited supplies of fossil fuels, than to ship in a commodity from afar.

Agriculture is in third place, far behind energy use and chlorofluorocarbon production, as a contributor to climate warming. For farms to play a significant role, changes in cropping practices must happen on a large scale, across large swaths of India and Brazil and China and the American Midwest. As Bender suggests, farmers will be able to shore up their defenses against climate change, and can make obvious reductions in their own energy use which could save them money.

But the lasting solution to greenhouse gas emissions and climate change will depend mostly on the choices that everyone else makes. According to the London-based NGO Safe Alliance, a basic meal—some meat, grain, fruits, and vegetables—using imported ingredients can easily generate four times the greenhouse gas emissions as the same meal with ingredients from local sources. In terms of our personal contribution to climate change, eating local can be as important as driving a fuel-efficient car, or giving up the car for a bike. As politicians struggle to muster the will power to confront the climate crisis, ensuring that farmers have a less erratic climate in which to raise the world's food shouldn't be too hard a sell.

Brian Halweil *is a senior researcher at Worldwatch Institute, and the author of* Eat Here: Reclaiming Homegrown Pleasures in a Global Supermarket.

World Population, Agriculture, and Malnutrition

Increases in food production, per hectare of land, have not kept pace with increases in populaton, and the planet has viturally no more arable land or fresh water to spare. As a result, per-capita cropland has fallen by more than half since 1960, and per-capita production of grains, the basic food, has been falling worldwide for 20 years.

David Pimentel and Anne Wilson

Entering the new millennium, stark contrasts are apparent between the availability of natural resources and the demands of billions of humans who require them for their survival. According to the Population Reference Bureau, each day almost a quarter-million people are added to the roughly 6.4 billion who already exist. Yet the stocks of natural resources that support human life—food, fresh water, quality soil, energy, and biodiversity—are being polluted, degraded, and depleted.

Global population has doubled during the last 45 years. If the present growth rate of 1.3 percent per year persists, the population will double again within a mere 50 years. Growth rates vary from one country or region to another. For example, China's present population of 1.4 billion, despite the governmental policy of permitting only one child per couple, is still growing at an annual rate of 0.6 percent. Although China recognizes its serious overpopulation problem and recently passed legislation strengthening the policy, its young age structure means that the number of Chinese will continue to increase for another 50 years. India, with nearly 1.1 billion people (living on approximately one-third the land of either the United States or China), has a current population growth rate of 1.7 percent per year. This translates to a doubling time of 41 years. Taken together, the populations of China and India constitute more than one-third of the total world population. In Africa, despite the AIDS epidemic, the populations of most countries also are expanding. The populations of Chad and Ethiopia, for example, are projected to double in 21 and 23 years, respectively.

But the problem is hardly confined to the developing world. The U.S. population—among the most heavily consuming in the world—is growing rapidly. Now standing at nearly 300 million, it has doubled during the past 60 years. The U.S. Bureau of the Census reported in 2003 that sustaining the current growth rate of about 1.1 percent per year will double the population to 600 million in less than 70 years.

Current United Nations estimates of population stabilization at about 9 billion people by 2050 are questionable, mainly because of the very young age structure of the current world population and the momentum it fosters. A large share of the population is concentrated within the 15-to-40 range, where reproductive rates are high. Even if all the people in the world adopted a policy of only two children per couple, it would take approximately 70 years before the world population would finally stabilize at about 12 billion, twice the current level.

Land

Many human beings already suffer from hunger and/or malnourishment. The United Nations Food and Agricultural Organization (FAO) reports that the quantity of food produced per capita has been declining since 1984, based on available cereal grains, which make up about 80 percent of the world's food supply. Although grain yields per hectare in both developed and developing countries are still increasing, the rate of increase is slowing. According to the U.S. Department of Agriculture, U.S. grain yields increased at about 3 percent per year between 1950 and 1980, but since then the annual rate of increase for corn and other major other grains has been only about 1 percent. Yet the World Health Organization estimates that more than 3 billion people are malnourished (deficient in intake of calories, protein, iron, iodine, and/or vitamins A, B, C, and D). This is the largest number and proportion of malnourished people ever reported.

At the same time, cropland resources are under severe strain. FAO Food Balance Sheets show that more than 99.7 percent of human food (calories) comes from the terrestrial environment, while less than 0.3 percent comes from the oceans and other aquatic ecosystems. Of the total of 13 billion hectares of land area on Earth, cropland accounts for 11 percent, pastureland 27 per cent, forested land 32 percent, and urban lands 9 per cent. Most of the remaining 21 percent is unsuitable for crops, pasture, and/or forests because the soil is too infertile or shallow to support plant growth, or the climate and region are too cold, dry, steep, stony, or wet.

In 1960, when the world population numbered only 3 billion, approximately 0.5 hectare of cropland per capita was available, the minimum area considered essential for the production of a diverse, healthy, nutritious diet of plant and animal products like that enjoyed widely in the United States and Europe. But as the human population continues to increase and expand its economic activity and related artifacts, including transport systems and urban structures, vital cropland is being covered and lost from production. Globally, available per-capita cropland is now about 0.23 hectare. In the United States, there is already about 0.4 hectare (1 acre) of land per person tied up in urban buildings and highways and the available cropland per capita has shrunk over the last 30 years or so to 0.5 hectare. In China, per-capita cropland has declined to 0.08 hectare from 0.11 hectare 25 years ago, due to continued population growth as well as extreme soil erosion and degradation. This relatively small amount of cropland provides the Chinese people a primarily vegetarian diet.

The United States produces 1,481 kilograms per year of agricultural products for each American, while the Chinese food supply averages only 785 kilograms per year per capita (mostly grains in both cases). Lester Brown of the Earth Policy Institute has suggested that by all available measurements the Chinese have reached or exceeded the limits of their agricultural system. The Chinese reliance on large inputs of fossil fuel-based fertilizers to compensate for shortages of arable land and severely eroded soils, combined with their limited fresh water supply, suggests severe problems looming ahead. Even now, China imports large amounts of grain from the United States (which also relies heavily on fossil inputs for agriculture) and other nations, and is expected to increase imports of grains in the near future.

The decline of per-capita cropland is aggravated by the degradation of soils. Throughout the world, current erosion rates are higher than ever. According to a study for the International Food Policy Research Institute, each year an estimated 10 million hectares of cropland worldwide are abandoned due to soil erosion and diminished production caused by erosion. Another 10 million hectares are critically damaged each year by salinization, in large part as a result of irrigation and/or improper drainage methods. This loss amounts to more than 1.3 percent of total cropland annually. Most of the additional cropland needed to replace yearly losses comes from the world's forest areas. The urgent need to increase crop production accounts for more than 60 percent of the massive deforestation now occurring worldwide.

Erosion losses are critical because topsoil renewal is extremely slow. It takes about 500 years for 2.5 centimeters (1 inch) of topsoil to reform under agricultural conditions. Soil erosion rates on cropland range from about 10 metric tons per hectare per year (t/ha/yr) in the United States to 40 t/ha/yr in China. During the past 30 years, the rate of soil loss throughout Africa has increased 20-fold. A 1996 study in India found that as much as 5,600 t/ha/yr of soil were lost under some arid and windy conditions.

Some crops can be grown under artificial conditions using hydroponic techniques, but the cost (in energy and dollars) is approximately 10 times that of conventional agriculture. Such systems are neither affordable nor sustainable for the future.

Water

The availability of adequate supplies of fresh water for human direct use and agriculture is already critical in many regions, especially the Middle East and parts of North Africa where low rainfall is endemic. Surface waters, for instance, are often poorly managed, resulting in water shortages and pollution, both of which threaten humans and aquatic biota. Groundwater—rainfall lying in underground aquifers—is another vital source of water for agriculture; it too is often used profligately. Aquifers recharge very slowly, usually at rates of 0.1 to 0.3 percent per year, according to the UN Environment Programme. At these rates, groundwater resources must be carefully managed to prevent overuse and depletion, but this wisdom is often ignored. For example, in Tamil Nadu, India, groundwater levels dropped 25 to 30 meters during the 1970s because of excessive pumping for irrigation. In Beijing, China, the groundwater level is falling at a rate of about 1 meter per year, while in Tianjin, China, it is dropping 4.4 meters per year. In the United States, groundwater overdraft is high, averaging 25 percent greater than replacement rates. The capacity of the Ogallala aquifer, which underlies parts of Nebraska, South Dakota, Colorado, Kansas, Oklahoma, New Mexico, and Texas, has decreased by 33 percent since about 1950. Withdrawal from the Ogallala is three times faster than its recharge rate. Aquifers in some parts of Arizona are being overpumped more than 10 times faster than the recharge rate.

Irrigation enables crop production in arid regions, provided there is an adequate source of fresh water and enough energy (generally fossil in origin) to pump and move the water. About 70 percent of the water removed from all sources worldwide is used solely for irrigation. Of this amount, about two-thirds is consumed by growing plants and is non-recoverable, i.e, lost to the hydrologic cycle via evapotranspiration. Irrigation is less water-efficient than rainfed watering of crops, and the limitations of surface and ground water resources for irrigation, its high economic costs, and the large energy inputs required will tend to limit future agricultural irrigation, especially in developing nations that cannot afford such expenditures.

Pollution is a major threat to maintaining ample fresh water resources. Although considerable water pollution has been documented in developed nations like the United States, the problem is of greatest concern in countries where water regulations are not rigorously enforced or do not exist. This is common in most developing countries, which (according to the World Health Organization) discharge 95 percent of untreated urban sewage directly into surface waters. For instance, of India's 3,119 towns and cities, only 209 have even partial sewage treatment facilities, and a mere eight possess full facilities. Downstream, the polluted water is used for drinking, bathing, and washing.

Energy

Humans have relied on various sources of power for centuries, beginning of course with solar energy—fundamental to nearly all natural ecosystems—and their own muscle power. Other sources have included animals, wind, tides, water, wood, coal, gas, oil, and nuclear energy. Since about 1700, increasingly abundant fossil fuel energy supplies have made it possible to augment agricultural production to feed an increasing number of humans, as well as improve the general quality of human life in many ways.

Since the fossil era began, the rate of energy use from all sources has grown even faster than world population. From 1970 to 1995, energy use increased at a rate of 2.5 percent per year (doubling every 30 years), compared with worldwide population growth of 1.7 percent per year (doubling about every 40 years). During the next 20 years, energy use is projected to increase by 4.5 percent per year (doubling every 16 years) and population by 1.3 percent per year (doubling every 54 years).

Although about half of all the solar energy captured by worldwide photosynthesis is used by humans, this amount is still inadequate to meet all human needs for food and other purposes. To make up for this shortfall, the world consumes a lot of fossil energy: about 345 quadrillion British thermal units (3.64×10 to the 20th joules) of it in 2001, with the United States alone accounting for 83 quadrillion Btu of fossil energy consumption that year. (Each year, in fact, the U.S. population uses twice as much fossil energy as all the solar energy captured by harvested U.S. crops, forest products, and other vegetation.) A great deal of this supplemental energy goes into agriculture. In China, for instance, while most fossil energy is used by industry, about one-quarter is used for agriculture and the food production system. Like some other developing nations with high rates of population growth, China is increasing fossil fuel use to augment agricultural production of food and fiber. Since 1955 Chinese agriculture has boosted energy use 100-fold for fertilizers, pesticides, and irrigation.

In general, however, in what may be a harbinger of an approaching crunch, the International Fertilizer Organization reports that fertilizer production has declined by more than 17 percent since 1989, especially in the developing countries, because of fossil fuel shortages and resulting high prices. In fact, the projected global availability of fossil energy resources for fertilizers, not to mention all other purposes is discouraging.

British Petroleum and other authorities have estimated that the world supply of oil would last approximately 50 years at current production rates. Perhaps somewhat optimistically, the global natural gas supply is considered adequate for about 50 years and coal supplies for at least 100 years. However, demand is not static, but rising dramatically. Moreover, even adequate production in one place may not translate into adequate supply elsewhere; natural gas supplies are already in short supply in the United States and U.S. reserves may be depleted in as little as 20 years, yet transporting natural gas in liquid form to the United States from places where it is abundant poses serious technical and financial challenges.

An even more sobering prospect is that of the imminent peak in production of oil and natural gas. The experience of the United States may portend the fate of global oil and gas production. Walter Youngquist, formerly an oil geologist with Exxon, reports that current oil and gas exploration drilling data have not borne out some of the earlier optimistic estimates of these resources yet to be found in the United States. U.S. oil production peaked around 1970 and has been declining ever since. Youngquist estimates that about 90 percent of U.S. oil resources already have been mined. A key consequence is that U.S. net imports of oil rose to about 53 percent of total consumption in 2002 and are still going up, placing the economy at risk from fluctuating oil prices and difficult political situations.

This scenario is likely to repeat globally. Predictions for the peak year of global oil production, for instance, range from 2005 to 2035, but in our view the most plausible of these estimates are those in the 2005-2010 range. Whenever the peak occurs, the impacts of rising energy prices on the economies of most nations will be profound. Modern agriculture, no less than other sectors of the economy, depends on large quantities of oil and natural gas (used in nitrogen fertilizer production) and higher energy prices are already having an impact on agricultural production.

Wild Facts

These facts speak for themselves. They starkly signal a rapidly approaching time of grave challenge for the agricultural system. During the 20th century, increased food production—supporting a period of unprecedented growth in the world population—depended on the availability of cheap fossil energy, primarily oil and natural gas. The consequent expansion of human needs and activities has been depleting the land, water, and biological resources that are essential for sustainable agricultural production. Already, more than 3 billion people in the world are malnourished, yet per-capita production of cereal grains, basic world foods, has continued to decline for the past 20 years, despite all the new biotechnologies.

As the world population continues to expand, all vital natural resources will have to be divided among increasing numbers of people and per-capita availability will decline to low levels. When this occurs, we believe that it will become quite difficult to maintain prosperity, a quality life, and even personal freedoms for those who already enjoy them, much less secure those benefits for the billions currently living without. Meeting this challenge will test humanity's resourcefulness and goodwill to the utmost.

David Pimentel is a professor in the College of Agriculture and Life Sciences at Cornell University. *Anne Wilson* is a research assistant at Cornell.

From *World Watch Magazine*, September/October 2004, pp. 22–25. Copyright © 2004 by Worldwatch Institute, www.worldwatch.org. Reprinted by permission.

UNIT 3

Energy: Present and Future Problems

Unit Selections

Key Points to Consider

- What is the nature of the legal concept of "mineral rights" that can have negative impacts on the environmental quality of farm and ranch land? Should the owners of mineral rights have unlimited access to fossil fuels, despite the damage they may produce to the land's surface?

- What are some of the major benefits of such alternate energy sources as solar power and wind power? Do these energy alternatives really have a chance at competing with fossil fuels for a share of the global energy market?

- Why should the hydrogen fuel cell be considered as an important component of energy systems of the future? What limiting factors have stood in the way of development of hydrogen as an alternative fuel source?

Student Website

www.mhcls.com/online

Internet References

Further information regarding these websites may be found in this book's preface or online.

Alliance for Global Sustainability (AGS)
http://globalsustainability.org/

Alternative Energy Institute, Inc.
http://www.altenergy.org

Energy and the Environment: Resources for a Networked World
http://zebu.uoregon.edu/energy.html

Institute for Global Communication/EcoNet
http://www.igc.org/

Nuclear Power Introduction
http://library.thinkquest.org/17658/pdfs/nucintro.pdf

U.S. Department of Energy
http://www.energy.gov

There has been a tendency, particularly in the developed nations of the world, to view the present high standards of living as exclusively the benefit of a high-technology society. In the "techno-optimism" of the post–World War II years, prominent scientists described the technical-industrial civilization of the future as being limited only by a lack of enough trained engineers and scientists to build and maintain it. This euphoria reached its climax in July 1969 when American astronauts walked upon the surface of the Moon, an accomplishment brought about solely by American technology—or so it was supposed. It cannot be denied that technology has been important in raising standards of living and permitting Moon landings, but how much of the growth in living standards and how many outstanding and dramatic feats of space exploration have been the results of technology alone? The answer is few—for in many of humankind's recent successes, the contributions of technology to growth have been no more important than the availability of incredibly cheap energy resources, particularly petroleum, natural gas, and coal.

As the world's supply of recoverable (inexpensive) fossil fuels dwindles and becomes, as evidenced by recent international events, more important as a factor in international conflict, it becomes increasingly clear that the energy dilemma is the most serious economic and environmental threat facing the Western world and its high standard of living. With the exception of the specter of global climate change, the scarcity and cost of conventional (fossil fuel) energy is probably the most serious threat to economic growth and stability in the rest of the world as well. The economic dimensions of the energy problem are rooted in the instabilities of monetary systems produced by and dependent on inexpensive energy. The environmental dimensions of the problem are even more complex, ranging from the hazards posed by the development of such alternative sources as nuclear power to the inability of developing world farmers to purchase necessary fertilizer produced from petroleum, which has suddenly become very costly, and to the enhanced greenhouse effect created by fossil fuel consumption. The only answers to the problems of dwindling and geographically vulnerable, inexpensive energy supplies are conservation and sustainable energy technology. Both require a massive readjustment of thinking, away from the exuberant notion that technology can solve any problem. The difficulty with conservation, of course, is a philosophical one that grows out of the still-prevailing optimism about high technology. Conservation is not as exciting as putting a man on the Moon. Its tactical applications—caulking windows and insulating attics are dog-paddle technologies to people accustomed to the crawl stroke. Does a solution to this problem entail the technological fixes of which many are so enamored? Probably not, as it appears that the accelerating energy demands of the world's developing nations will most likely be met first by increased reliance on the traditional (and still relatively cheap) fossil fuels. Although there is a need to reduce this reliance, there are few ready alternatives available to the poorer developing countries. It would appear that conservation is the only option. But conservation requires social and economic change in

order to solve the environmental problems related to fossil fuel use, as do alternate strategies for acquiring energy.

Alternate approaches—technological, social, and economic—to the dominant forms of the extraction and use of energy form the basis for nearly all the articles in this section. In the first article, science writer Keith Kloor suggests in "Powder Keg" that increased natural gas development in Wyoming's Powder River Basin, while producing cleaner and more-efficient energy, has environmental and social consequences. Here, large energy companies own the mineral rights under the land owned and operated by livestock ranchers. The disjunctions between what the energy companies are allowed to do at the surface because they own the subsurface, and the best uses of the surface of the land by ranchers, are profound. A three-way struggle between big energy, the federal and state governments, and environmental advocates puts ranchers in the unusual position of siding with environmentalists. In "Personalized Energy," energy expert Stephen Millett suggests that, since fossil fuels have so many negatives for the environment and since alternative energy systems such as wind and solar are still a long ways from being capable of satisfying current demands, a new approach to energy shortages is needed. This new approach actually represents a return to an older system of energy acquisition and use in which local communities and individual households played a more important role in providing their own energy. He points to intriguing new developments in fuel cell technologies, for example, as mechanisms through which energy production may be redistributed from the huge producing plants to much smaller ones serving localities.

The next two selections in the section deal with some of the most promising renewable or alternative energy strategies. In "Wind Power: Obstacles and Opportunities," Martin J. Pasqualetti of Arizona State University suggests that one of the world's first energy sources to be harnessed for human use is also one of the most promising alternative energy sources. While the most traditional use of wind power has been to pump water, the same kinds of environments that produce efficient windmill water pumps also produce efficiency wind turbines that "pump energy" rather than water. The use of wind-driven turbines to produce electricity, while still a developing technology—and one that often encounters strident local opposition when it interferes with land use or aesthetics of scenery—holds great promise for the future. Particularly in the Great Plains of the United States, the dedication of areas of land to electrical energy production could slow a decades-long trend toward farm abandonment and consolidation and restore the economic vitality of small, former agricultural communities. In "Hydrogen: Waiting for the Revolution," freelance writer Bill Keenan argues that hydrogen may well be the fuel of the future and that it has enormous benefits for both the economy and the environment. As ideal as hydrogen seems to be as a fuel source—and as technologically feasible as scientists suggest—there still is little concrete evidence that delivery of hydrogen as a cheap and readily available, non-polluting energy source is around the corner. This will not happen until large corporations, still linked to the fossil fuel energy system, become truly interested in the development and use of hydrogen and put R&D money into a hydrogen-based fuel system.

POWDER KEG

THE GAS INDUSTRY HAS BEEN BUSY IN WYOMING'S PRAIRIES AND GRASSLANDS, BUILDING THOUSANDS OF MILES OF ROADS AND SINKING MORE THAN 10,000 WELLS IN THE PAST THREE YEARS. BUT IN THE POWDER RIVER BASIN, RANCHERS ARE JOINING ENVIRONMENTALISTS TO TRY TO STILL THE DRILLS.

BY KEITH KLOOR

Eᴅ SWARTZ DOES NOT SEEM LIKE THE KIND OF guy you would threaten with bodily harm. He has spent almost all of his 62 years in a windswept corner of northeastern Wyoming, herding cattle, baling hay, and building waterlines to keep his ranch from going dry. He still works as long as the daylight lasts, even after suffering a near-fatal heart attack in 1995. But shortly after his ranch started dying in 1999, he stood up on a bus full of state officials from Wyoming and Montana and started fuming about his dried-up meadows and polluted creekbeds. Most audiences would have sympathized with Swartz. But this particular bus carried coal-bed methane executives and various state officials, including the governors of Wyoming and Montana. They were touring drilling sites in the Powder River basin; the Wyoming officials wanted to impress their counterparts in neighboring Montana with how smoothly the booming development was going.

Ed Swartz's property, however, which sits in the coal-rich basin, has been inundated with coal-bed methane discharge water from drilling sites near his ranch, contaminating his creek and preventing him from irrigating his alfalfa, the cattle ranch's lifeline. On this day he just wanted to make sure everyone on the bus was aware of it. And that he was not the only rancher with this problem. John Kennedy, a local coal-bed methane operator, was furious. "You're a liar!" he yelled at Swartz, cutting him off. Swartz, a strapping, leathery, third-generation Wyomingite whose grandfather homesteaded the family ranch in 1904, kept talking, unperturbed. "I'm going to hurt you!" Kennedy warned. Swartz said to go ahead and try. Inflamed, Kennedy again thundered, "I'm going to hurt you!" But the blustery driller never left his seat. Finally, Montana's governor, Judy Martz, settled things down, so the tour could continue.

Sitting around his kitchen table, chain-smoking low-tar cigarettes, Swartz tells me the story over a pot of turbo-charged coffee, a glinty grin on his face. Kennedy, I learn, has a reputation for harassing similarly aggrieved ranchers at public meetings and over the phone. Swartz, who has long been an active, bedrock Republican in what is a virtual one-party—Republican—state, is now considered a pariah for rocking the boat. "People think I'm just a rabble-rousing rancher around here," he says defiantly. That, undoubtedly, is because of the Clean Water Act lawsuit he has filed against Wyoming state officials and the drilling company (Denver-based Redstone Inc.) he asserts is responsible for the damages to his land.

Coal-bed methane development is a relatively new form of natural gas extraction that has exploded in the Powder River basin since 1997, with more than 10,000 gas wells already sunk on private and state lands. It is like mother's milk to state officials, because it produces both tax revenues and campaign contributions from the energy companies. "They are so oriented to the energy industry that they could give a red rat's ass about a rancher," Swartz bristles. "I'm as serious as I can be about that. There's not one of them that is concerned about a rancher around here."

ALONG THIS STORIED AND BLOODIED FRONTIER, INDIAN BATTLES RAGED, RANGE WARS WERE FOUGHT, AND BUFFALO BILL AND BUTCH CASSIDY SEALED THEIR MYTHS.

Tʜᴇ 8-MILLION-acre Powder River basin straddles the Wyoming–Montana border, offering a snapshot of the quintessential Old West, with its rolling hills and prairies nestled between the Bighorn Mountains and the Black Hills, 100 miles to the east. Along this storied and bloodied frontier, Indian battles raged, range wars first erupted between homesteaders and cattle barons, and Buffalo Bill and Butch Cassidy sealed their myths. It is the place, many claim, where the American cowboy was born.

The Powder River region is also where the Great Plains meet the Rocky Mountains, a bare, reddish, lunarlike landscape that skips between parched, stumpy buttes, green meadows, and unspoiled streams. It is a vast, mixed ecosystem of grassland and sagebrush, where, in that timeless Darwinian race, prairie dogs burrow deep and antelope hurtle fast and high, coyotes and mountain lions quick on their heels. The Powder River—described by settlers as "a mile wide and an inch deep, too thin to plow and too thick to drink"—is a 375-mile tributary of the Yellowstone River. Plying its shallow, muddy waters are the globally imperiled sturgeon chub, the channel catfish, and 23 other native fish.

During the 20th century the long boom-and-bust cycles of strip mines and oilfields left their own indelible imprint on the landscape, in the form of sawed-off hilltops and abandoned "orphan" wells. Even so, ranchers and environmentalists have always taken solace that the deep and large gashes scarring the land were mostly confined to a few areas.

No longer. The latest boom to hit the Powder River basin has spread out in a chaotic patchwork, pockmarking the historic landscape with thousands of miles of powerlines, pipelines, roads, compressor stations, and wellheads. Methane is a natural gas found in the region's plentiful coal seams. Water pressure holds the gas in the coal; pumping the water out in large volumes releases the gas. The process also produces wastewater laced with sodium, calcium, and magnesium—too saline to be used for irrigation, too tainted to be dumped in waterways. So in a semi-arid region where water is precious, energy companies are forced to store the methane water in "containment" pits, from which it often runs into water wells and into the tributaries of the Powder River.

"IT'S SO DAMN DISCOURAGING. EVERYTHING I WORKED FOR, THAT MY GRANDFATHER WORKED FOR, AND THAT MY SON IS WORKING FOR IS BEING WIPED OUT."

The resulting environmental damage in the Powder River basin has hit ranchers and the land equally hard. "It's so damn discouraging," Swartz tells me in a craggy voice tinged with resignation. "Everything I worked for, that my grandfather worked for, and that my son is working for is being wiped out." Over the years the family has endured many droughts (including the one the region is suffering today), its share of machinery breakdowns, and several diseases afflicting their cattle. But nothing compares with the poisonous runoff that is killing the ranch's vegetation. "They're [Redstone] using my place as a garbage dump," Swartz fumes.

About a year and a half ago, at the boom's peak, drillers were pumping 55 million gallons of water to the surface every day. Underground aquifers were being depleted and cottonwood trees flooded. The massive runoff of the methane-tainted water

has become so alarming—polluting creeks and streams and altering natural river flows—that earlier this year the conservation group American Rivers named Wyoming's Powder River as one of America's Most Endangered Rivers for 2002. "There could be 139,000 coal-bed methane wells in the Powder River basin by the end of the decade," says Rebecca Wodder, president of American Rivers, referring to energy-industry and government estimates. "Despite this, federal and state agencies have yet to formulate an adequate plan for minimizing the environmental consequences of drilling in the Powder River basin."

And they weren't about to until the U.S. Environmental Protection Agency (EPA) issued a report last May, slamming the Bureau of Land Management (BLM) for failing to assess the fallout from coal-bed methane development. At the time, drillers were already having their way on Wyoming's private and public lands, owing to lax state and federal environmental safeguards. They had just set their sights on a mother lode of rich coal-seam deposits on BLM lands in the Powder River region. Then, with the bureau poised to give the operators a quick go-ahead, the EPA's report stopped them in their tracks, throwing into doubt the development of gas wells on 8 million public acres in Wyoming. In particular, the EPA cited concerns about air quality from dust and compressor emissions, and the impact on wildlife and water quality. The EPA report forced the BLM to redo its environmental-impact statement on 51,000 new gas wells slated for development on federal lands.

The reassessment, scheduled to be released in January, stands to reverberate through out the Rocky Mountain West, where a gold rush mentality has taken hold. Energy officials have called the area the "Persian Gulf of natural gas." Gas companies have already struck hard and fast in Colorado's San Juan basin; now they're waiting on the green light on federal lands to expand there and across Montana and Wyoming.

Moreover, coal-bed methane development in the West is the cornerstone of President Bush's proposed domestic energy plan, which claims the area has enough natural gas to supply the energy needs of the United States for seven years. "The region is enormously rich in minerals," Ray Thomasson, a Colorado-based energy consultant, told a recent Denver gathering of energy experts and industry officials. "We just have to find out where the sweet spots are."

IN WYOMING, AS IN MOST OF THE WEST, SUBSURFACE RIGHTS SUPERSEDE SURFACE RIGHTS, SO UNLESS A LANDOWNER OWNS BOTH—FEW DO—WHOEVER OWNS THE MINERAL RIGHTS UNDER THE LAND HOLDS THE TRUMP CARD.

Despite their proud bearing, Nancy and Robert Sorenson can't mask the sorrow and anguish in their voices as they describe what it's like to live in the first "sweet spot" of the coal-bed methane boom. Both natives of Wyoming, the Sorensons have spent the past 30 years—almost half their lives—on a 3,500-acre cattle ranch in the Powder River basin, 20 miles north of Ed Swartz's spread. They, too, have had methane-contaminated water run off onto their property, flooding their soil and boxwood elder trees.

The Sorensons have graciously invited me into their home for a lunch of stir-fried chicken and homemade nut bread and a discussion of their unwanted quandary. Robert is taciturn yet direct, and casts a sharp gaze behind his full mustache; Nancy has a warm, open smile but speaks softly and deliberately, always searching for the right words. Recently, rich coal seams have been discovered under their land. But since the Sorensons own only partial mineral rights, when the drillers came knocking, they had no power to turn them away. In Wyoming, as in most of the West, subsurface rights supersede surface rights, so unless a landowner owns both—few do—whoever owns the mineral rights under the land holds the trump card.

Though the Sorensons stand to collect royalties, they are agonizing over the repercussions. "I worry about our neighbors downstream, who will be hit hard by this," says Nancy, a retired schoolteacher and an active board member of the Powder River Basin Resource Council, a grassroots group that has united ranchers and local environmentalists. She also worries about the fate of the wildlife she sees every day on her morning walks around the ranch, such as antelope, mule deer, and foxes. (" I've seen a mountain lion twice, which was quite a thrill!") Living on the land all their lives, the Sorensons have become attuned to its natural rhythms, mindful of even the subtlest changes in predator-prey relationships. Over the years the couple has watched, with admiration, as ecological forces exerted their own balance. "When the rabbits become too much of problem, the bobcats take care of it," says Nancy. But, she adds in a doleful whisper, "I hate to see what happens to all these animals once this drilling starts, because you are fragmenting their environment."

Overall, the Powder River basin is home to more than 157,000 mule deer, 108,000 pronghorn antelope, and almost 12,000 elk. Even the BLM admitted in its first—albeit inadequate—environmental assessment that the proposed coal-bed methane development "may result in loss of viability of federal lands… and may result in trends toward federal listing" under the Endangered Species Act for 16 species, including the white-tailed prairie dog, the burrowing owl, and the Brewer's sparrow.

As clouds darken the early afternoon and a light, intermittent drizzle begins to fall, the Sorensons mourn the transformation—almost overnight—of their rural community into an industrial zone. "It's a total change of a way of life," says Robert, whose family homesteaded in the Powder River basin in 1881.

He's right. For two days I have zigzagged hundreds of miles west from the city of Gillette—the satellite home base of energy companies—to Sheridan, another energy outpost. In between antelope sightings, it seemed that for every 20 miles I rode along Highway 14-16, new dirt roads were being bulldozed in the rolling foothills for pipelines, and open pits were being dug for wastewater containment ponds. What's more, this was a quiet period, because the unseasonably wet, cold weather hampered drilling. In warmer temperatures, for instance, coal-bed methane operators use an atomizer to spray methane water at high pressure, so that most of it will evaporate—although the salt and minerals still coat the ground.

"IMAGINE IF YOU LIVED ACROSS THE STREET FROM A POWER PLANT. THAT'S WHAT IT'S LIKE FOR PEOPLE THAT LIVE NEAR A COAL-BED METHANE FIELD." SOME RANCHERS LIKEN THE SOUND TO THAT OF 747S TAKING OFF—CONSTANTLY.

During construction of a particular site, it's not uncommon, the Sorensons say, to have a hundred trucks trundling in 24 hours a day. Once the wellheads are sunk and the compressor stations built alongside—sheltered in houselike structures—the noisy, whirring process of methane gas extraction runs all day and all night. "Imagine if you lived across the street from a power plant," says Robert. "That's what it's like for people that live near a coal-bed methane field." Some ranchers liken the sound to that of 747s taking off—constantly.

Yet as Nancy admits, somewhat awkwardly, not everyone in the community is put off by the development. Though Wyomingites are deeply proud of their ties to the land and the ranching tradition, it's not an easy life. "I'm aware of neighbors who welcomed this [the gas development] with open arms because they were so far at the end of their rope," she says. "And because this is going to keep them on the land a little while longer, I'm absolutely understanding of their position." Her voice trails off, a faraway look in her eyes. "You know," she continues, a few seconds later, "we've been able to make a living off this place, which is quite unusual. Robert has made some smart moves, and we've both worked hard, but we've also been lucky in that we haven't had a major illness. So I can't blame other people for how they feel."

Like Ed Swartz, the Sorensons direct their anger at state officials, who they believe have permitted the coal-bed methane operators to run roughshod over ranchers and the land. For their ranch, the Sorensons signed 13 separate lease deals, like pipeline and powerline rights-of-ways, with energy companies. "Not once did we have a choice," says Nancy. What's more, Wyoming, unlike neighboring Montana and most states, doesn't have laws mandating that proposed industrial development—such as gas extraction—on nonfederal lands be assessed beforehand for environmental impact.

Shortly after the widespread environmental damages became evident several years ago, the Sorensons, Ed Swartz, and many other ranchers met with their state representatives to plead for tighter regulations on the drilling and for a set of rancher rights—to no avail.

"They've all just been bought and sold," says Robert. "Really, they all have." (Nearly 70 percent of all campaign contributions to Wyoming's state legislators come from the oil and gas industry.)

"They all know where the paychecks come from," Nancy chimes in. "There isn't anybody in local government that is going to fight this. And neither am I. We just need some tougher regulations."

The ranchers' biggest concern is the depletion of their underground aquifers. In Colorado, where coal-bed methane development has taken off in a number of areas, drillers are required by state law to clean the discharged methane water of all pollutants and "reinject" it into the ground. Not so in Wyoming. Drillers say that requirement would make their operations less cost-effective. Many ranchers, joining forces with the formidable Powder River Basin Resource Council, have also been petitioning state and federal lawmakers for a mandatory Surface Owner Agreement—which would allow landowners to have a say about where pipelines and roads are built. (The government broke its promise that powerlines would be buried and kept to minimum.) Above all, what pains the ranching community is the lack of adequate bonding—money the energy companies pay to cover land damages and reclamation. Just look around, Nancy says, at all the abandoned wells from previous booms. Undoubtedly, history will repeat itself, she asserts, if energy companies aren't required to pony up much more than the paltry $25,000 bond they pay for unlimited wellheads on Wyoming's federal lands (it's $75,000 for state and private lands).

Given this history, I'm surprised to hear the Sorensons say they are not against coal-bed methane development in principle (and that goes for Ed Swartz and the other Powder River basin ranchers I spoke with).

"No," Robert answers resolutely. "We just want to see it done right."

A FEW MONTHS AFTER MY TRIP TO THE Powder River basin last May, the bottom fell out of the natural gas market, dropping prices to less than a dollar for every thousand cubic feet, from a high of $12 two years earlier, a price that fueled the drilling frenzy. Some industry experts and plenty of environmentalists attribute the Powder River boom to the manipulation of energy prices in California by Enron and other energy companies. Whatever the reason, the plunging price has slowed development in the Powder River basin and given ranchers—and wildlife—some breathing room.

"We have a window of opportunity to get the problems addressed," says Jill Morrison, an organizer with the Powder River Basin Resource Council. "It's good we have this slowdown. Maybe now we can get better planning and development." Morrison cautions it could merely be a lull in a volatile energy market, and that in the meantime, drillers might use the depressed prices as an excuse to pay less money to repair damages to the land.

No matter the outcome, Ed Swartz, the Sorensons, and other Powder River ranchers under siege aren't counting on their politicians for help. In August, three months after the EPA released its critical report of the BLM's environmental assessment, Wyoming's governor, Jim Geringer, lambasted the EPA for its "aggressive and ill-informed approach" to coal-bed methane development in the Powder River region. And in September, perhaps to shore up their flagging spirits, Montana Governor Judy Martz told a roomful of oil and gas lobbyists that they were "the true environmentalists"—as opposed to, as one meeting attendee said, the "radical" critics of energy development.

After I heard both comments, I figured these officials had never seen Ed Swartz's polluted stream or had lunch with the Sorensons. "We are very environmentally conscious," Nancy told me during my visit, "and most ranchers we know are." Her tone was gentle but firm. "I'm not opposed to having development, but there should be some equilibrium."

It never occurred to me then, or months later, when Governor Martz and the gas-industry proponents made their barbs, that the Sorensons or Ed Swartz were "radicals." But I sure do know who the "true environmentalists" are in Wyoming.

WHAT YOU CAN DO

For more information, call the Powder River Basin Resource Council at 307-672-5809 or log on to www.powderriverbasin.org. You can also call or e-mail the Wyoming state office of the BLM at 307-775-6256; state_wyomail@blm.gov.

Personalized Energy
THE NEXT PARADIGM

In the future, energy will be more in the control of neighborhoods and home-owners. But for that to happen, new technologies need to be developed that bring efficiency and reliability up and costs down, says a technology futurist.

By Stephen M. Millett

Consumers want more control over their energy, and they want it cheaper, cleaner, more convenient, and more reliable. They want energy, like other commodities, to be more personalized. Technological improvements will focus on meeting those demands, but they won't happen quickly. Current forecasts for energy supply and demand are limited by a mind-set stuck in the past. But new technologies and new consumer imperatives will spawn new ideas about energy that could get us off the grid and bring power generation into neighborhoods and even into homes.

The primary question we need to ask about the future of energy is whether the old supply-and-demand paradigm of fossil fuels still applies. In that case, the key to solving our energy woes lies in finding ways to increase production of traditional hydrocarbon fuels (such as oil, natural gas, and coal) and promote consumer conservation. But if the old paradigm is out, there may be a whole new paradigm emerging, where new technologies, for in-stance, could change the whole energy picture.

Right now, we hear too many discussions about drilling more oil, conserving energy, and other actions based on old-paradigm thinking. Indeed, statistics show a big gap between projected energy demand and supplies in the United States: Oil and natural gas consumption are going up and available quantities are going down, so we're going to have a big projected shortfall.

The biggest jump in American energy consumption in the twentieth century was the use of petroleum, and that's almost exclusively transportation. Transportation relies on petroleum to meet 95% of its energy needs, according to the U.S. Bureau of Transportation Statistics. The story on coal is a little bit different. Developed economies such as the United States used to use coal in homes for heating, but that's done almost nowhere anymore. Americans are using more and more coal, but it's to generate electricity in large power plants. Coal is now rarely used at the individual level.

The 2001 report of the president's National Energy Policy Development Group stated, "Renewable and alternative fuels offer hope for America's energy future but they supply only a small fraction of present energy needs. The day they fulfill the bulk of our [energy] needs is still years away. Until that day comes we must continue meeting the nation's energy requirements by the means available to us."

This assertion assumes no changes to the existing energy paradigm: no new technological breakthroughs, no shifts in people's values or consumers' demands, no surprising events—natural or manmade—to alter the energy picture. But this paradigm-blinder limits our thinking—and our forecasts. Paradigms and social systems are rarely permanent, and new technology often drives a transition to other paradigms.

My thesis is that we have just begun the shift away from what I call the "carbon-combustion paradigm" to a new "electro-hydrogen paradigm." The shift is going to be very dramatic in the next 20 years, but the

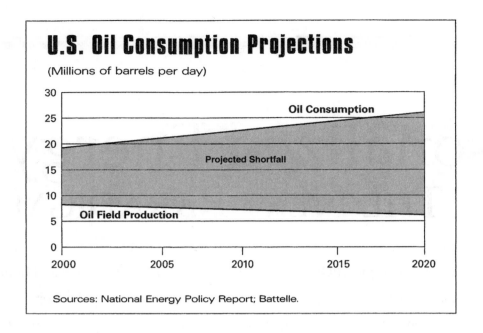

U.S. Oil Consumption Projections

(Millions of barrels per day)

Oil Consumption

Projected Shortfall

Oil Field Production

Sources: National Energy Policy Report; Battelle.

full integration is going to take easily a hundred years. We're going to see a lot of exciting technology innovations in the laboratory and in prototype systems in the next 10 to 20 years, but to go from our current paradigm and all its infrastructure to a new paradigm and all of its infrastructures is going to take a very long time.

Energy and the Consumer

Changes in consumer behavior are driving many trends. In the U.S. market, baby boomers seek convenience, while the elderly put heavy emphasis on the reliability and affordability of power. The question for policy makers and the energy industry is how reliable the electric grid will be in the future. Both baby boomers and Generation X'ers value customization—the personalization of products, especially computers and cell phones. Consumers also want more mobility and longevity in their products. And of course we all want inexpensive energy.

Along with consumer behavior, there are marketplace trends working toward this paradigm shift, including the effects of more-stringent environmental-quality regulations. The bad news is that no energy system will ever be 100% environmen-

tally friendly; the good news is that the next paradigm will be a lot friendlier than the last one was.

U.S. energy policy calls for greater energy self-sufficiency. An additional need, a new element since September 11, 2001, is security of the energy infrastructure. Before then, people didn't even worry about infrastructure security, but now it's a major issue. The electric grid system is absolutely vulnerable to weather and potential security compromises.

Another marketplace issue is the need for energy-cost stability and continued necessity for economic growth. If you take a strictly conservationist approach to this issue and say this cost stability and continued economic growth can be realized by voluntary simplicity, then you are limited by the old energy paradigm. This low-growth scenario has a low probability of occurring. We need more energy for continued economic growth, and that is still what most people in the world want.

Also impacting marketplace trends are emerging technologies such as those behind a gradual shift now beginning from central-station generation to decentralized, or distributed, generation of power from local sources. The paradigm shift to distributed generation, or distributed re-

sources, parallels the paradigm shift from fossil fuels to hydrogen. For example, one of the biggest needs we have is gasification of coal, though surprisingly little work is yet being done in this area. The current gasification technology has been in existence for at least 20 years, it isn't really very good, and it's very expensive. It's just not competitive.

The Real Hydrogen Future

It's easy to speculate about the hydrogen economy's potential and to go off into science-fiction scenarios. But because Battelle deals with the real-world challenges of governments and corporations, we spend a lot of time separating science from science fiction. So here's a little reality check: All forms of energy are going to have some negative environmental consequences. We need to recognize that fact and then try to make the energy system a lot better in the future than it is today. Another reality check is that no fuels will be free: There will always be costs for both the fuels and their infrastructure of production and distribution. The challenge is to find the new ways that improve the value (benefits/costs) relationship.

The challenge in this transitional period to the hydrogen future is to

About Battelle

The Battelle Memorial Institute was established in 1929 in Columbus, Ohio, and now manages or co-manages four of the 16 U.S. national labs. We are in the business of technology development, management, and commercialization. We do mostly government work, but we also have industrial clients. We're independent, meaning we have stakeholders but no stockholders. We are technically not for profit but, as we're reminded daily by our CEO, we are not for loss. We do more than $1.7 billion in business a year, and that's a lot of R&D. Battelle scientists have contributed to a wide range of breakthroughs, such as copy machines, optical digital recording, and bar codes; Battelle's R&D yields between 50 and 100 patented inventions a year.

All corporations and organizations face the challenge of keeping up with and anticipating change. At Battelle, we do futuring. I like to use the word *futuring* because the participle adds the action of making or doing something. We do trend analysis, expert focus groups, and expert judgment, and we have our own process of scenario analysis based on cross-impact analysis. We also have our own scenarios software, which we've been using for 20 years for our work with corporations.

Battelle studies "consumer value zones," where marketplace trends, new customer demands, and emerging technologies all converge. The study of energy's consumer value zone leads us to conclude, for example, that the future of energy is personal; that is, energy production will increasingly move from large, centralized power plants to distributed power.

Among its outreach projects, Battelle does an annual technology forecast and maintains a separate Web site for our scenarios and trends: www.dr-futuring.com.

For more information, see Battelle's Web site, www.battelle.org.

— *Stephen M. Millett*

extract hydrogen from hydrocarbon sources in an affordable way. There are many potential avenues being explored today. The approach at Battelle is to develop a universal reformer for the fuel cell, where we would take methane, methanol, and even gasoline and convert it into hydrogen at the point of burning it. The ability to extract sufficiently pure hydrogen from methane, methanol, or gasoline means that we could continue to use the existing infrastructure (such as all of those gas stations) to distribute safe liquid fuels without the expense and hazards of storing hydrogen today. Avoiding new infrastructure costs in the short run would greatly help the transition to the electrohydrogen paradigm of fuel cells for both transportation and stationary power generation applications.

Economics really favor the current, fading energy paradigm; economics do not yet favor the next one. We're going to have to see a lot of economic and regulatory changes, as well as technological changes. The challenge is cost. We can make fuel cells, we can produce hydrogen, but we can't do it at competitive prices relative to the existing hydrocarbon system, the carbon-combustion system. Electric utilities use a benchmark of $1,500 per kilowatt capacity; anything costing more than that is simply not competitive enough. Researchers at United Technologies, for example, are getting the cost of the fuel cell down below $3,000 per kilowatt—it has been as high as $15,000—but the price is still too high for general commercialization.

And what about solar cells? There's no question that there's a market today for solar cells, but it's largely a vanity technology for people who put it on their houses. If you take a pocket calculator that has solar panels, generating power measured in milliwatts, and normalize that to a kilowatt, the cost might be as high as $17,000. So, clearly, there's a long way to go before solar technology can replace the carbon-combustion system.

Alternative fuels like wind power, which is now growing, have been attractive because of a number of government incentives and subsidies at both the federal and the state levels.

Technologies to Watch

The technologies to watch include:

- Innovations in materials for batteries and fuel cells, especially PEM (polymer electrolyte membrane) and solid-oxide fuel cells.
- Breakthroughs in reducing diesel emissions.
- Innovations for reconfiguring backup and emergency power generation into distributed generation systems.
- Biofuel development.
- New approaches to the gasification of coal.
- Global warming and carbon-dioxide management.

For fuel cells and batteries, the biggest challenge is in materials development. Exciting new developments in battery technology include the sodium sulfur battery, which the American Electric Power Company is working on in Columbus, Ohio.

For PEM and solid-oxide fuel cells, the name of the game is the materials and getting their costs down. For instance, the current membrane material used in PEM fuel cells now costs as much as $800 per square meter. We need to get the cost down to $8 to make this transition to the next paradigm. So we need technology breakthroughs that bring costs down. In addition, the current use of platinum as the catalyst is obviously very expensive and needs to be changed.

Reducing diesel emissions is another significant area of research at Battelle and other institutions. Biofuel blending with diesel and other fuels is a very exciting growth area.

Bringing diesel emissions down will promote the transition of current backup generation to distributed resources coordinated with the power grid. Most utilities now dismiss customer-driven backup generation as simply being irrelevant to the grid, but if you can make all of those diesel generators environmentally compliant, and if you can coordinate them with the grid, then you've got a prototype distributed-generation system already in place.

Biofuel development, not just bioblending, is another breakthrough area. The DNA revolution in agriculture is very exciting because we could design plants—not just corn, but also chickweed or garbage grass, for instance—that could be engineered for high-starch content to be more easily converted into methanol.

Gasification of coal is a huge area of research and development. Affordable and efficient coal gasification would enable us to break down the constituent parts of coal and get the hydrogen atoms out of it. In an ideal process, we would be able to separate sulfur and other undesirable constituents out of the coal and extract pure hydrogen. We could also separate out the carbon content that produces carbon-dioxide emission from stacks. An innovative coal gasification technology would be a tremendous boon for the American economy—and the economies of Germany, Russia, China, and India, to mention just a few others. Hydrogen from coal would be a major step in the transition to fuel cells.

The new energy paradigm is also about the environment, and we at Battelle are concerned about global climate change and carbon-dioxide management. To that end, Battelle is actively pursuing approaches like carbon sequestration. We are currently working with the U.S. Department of Energy to evaluate how to capture carbon dioxide and store it underground so that it cannot escape into the atmosphere.

Toward a Distributed Power System

We're not suddenly going to do away with our coal-burning plants, but there are emerging opportunities to use large fuel cells and batteries in conjunction with central generation. This could produce emergency and peak power at the generation site as well as provide supplementary power at distributed sites. By "distributed," I don't mean we're going from the big power plant to the home all in one jump. Energy will be distributed first at the level of neighborhoods and districts, and then we'll work it on down to homes generating their own power. It'll go step by step, but the trend favors personalized energy.

We're going to see some exciting technologies developed in the next 10 years, but it's going to be a slow process toward full-blown commercialization. If we're on a low technology-development trajectory, it will take more time. If we get a couple of breakthroughs in technology or some regulatory changes, then we can be on a faster track, but no sooner than 2008 or even 2010 at the earliest unless a desperate need for power drives the trends faster. Slow progress favors the "tracker" and "adapter" companies and organizations, while fast progress favors the early innovators. Many companies are now agonizing over whether to be the progress leaders or the followers (or fast followers).

Who's going to lead this energy paradigm shift? Who really is going to provide the thought leadership and the breakthroughs? The Japanese are clearly ahead of the Americans in fuel cells. Honda and Toyota are ahead of the Big Three auto manufacturers in Detroit on energy breakthroughs for transportation. Honda in particular is the world leader in thinking through distributed power generation.

As for regulatory leadership, the question is who is going to provide the standards. There's a dearth of leadership for the new energy paradigm in the United States. Neither the federal government nor the states are showing signs of leadership, and there are very few progressive electric and gas utilities out there in the United States. Wherever the leadership comes from for the new energy paradigm, that's who will likely succeed and capture the largest market share.

But for now, those who should lead seem to be saying, "Change is good. You go first."

Stephen M. Millett is a thought leader at Battelle and co-author of *A Manager's Guide to Technology Forecasting and Strategy Analysis Methods*. His address is Battelle, 505 King Avenue, Columbus, Ohio 43201. E-mail milletts@battelle.org; Web site www.battelle.org.

Originally published in the July/August 2004 issue of *The Futurist*, pp. 44-48. Copyright © 2004 by World Future Society, 7910 Woodmont Avenue, Suite 450, Bethesda, MD 20814. Telephone: 301/656-8274; Fax: 301/951-0394; http://www.wfs.org. Used with permission from the World Future Society.

Wind Power

Obstacles and Opportunities

Martin J. Pasqualetti

To know the wind is to respect nature. You ride with the wind when it fills your sails, but pay its power no heed and risk inconvenience, expense, even death. Drive through calm air in Los Angeles one moment only to encounter 30 minutes later Santa Anas whipping wildfires across mountaintops and pushing tractor-trailers into ditches. Lounge on the beach on Kauai one day, but find yourself huddling for protection the next day as a hurricane levels entire forests.[1] Sit on the porch during a quiet and muggy Oklahoma night when suddenly a mass of debris, once a house, swirls past, before dropping nearby as a pile of kindling and shattered dreams. More than any other force of nature, we have little defense against the wind. The wind keeps us on our toes.

If we cannot control the wind, perhaps we can put it to our use. It is a challenge with which we have had some success. Historically, we have used the wind to help us with work that would otherwise fall heavily upon our own backs. The wind helped humans explore the world; they had no other energy source. It continues to help us prepare foods and pump water. In some places, the wind is such a part of daily life that in its absence, silence blankets the landscape and puts us out of sorts. When it picks up again, flags flutter, well water rises, and grains are again ground to flour.

The wind machines humans developed were among the earliest icons of civilization. We can see them in early sketches from the Orient, scrolls from Persia, paintings from the Low Countries, photographs from the Dust Bowl, and even movies from Hollywood. Putting the wind to work was our first conscious use of solar power.

Perhaps the most widespread use of wind machines, at least in the United States, has been to pump water. Dotting the Great Plains by the hundreds of thousands, farm windmills—along with grain silos—were once as characteristic of the landscape as coal spoils were of Appalachia. Spinning whenever air moved, they brought to the surface the water that allowed ranches and settlements to flourish in an area otherwise too dry for either to exist for long. Most of these ingenious whirling devices eventually gave way to powerful compact motors that ran on fossil fuels, and as a result, wind energy landscapes largely disappeared. Before they all were removed, some folks preserved a few of them, drawn to their quaint beauty and the nostalgia they evoked as symbols of a Great Plains lifestyle. By then, however, most people considered the era of wind machines dead.

As it has happened, the epitaphs were premature. Today, wind machines are back. It has not been a quiet resurrection but rather one with substantial notoriety and publicity, plus a controversial mix of support and resistance. The new devices look and act little like their ancestors: Instead of the creaking, wooden machines of the past, those of the new species are made of metal and fiberglass—and are bigger, quieter, sleeker, and more powerful than ever. Instead of pumping water, the moving blades spin generators housed with an assemblage of gears in the nacelle, which is located behind the hub where all the blades meet. Instead of a stream of water, modern wind machines are pumping a stream of electrons, a product proving to be a valuable asset to farmers who are trying to address present day economic realities of living off the land.

The new appearance and mechanics of wind machines reflects their different role. Instead of producing mechanical power for the purposes of pumping and grinding, the new machines convert mechanical energy into electricity. Instead of being erected here and there in splendidly independent isolation, many are being clustered in symmetrically interdependent neighborhoods, designed to work together as parts of a larger organism. Nor are they just generating electricity: Unexpectedly, modern wind machines are prompting us to consider how best to weigh the energy we need against the environmental quality we want. All the while they are continuing their transformation from public indifference to public curiosity, from an overlooked energy supplier to alternative energy's "holy grail," one possible way to get most of what we want and little of which we do not.

Table 1. Historical wind turbines							
Turbine, Country	Date in service	Diameter (meters)	Swept area (meters)	Power (kilowatts)	Specific power (kilowatts per square meter)	Number of blades	Tower height (meters)
Poul la Cour, Denmark	1891	23	408	18	0.04	4	—
Smith-Putnam, United States	1941	53	2,231	1,250	0.56	2	34
F. L. Smidth, Denmark	1941	17	237	50	0.21	3	24
F. L. Smidth, Denmark	1942	24	456	70	0.15	3	24
Gedser, Denmark	1957	24	452	200	0.44	3	25
Hütter, Germany	1958	34	908	100	0.11	2	22

SOURCE: P. Gipe, *Wind Energy Comes of Age* (New York: John Wiley & Sons, 1995), 78.

An Old Resource with a New Mission

Compared to the variety of uses that stretch back millennia, converting wind energy to electricity is a recent application. Although a few people were trying to accomplish this at the same time Thomas Edison opened his coal-fired Pearl Street generating plant in the latter years of the nineteenth century, it would be another 80 years before such proof-of-concept machines would evolve into the commercial generators that started sprouting in the California landscape in 1981 (see Table 1). Indeed, the beginning of the modern era of wind power bore few similarities to earlier water pumping. The vision of modern wind power was much grander in scale, one that has evolved to row upon row of machines spreading over hundreds of acres, contributing enough electricity to power an entire city but—and this is the big difference—without undesirable side effects that accompanied the use of conventional resources.

Obviously, the machinery of the late twentieth century differs both in form and function from the equipment that nineteenth century ranchers and farmers developed to help them wrest a living from the dry lands that predominate west of the 100th meridian. For them, it was enough that machines were turning when the air was on the move. In the new era, such simple fulfillment is not enough; wind power today is viewed less from the living room and more from the boardroom. Wind power is big business, and the managers of that business must be sophisticated not just in the ways of making money, but in several disciplines that lead to success. Even something as seemingly innocent as turbine placement can no longer be considered just from the perspectives of convenience, necessity, or whim. Instead, the new wind barons must understand meteorology, metallurgy, physics, aerodynamics, capacity factors, land ownership, planning, zoning, and the influence of public perception.

Wind power's popularity is widespread and growing, a result of its increasing profitability and the perceived environmental benefits it engenders. It also results from the simple fact that, unlike fossil and nuclear fuels, wind is a widely available, familiar element of the environment. The first step is seemingly the easiest: finding it.

Wind and the Family Farm

The initial step in developing any resource, be it gold or wind, is locating it. While it is often an uncomplicated step, not every place is attractive. Just like gold, wind is not evenly distributed in the richness of its product. Subtropical deserts, such as the Sonoran Desert surrounding Phoenix, are created by persistent high pressure and are often unsuitable for wind development. For obvious reasons, forested areas are unattractive for wind turbines, as are equatorial areas with their characteristically light and variable winds. The rest of the world is more promising, although detailed data collection must precede full-fledged capital investment.

Finding windy places is a relatively easy step: Unlike fossil or nuclear fuels, it does not require drilling rigs, seismic gear, or Geiger counters. Wind power, in most places, is simply "out there." This means that the most obvious early task of wind prospectors is to determine where winds are strong enough. Once identified, such areas reveal several common characteristics, including exposed terrain, colliding air masses, and, in particular circumstances, topographic funneling, as through mountain passes. In the United States, several areas meet such criteria, including sites in California, southeastern Washington, central Wyoming, east-central New Mexico, and most notably the Great Plains (see Figure 1). In Europe, strong winds are significant along the west coast of Ireland, Great Britain, the eastern North Sea, the southern Baltic Sea, the Pyrénées Mountains, and the Rhone Valley and (see Figure 2). Many of these areas are "nuggets" that

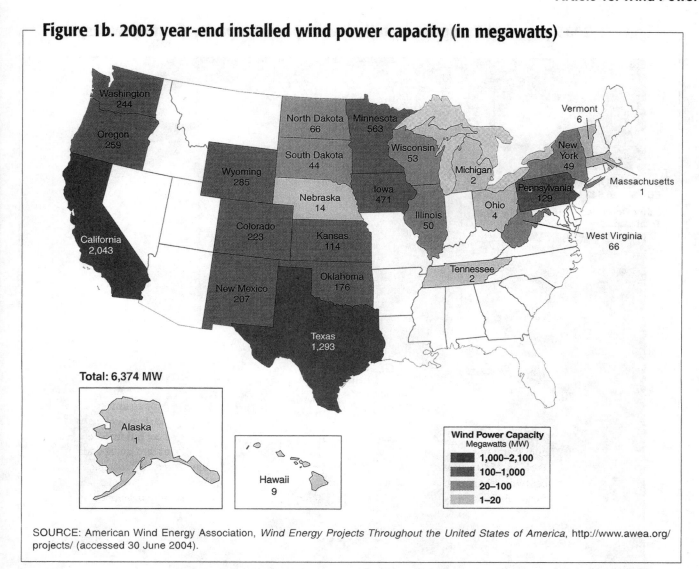

Figure 1b. 2003 year-end installed wind power capacity (in megawatts)

Washington 244
Oregon 259
California 2,043
Wyoming 285
North Dakota 66
South Dakota 44
Nebraska 14
Colorado 223
New Mexico 207
Kansas 114
Oklahoma 176
Texas 1,293
Minnesota 563
Iowa 471
Wisconsin 53
Michigan 2
Illinois 50
Ohio 4
Tennessee 2
Pennsylvania 129
New York 49
Vermont 6
Massachusetts 1
West Virginia 66

Total: 6,374 MW

Alaska 1

Hawaii 9

Wind Power Capacity
Megawatts (MW)
1,000–2,100
100–1,000
20–100
1–20

SOURCE: American Wind Energy Association, *Wind Energy Projects Throughout the United States of America*, http://www.awea.org/projects/ (accessed 30 June 2004).

we are plucking first, and they have been stimulating further prospecting and the development of grand plans for the future. By the end of 2003, the total installed capacity in the European Union was 28,440 megawatts (MW).[2]

Like the gold rush of the 1850s, the modern wind rush started in California. California still leads the way, with 2,042 MW installed by January 2004, principally in four locations: Altamont Pass, on the edge of the Central Valley east of San Francisco; Tehachapi Pass at the southern end of the Sierra Nevada; San Gorgonio Pass, near Palm Springs; and in the rolling hills between San Francisco and Sacramento. Although first, the primacy of California is certainly temporary: With a potential for 6,770 MW, it ranks only seventeenth among the 50 states in potential (see Table 2).[3]

Among other states attracting interest, Texas has been the pacesetter with more than $1 billion in new wind investment and 1,293 MW installed, mostly in the western part of the state near such communities as Big Springs

and McCamey. Coincidentally, many of these developments are positioned among the now derelict oil equipment that helped bring great wealth to this part of the state and has underpinned many of its towns and cities. At the end of January 2004, California, Texas, and 24 additional states held within their borders an installed capacity of 6,374 MW.[4]

Another 2,000 MW has been proposed for development in the near future, with some of the largest projects planned for the states of Washington and Massachusetts. However, the greatest potential remains where wind machines once so dominated the landscape—midway between these extremes in the Great Plains.

The wind of the Great Plains is as obvious as is its treeless expanse. When Francisco Vásquez de Coronado and his men crossed this region searching for the Seven Cities of Cíbola in 1540, they found no gold but two other resources instead. The most obvious and most useful to Coronado were the great herds of bison—totaling per-

Figure 2. 2003 year-end European installed wind power capacity (in megawatts)

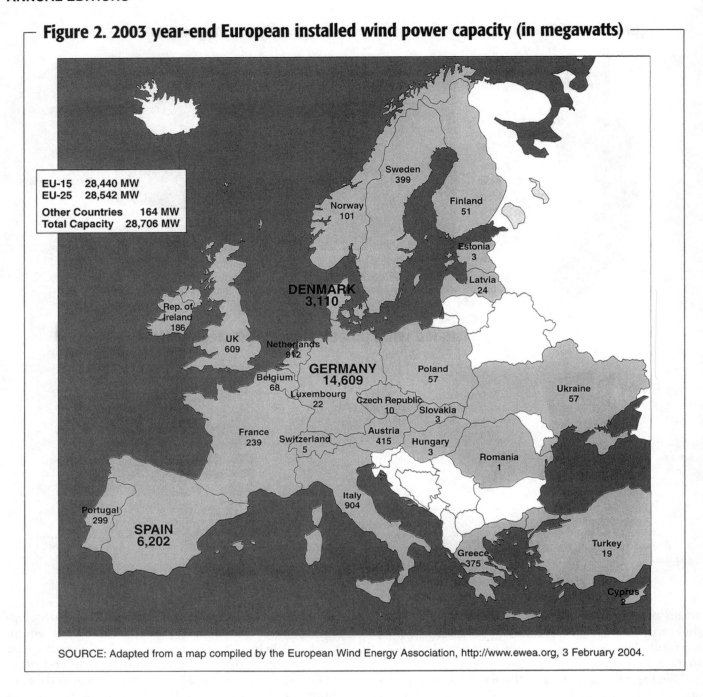

EU-15   28,440 MW
EU-25   28,542 MW

Other Countries   164 MW
Total Capacity   28,706 MW

Sweden
399

Norway
101

Finland
51

Estonia
3

Latvia
24

DENMARK
3,110

Rep. of
Ireland
186

UK
609

Netherlands
912

Poland
57

Ukraine
57

GERMANY
14,609

Belgium
68

Luxembourg
22

Czech Republic
10

Slovakia
3

France
239

Switzerland
5

Austria
415

Hungary
3

Romania
1

Portugal
299

Italy
904

SPAIN
6,202

Greece
375

Turkey
19

Cyprus
2

SOURCE: Adapted from a map compiled by the European Wind Energy Association, http://www.ewea.org, 3 February 2004.

haps 50 million head—that were scattered across a million square miles of grassland. Not only did they provide food, but their droppings helped guide the expedition across otherwise indistinct landscapes. The other resource was the wind, but three centuries would pass before it was appreciated.

By the late 1800s, the general pattern had reversed; bison were being decimated for sport, and wind power was lifting water for irrigation. Today, only this second resource remains, yet it is being used for a different mission: to generate electricity and make money.[5] It is a realistic ambition: The winds of the Great Plains are so abundant

that the energy potential from just three states (North Dakota, Texas, and Kansas), were they fully developed, would match the electrical needs of the entire country. These and several other Great Plains states hold the largest expanse of class 4 (400–500 watts per square meter) lands in the country (see the box).[6]

Although weather on the Great Plains is often viewed as being inhospitable—farming families there endure swirling snow in winter and blowing dust in summer—attitudes toward the frequent tempests are lately bending in a new direction. Always alert for new sources of income to ease their financial volatility, locals are turning to

Table 2. Top 20 states for wind energy potential (measured by annual energy potential in billions of kilowatt hours)		
1	North Dakota	1,200
2	Texas	1,190
3	Kansas	1,070
4	South Dakota	1,030
5	Montana	1,020
6	Nebraska	868
7	Wyoming	747
8	Oklahoma	725
9	Minnesota	657
10	Iowa	551
11	Colorado	481
12	New Mexico	435
13	Idaho	73
14	Michigan	65
15	New York	62
16	Illinois	61
17	California	59
18	Wisconsin	58
19	Maine	56
20	Missouri	52

NOTE: As of July 2004, reevaluation of the wind potential has been done for 28 states by the U.S. Department of Energy's Windpowering America program; reevaluation of additional states is still in process. Once complete, the numbers for each state might change, as might the relative rankings. In many cases, the potential will increase. The Great Plains will continue to dominate the rest of the country in terms of potential.

SOURCE: Pacific Northwest Laboratory, *An Assessment of the Available Wind Land Area and Wind Energy Potential in the Contiguous United States* (Richland, WA: Pacific Northwest National Laboratory, 1991).

wind developers with equanimity and even enthusiasm. They are finding that the same winds that strip soil from the fields and bury houses in snow can fuel rural economic development.[7]

Construction of a typical 100 MW wind farm produces more than 50,000 days (approximately 419,020 manhours) of employment. In Prowers County, Colorado, the recently completed wind development is each year providing $764,000 in new revenues, $917,000 in school general funds, $203,000 in school bond funds, $189,000 to the Prowers Medical Center, and $189,000 in additional revenue to the county tax base.[8] Meanwhile, a 250 MW project in Iowa is providing $2 million in property taxes and $5.5 million in operation and maintenance income. The leases on offer to farmers in this area commonly provide yearly royalties of more than $2,000 per turbine: For a single Iowa project, local farmers are receiving $640,000 annually. Other projects return $4,000–$5,000 per turbine. In some cases, a one-megawatt turbine could generate revenues for the owner of $150,000 per year once the debt for purchase is repaid.[9] Though appreciable at the individual level, such royalties are only a small part of the variable costs for wind

developers (see Figure 3). To them, such largesse seems good business. To farmers, such revenues can mean the difference between bankruptcy and prosperity.

In spite of a weighty list of environmental attributes, wind power carries some unexpectedly heavy baggage.

In spite of promising trends, it would be an overstatement to claim that wind power offers an energy panacea or a reversal of the national trend toward increasingly concentrated generation of electricity: It is, at least so far, a relatively small enterprise. Taken together, all the wind developments in the country contribute less than 1 percent of our current needs.[10] However, there is a real attraction to wind power's promise for the future: Its estimated generating potential in the United States alone is 10,777 billion kilowatts per hour (kWh) annually, or three times the electricity generated in the entire country today.[11] In recognition of such potential for pollution-free electricity, the U.S. government is sponsoring a program called Wind Powering America (WPA) to tap more deeply this vast natural resource. WPA's new goal is to increase to 30 the number of states with more than 100 megawatts of wind-generating capacity by 2010.[12] The program also aims to increase rural economic development and, to some degree, local energy independence.

The Environmental Irony

Wind power attracts many adherents from the environmental community. These organizations focus on its solar roots, emphasizing that it requires no mining, drilling, or pumping, no pipelines, port facilities, or supply trains. It produces no air pollution or radioactive waste, and it neither dirties water nor requires water for cooling. Wind power is relatively benign, simple, modular, affordable, and domestic. It is, in short, an environmental golden goose.

However, in spite of such a weighty list of attributes, wind power carries some unexpectedly heavy baggage. In England, anti-wind epithets have been particularly colorful: Developers have suffered their machines being called everything—including "lavatory brushes in the air" for their busy top ends. This leads us to an irony of wind power: While we usually consider wind power environmentally friendly, most of the objections to its expansion have had environmental origins.

We can follow the thread of such reactions most clearly to Palm Springs, California, in the mid-1980s. Soon after installing thousands of turbines in windy San Gorgonio Pass just north of the city limits, developers were battered with complaints that the machines interfered with television reception, produced annoying and inconsistent noise, posed risks to wildlife and aircraft, and represented incompatible land-use practices.[13] The most trou-

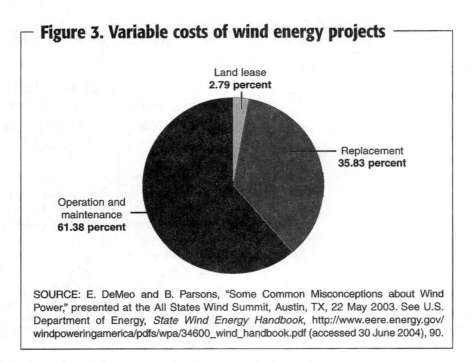

Figure 3. Variable costs of wind energy projects

Land lease
2.79 percent

Replacement
35.83 percent

Operation and
maintenance
61.38 percent

SOURCE: E. DeMeo and B. Parsons, "Some Common Misconceptions about Wind Power," presented at the All States Wind Summit, Austin, TX, 22 May 2003. See U.S. Department of Energy, *State Wind Energy Handbook*, http://www.eere.energy.gov/windpoweringamerica/pdfs/wpa/34600_wind_handbook.pdf (accessed 30 June 2004), 90.

WIND CLASSES

Developers need to know the average wind speed at a particular site to design and build the most appropriate turbines. It would be no more prudent to size the turbines for the slowest speed than it would be to size them from the fastest, but infrequently occurring speed. One can get a good idea of this relationship by using the Weibull distribution, a plot of frequency against speed. This distribution helps identify various classes of wind (see the table below). The higher the number, the stronger the average speed. A good wind speed is 7 meters per second (mps); 20 mps may be excessive and cause damage to equipment. Turbine manufacturers have a "rated wind speed" for all models and sizes of turbines they sell. Typical rated wind speed requirements are in the range of 8-13 mps, but many machines will produce some power with much slower speeds.[1] Currently, developers are concentrating on class 4 and above as the most promising areas.

34. For an excellent and detailed description of these and other principles, see P. Gipe, *Wind Energy Comes of Age* (New York: John Wiley & Sons, 1995).

WIND POWER CLASSIFICATION

Wind power class	Resource potential	Wind power density at 50 meters (in watts per square meter)	Wind speed at 50 meters (in meters per second)	Wind speed at 50 meters (in miles per hour)
2	Marginal	200–300	5.6–6.4	12.5–14.3
3	Fair	300–400	6.4–7.0	14.3–15.7
4	Good	400–500	7.0–7.5	15.7–16.8
5	Excellent	500–600	7.5–8.0	16.8–17.9
6	Outstanding	600–800	8.0–8.8	17.9–19.7
7	Superb	800–1600	8.8–11.1	19.7–24.8

bling, bitter, and outraged complaint, however, was that the wind machines destroyed the aesthetic appeal of the landscape, thereby threatening the very attribute that most attracts tourists to the area's fancy resorts.[14]

Developers and bureaucrats of the day, all of whom had expected a warmer welcome, were startled by such reactions. It was apparent to everyone with hopes for contributions from wind power that any future success would have to rest on greater environmental compatibility and a more complete respect for public attitudes and opinions. The initial experience provoked industry musing as to what actions might better attract support. Soon, manufacturers began making improvements to design and engineering. Locally, the concerns over wind impacts led to stricter planning rules and more uniform standards. These adjustments softened the problems, but they could not eliminate them. Turbines remained unavoidably visible and the center of a classic example of incompatible land use. The very characteristic that had long kept residential development in the San Gorgonio Pass minimal—the strong wind—was the same characteristic that prompted developers to fill it with machines. There was very little compromise potential.

So strong was the backlash against wind development that the City of Palm Springs sued the U.S. Bureau of Land Management and the County of Riverside, claiming that developers had not followed proper environmental procedures. Although the suit was eventually abandoned, it was not before the local jurisdictions, including Palm Springs and Riverside County, enacted a long list of required adjustments, stipulating (for example) height limitations, the use of nonglinting paint, reporting mechanisms for endangered species, and the establishment of decommissioning bonds.

A few years later, in an unpredictable turnaround, attitudes changed. This happened once Palm Springs, led by its mayor Sonny Bono, began eyeing wind machines as generators of tax revenue, as well as electricity. With a financial windfall in mind, the city annexed several square miles of land in the middle of the windiest part of the pass, thereby enlarging the city limits and sweeping additional tax revenues into the municipal treasury. Also, counter to the early intuition and opposition of city officials, the wind turbines have become something of a tourist attraction. Organized tours are available, images of wind farms adorn many local postcards, and brochures advertise the Palm Springs wind industry. Even Hollywood producers have incorporated the striking wind energy landscapes in movies and advertisements. These changes reflect the progression in public attitudes toward greater acceptance, although a closer look still finds disgruntled residents who have the original objections. As they point out, we can paint them, size them, sculpt them, and engineer them to a fine edge, but we cannot make them disappear.

The Aesthetic Core

Reactions to wind power tend to be both quick and subjective. While one group fights intrusion, another is organizing visits for enthusiastic tourists. Where one person loathes turbines, that person's neighbors find them fascinating. Whichever reaction prevails in any given location, wind turbines cannot be ignored, for they do not fit naturally upon the land. They are, to apply Massachu-

setts Institute of Technology historian Leo Marx's famous phrase, "machines in the garden."[15]

The spread of wind power encounters the most strident opposition where it interferes with local land use.

Wind power's development contained a surprise: Among its corps of supporters, no one anticipated the need to defend wind projects. Why did no one foresee objections? We can only speculate, but it seems that the advantages were considered by adherents to be so obvious, especially when compared to nuclear power, that developing a defensive strategy for this new technology seemed superfluous. Supporters failed to recognize how opposite is the signature between the two: Nuclear power is compact and quiet, whereas wind power is expansive and obvious. Reflecting on this difference, resistance to nuclear power accumulated slowly only after a long educational process that culminated with accidents at Three Mile Island and Chernobyl, while resistance to wind power was immediate and instinctive.

Although this difference suggests the heft of visual aesthetics in shaping public opinion, it masks two other ingredients of equal importance. One is the immobility of the resource: Wind moves, windy sites do not. In this way, wind differs from coal and most other fuels, because its nature does not allow it to be extracted and transported for use at a distant site. For wind power to be successful, turbines must be installed where sufficient wind resources exist or not at all. Thus, just like two other resources— geothermal energy and hydropower— the site-specific nature of wind developments intrinsically invites conflicts with existing or planned land uses. This is true even in deserts, the common dumping ground of society.[16]

The second ingredient in helping form public attitudes toward wind power is the landscape itself. Simply put, some landscapes are more valued than others. Place turbines in sensitive areas—perhaps along the coast or in a national park—and prepare for an uproar. Place them out of view or in low-value areas—sanitary landfills, for example—and opposition diminishes.

These characteristics produce wind power's most intractable challenges. First, owing to resource immobility and the subjectivity of its aesthetic impact, total mitigation is impossible. Second, because environmental competition changes from place to place and from one time to another, generic solutions are few and elusive. Third, because nothing can make turbines invisible, little we do will make them more acceptable to those perceiving land-use interference. There is no escaping the essence of wind turbines: They will always be spinning, pulsing, exoskeletal contraptions that naturally attract the eye.

The foregoing notwithstanding, the future of wind power remains both promising and substantial, if we can identify and follow the appropriate path. Two general strategies suggest themselves: work to bend public opin-

LARGER AND LARGER TURBINES

Wind turbines are getting larger and larger. What is driving this trend? To answer this question, we need to know that movement obtains its impetus from the sun; as solar energy strikes the surface of the Earth, it creates differences in pressure. The wind, in turn, moves "downhill" along the pressure gradients that are formed, from higher to lower pressure. Speed increases as the horizontal distance between different pressures is shortened. Wind also typically accelerates when it is constricted, as when it moves through a mountain pass. The faster the wind moves, the more energy it carries, but it is not in a linear function. Rather, it increases with the cube of the wind speed, usually written x^3. This means that a wind speed of 8 meters per second (mps), yields 314 (8^3) watts for every square meter exposed to the wind, while at 16 mps, we get 2,509 (16^3) watts per square meter, again eight times as much. This relationship puts a premium on sites having the strongest winds. This relationship also explains why the area "swept" by the turbine blades is so important and why the wind industry has been striving fervently to increase the scale of the equipment it installs. A one-megawatt turbine at a typical European site would produce enough electricity annually to meet the needs of 700 typical European households.

ion in favor of wind power, or install the turbines out of view. The first approach is under way but slow. The second approach can be quicker and would seem to hold promise, but it is being met with mixed results, especially when projects are proposed for offshore locations—the newest tactic to avoid public criticism and maximize profits.

Moving Offshore

The spread of wind power encounters the most strident opposition where it interferes with local land use. Tourism, recreation, entertainment, resorts, and a host of other outdoor activities create most of the challenge because their function is to help people escape reality. For relief from this dilemma, developers are looking for sites offshore, and they have been finding them, especially in the shallow, wind-swept waters of the Irish Sea and North Sea. Denmark, which is characteristically leading the way with this strategy, has already installed and activated several fields of this type. Other projects are in place or planned off Ireland, the United Kingdom, the Netherlands, and several other countries.

In addition to prohibiting wind projects in populated areas, positioning them offshore offers several operational advantages. For example, winds passing over water tend to be stronger than those passing over land; offshore placement removes no land from existing or planned uses; any noise produced at sea is muffled by that of the surf; road use is largely a moot issue; and negotiation with multiple landowners is unnecessary. Nonetheless, offshore placement requires some tradeoffs. For example, offshore equipment is more costly to construct and maintain, and it inherently increases the potential for conflict with any recreational use of the seashore. It also tends to encourage the installation of larger turbines (see the box on this page).

Strictly from a public perspective, offshore placement has the presumed advantage of mitigating complaints about aesthetic intrusion. It has not, however, turned out to be the expected universal remedy. Indeed, moving offshore is increasing rather than diminishing the enmity of wind power in some quarters, especially in the northeastern United States.

Tempted by the strong offshore winds of coastal Massachusetts and responding to the hostility to wind developments witnessed in California, several entrepreneurs advocated placing wind turbines on the shallow offshore banks. The proposal itself may go down as the most foolhardy miscalculation in renewable energy history. The problem, as usual, is incompatible use of space. Called Cape Wind, the project is proposed for Nantucket Sound, a site between the popular vacation spots on Cape Cod and the exclusive holiday retreats of Martha's Vineyard and Nantucket Island. Like development near Palm Springs, Cape Wind is colliding with the wishes of a prosperous and politically astute residential corps bent on protecting existing scenic and recreational qualities that it has come to cherish.

Riding a Roller Coaster

Wind energy has experienced a wild ride over the past 20 years, one where initial enthusiasm soured quickly with the perception that generous incentives and lax oversight were allowing virtually any wind farm development, no matter how carelessly designed or operated, to be financially tenable. An undertow that quickly started to pull against the early currents of promise was a perception that wind developments were being installed without sufficient public notification, due consideration, or individual benefit. By the late 1980s, it was clear that improved turbines and business situations were going to be necessary if wind power was to develop a significant position in the alternative energy mix in the United States or abroad.

Some of the earliest advances first came into view in Europe, where even casual inspection spotted substantial differences from early installations in California. For example, instead of large clusters of turbines spread haphazardly upon the land, deployment of European turbines was more sensitively organized into smaller groupings that were carefully integrated into the landscape. This was partly a result of a higher sensitivity to ex-

THE CAPE WIND PROJECT: A WIND POWER LIGHTNING ROD

Cape Wind is a proposed $500-$750 million wind development project for Horseshoe Shoal in Nantucket Sound. If approved, the turbines will come within 5 miles of land, spread over an area of 24 square miles, and consist of 130, 417-foot wind turbines connected to a central service platform that includes a helicopter pad and crew quarters. Each turbine blade will be 164 feet long with a total diameter of 328 feet. Each turbine will have a base diameter of 16 feet and an above-water profile taller than the Statue of Liberty.

The proposal has become a lightning rod for the wind industry. The Alliance to Protect Nantucket Sound (the Alliance),[1] which strongly opposes the project, has been accumulating arguments against it. They point out that

- each turbine will have about 150 gallons of hydraulic oil, and the service platform will have at least 30,000 gallons of dielectric oil, and diesel fuel;
- the project will be within the flight path of thousands of small planes; and
- the turbines will pose a navigation hazard to the commercial ferry lines in the area.

These and other objections, however, take a secondary position to the Alliance's primary objection, that of aesthetic intrusion. The Alliance claims that the turbines will be visible for farther than 20 miles, that they will be lighted at night, and that they will flicker with changing sun angle. The Alliance has developed many computer visualizations of how they would appear. (Some proponents might point out that the Visualizations illustrate how inconsequential the turbines would appear from the beach.)

The pro-development side has not been idle. Cape Wind has its own Web site,[2] which identifies the many benefits of the project, including that it offsets the need for 113 million gallons of oil yearly and creates approximately 600–1,000 new jobs. They also refer to many studies attesting to the benefits of such projects as Cape Wind. One of the most recent references the positive impacts on sea creatures around the wind turbines off the southern Swedish coast.[3] Other Web sites provide many testimonials to the good sense of offshore wind power.[4] The controversy over Cape Wind's offshore proposal is just the beginning of many other anticipated projects along the East Coast, such as off Long Island.[5]

1. http://www.saveoursound.org.
2. http://www.capewind.org.
3. L. Nordstrom, "Windmills off Swedish Coast are Providing Unexpected Benefit for Marine Life, Scientists Say," *Environmental News Network*, 11 February 2004, http://www.enn.com/news/200402-11/s_13011.asp.
4. windfarm@cleanpowernow.org; http://www.safewind.info.
5. See the Safe Wind Coalition's Web site, http://www.safewind.info/wind_farms_where.htm; and the Long Island Offshore Wind Initiative's site at http://www.lioffshorewindenergy.org/.

isting conditions and partly a measured response to the experiences in California that had dulled the promise of wind power and threatened its future.[17]

Reflecting improvements and continued support, wind power's trajectory is once again upward. Today it is the fastest-growing renewable energy resource in the world.[18] Wind power is especially popular outside the United States: In countries like Germany, it is welcomed, encouraged, and promoted as one way to reduce greenhouse gas emissions. In Denmark, the value of wind power to the economy now exceeds that of its economic mainstay, ham. Spain's development of wind power is currently growing at a faster pace than it is in any other country.

The roller coaster ride is not over, however: Even amid news of improvements and quickened growth, wind power continues to have its critics. The more determined of these opponents work to keep wind machines from their property and out of their view. They hire public relations experts, make abundant use of the Internet to promote their view and attract adherents, and invite the support of prominent citizens to their cause. The group Save Our Sound is perhaps the most visible example of such techniques.[19] Such determined resistance was never envisioned when the champions of wind power came calling more than two decades ago. Today, despite progress in assuaging public apprehensions, a measure of uncertainty still hangs over wind's future.

From Incentives to Independence

What is to be made of the many incentives that wind power enjoys? Tax incentives, utility portfolio standards, feed-in laws,[20] and many other aids currently help make it an economically viable alternative energy provider. Some would say that the requirement for these incentives demonstrates that wind power is not a legitimate competitor for our energy dollars. Others might argue that the mere existence of these aids suggests how narrow the economic gap is between a present need for subsidies and independent viability. While its increasingly competitive status results partly from a rising cost of conventional energy, it also reflects the declining costs of all alternatives,

WIND POWER AND BIRD MORTALITY

There is a persistent public impression that birds and windmills don't mix very well. Particularly for the smaller turbines, the spinning blades are hard to see during the day and are invisible at night. Many of those who campaign against wind power expansion cite this concern as part of their argument.

Concerns about turbine-related bird mortality stem largely from the experience at Altamont Pass, California, where approximately 7,000 wind turbines are located on rolling grassland 50 miles east of San Francisco Bay.[1] Between 1989 and 1991, 182 dead birds were found in study plots associated with wind turbines, including approximately 39 golden eagles killed per year by the turbines.[2] Golden eagles, red-tailed hawks, and American kestrels had higher mortality than more common American ravens and turkey vultures.[3] Deaths of eagles and potential danger to endangered California condors are the biggest issues at Altamont Pass. Bird mortality at comparably sized wind facilities has been reported as being similar or lower than those at Altamont Pass.[4]

While such fatalities are regrettable, there is serious question as to whether they are sufficient to slow or halt the use of wind power. One environmental group, The Defenders of Wildlife, recommends that bird mortality should be "kept in perspective."[5] For comparison glass windows kill 100–900 million birds per year; house cats, 100 million; cars and trucks, 50–100 million; transmission line collisions, up to 175 million; agriculture, 67 million; and hunting, more than 100 million.[6] Clean Power Now, an advocacy group encouraging wind development in Nantucket Sound, answers the question "Do wind turbines kill birds?" by stating "Very few and not always."[7] Altamont Pass, where much of the concern for avian safety originated, appears to be more of the exception than the rule. Data show the actual numbers killed in the pass do not exceed one bird per turbine per year, and for raptors, reported kill rates are 0.05 per turbine per year.[8] Nevertheless, the wind power industry has made several adjustments. For example, perch guards are being

installed and a program to replace the old machines with modern turbines on high monopoles is ongoing (One modern turbine replaces seven older machines).[9] More study on this matter would be welcome.

1. W. G. Hunt, R. E. Jackman, T. L. Hunt, D. E. Driscoll, and L. Culp, *A Population Study of Golden Eagles in the Altamont Pass Wind Resource Area: Population Trend Analysis 1997*, report prepared for the National Renewable Energy Laboratory (NREL), Subcontract XAT-6-16459-01 (Santa Cruz, CA: Predatory Bird Research Group, University of California, 1998).
2. S. Orloff and A. Flannery, *Wind Turbine Effects on Avian Activity, Habitat Use, and Mortality in Altamont Pass and Solano County WRAs*, report prepared by BioSystems Analysis, Inc. for the California Energy Commission, 1992.
3. C. G. Thelander and L. Rugge, *Avian Risk Behavior and Fatalities at the Altamont Pass Wind Resource Area*, report prepared for the National Renewable Energy Laboratory: SR-500-27545, (Santa Cruz, CA: Predatory Bird Research Group, University of California, 2000).
4. M. D. McCrary et al., *Summary of Southern California Edison's Bird Monitoring Studies in the San Gorgonio Pass*, unpublished data; and R. L. Anderson, J. Tom, N. Neumann, J. A. Cleckler, and J. A. Brownell, *Avian Monitoring and Risk Assessment at Tehachapi Pass Wind Resource Area, California* (Sacramento, CA: California Energy Commission, 1996).
5. Defenders of Wildlife, *Renewable Energy: Wind Energy Resources, Principles and Recommendations*, http://www.defenders.org/habitat/renew/wind.html.
6. Curry & Kerlinger, *What Kills Birds?* http://www.currykerlinger.com/birds.htm.
7. Clean Power Now, *Do Wind Turbines Kill Birds?* http://www.cleanpowernow.org/birdkills.php.
8. P. Kerlinger, *An Assessment of the Impacts of Green Mountain Power Corporation's Wind Power Facility on Breeding and Migrating Birds in Searsburg, Vermont, July 1996–July 1998*, report prepared for the Vermont Department of Public Service, NREL/SR-500-28591 (Golden, CO: NREL, March 2002), page 64.
9. R. C. Curry and P. Kerlinger, *Avian Mitigation Plan: Kenetech Model Wind Turbines, Altamont Pass WRA, California*, report presented at the National Avian Wind Power Planning Meeting III, San Diego, CA, May 1998, page 26.

including wind. The message is this: Even without incentives, wind power has been moving toward economic independence, and it seems destined to reach parity with conventional sources soon.

It is often the smallest margin of help that wins the day for an emerging technology. One way to demonstrate this is to examine the impact of higher conventional energy cost. In one study of 12 Midwestern states, where electricity sold at 4.5 cents per kWh, the regional potential for cost-effective wind power was about 7 percent of current total generation in the United States.[21] If the market would support a price of 5.0 cents per kWh, however, the

potential would grow to 177 percent of current generation. If one additional penny is added to the price, the potential blossoms to 14 times current levels.[22]

Until conventional energy makes this inevitable jump, wind operators need another way to bridge the gap. This brings us to the U.S. Production Tax Credit (PTC). This credit originally provided for an inflation-adjusted 1.5 cents per kilowatt-hour for electricity generated with wind turbines. With PTC now at about 1.9 cents, wind projects are economically favored. In its absence, however, development of new projects virtually ceases. This occurred, for instance, when the credit expired at the end of 2001, before it was reinstated some months later. PTC lapsed again on 31 December 2003, and discussions in Congress are once again under way as to whether to extend it for a five-year period.[23]

Without such a credit, the U.S. wind industry will suffer. According to Craig Cox, executive director of the Interwest Energy Alliance, "The lapse of the PTC has created uncertainty in the wind energy marketplace, and interest in new developments has slowed."[24] Renewal of the credit is part of the $31 billion energy bill that stalled in Congress at the end of 2003, again putting the wind energy industry back on its "roller coaster" in the United States. The world's major wind turbine manufacturer, Vestas Group, delayed its decision to build a wind turbine plant in Oregon because of the uncertainty of the credit. Ultimately, such uncertainty spreads to all phases of wind energy development, not just deployment of turbines. "We've been looking to establish a manufacturing facility in the U.S. but have not done that only because of the boom and bust cycle of the wind energy industry in the U.S.," Scott Kringen of Vestas told Reuters.[25] Other spokespersons have made similar observations: "Today, a wide range of U.S. companies are interested in the wind industry, but many are staying on the sidelines because of the on-again, off-again nature of the market produced by frequent expirations of the PTC," said Randall Swisher, executive director of the Washington, DC-based American Wind Energy Association.[26] Most countries offer more stable, longer-term policy support for wind than does the United States, and they use mechanisms that are inherently more pluralistic and egalitarian. This helps explain why wind power is on such a fast track in countries such as Germany, Denmark, and Spain.

Also playing an important role in helping wind power gain a competitive advantage are Renewable Energy Credits, or "green tags."[27] These tags result from laws currently in force in 13 states that require electricity providers to include a prescribed amount of renewable electricity in the electric power-supply portfolio they offer to their customers. Electricity providers meet this requirement through several possible approaches. They can generate the necessary amount of renewable electricity themselves, purchase it from someone else, or buy credits from other providers who have excess. The green tags rely almost entirely on private market forces. Taken together with production tax credits and various industry improvements, they are helping wind power continue its trend toward independent profitability. Such status, coupled with reduced public resistance, will move wind power from the realm of alternative to the position of mainstream energy resource. This might be possible if we can move from NIMBY to PIMBY.

From NIMBY to PIMBY

When plans encounter resistance, developers usually make amended suggestions to attract greater support.

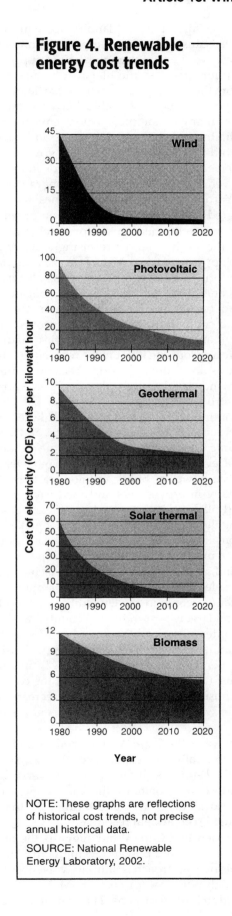

Figure 4. Renewable energy cost trends

NOTE: These graphs are reflections of historical cost trends, not precise annual historical data.

SOURCE: National Renewable Energy Laboratory, 2002.

ers, but it was not a part of planning for wind power in the mid-1980s. Instead, a naive impression prevailed that wind power would attract unquestioned support. However, the public resisted the blatant placement of wind turbines on the landscape. It was a particularly unexpected experience because California was known as a state where many of the most ardent environmentalists held forth. Instead of receiving congratulatory handshakes, wind developers (and various government officials) received notices of lawsuits for their trouble. NIMBY, even then a battlescarred acronym (for "Not In My Back Yard"), emerged in the headlines.

The ingredients mixed in the cauldron of subsequent wind power development made for a rich and complex brew. On the positive and promising side, developers learned to appreciate the power of public opinion and to work to inform it more completely. This also applied to regulators and policymakers. All parties began ascribing primacy to cooperation over imposition. By the mid-1990s, wind companies successfully improved efficiency and design, and jurisdictional authorities made zoning codes more appropriate if not more restrictive. The use of focus groups and public hearings became common elements in wind development planning procedures. As a result, controversy and press attention subsided, and projects continued to come on line with little fanfare or public notice in Iowa, Kansas, Minnesota, and Texas.

Then came Cape Wind, and much of the old debate began anew. Developers who had forgotten or never fully appreciated the power of public opinion started retreating. Despite many improvements and increasing experience, Cape Wind planners had devoted insufficient attention to considering the combination of factors that make one place unique from another. They reasoned that if offshore installations were meeting with success in Europe, why should they not find acceptance in "green" Massachusetts? However, they failed to realize the poor comparability between the mindset of people in the United States, who live in a spacious and largely post-industrial country, and their Europeans contemporaries who have been living with industrial landscapes, greater population density, and much less personal space for centuries. In making their calculations, they neglected to note that the coastal areas of Massachusetts, heavily utilized for recreation, is not comparable with the lightly settled coastal areas of Europe.

In many ways, the Cape Wind episode is an East Coast version of the California experience 20 years earlier. Admittedly, the setting is different—desert versus ocean—but the underlying problem is the same: wind turbines—immovable and numerous—interfering with the aesthetics of a valued recreational resource.

The experience of Cape Wind suggests the need for a fresh approach to wind power development. The key element of this new approach is simplicity itself: Avoid sites having a high potential for conflict. Making this assessment would involve two steps. The first step would be to assign sites "compatibility rankings," starting with the most compatible sites.

• Rank #1 properties would be those where it is not only suitable but overtly requested for wind development, such as farms in Iowa or Kansas.

• Rank #2 properties would likely be acceptable, such as in southeastern Washington.

• Rank #3 properties might be acceptable in certain circumstances, such as near Palm Springs.

• Rank #4 properties would be completely off-limits, for example, on the top of Mt. Rushmore.

Ranks would be determined according to points assigned to site-specific characteristics, including lines of site, type and color tone of terrain, ownership, bird flyways, endangered species, competitive economic value, transmission lines/corridors, protected status (such as national parks), economic development, energy security, and so forth. This should be part of the process of environmental impact assessment, and it should be initiated at any location with a strong, class 4 or above wind resource. Without such rankings, the current ad hoc and contentious approach will continue. It would be akin to a general plan for a city: Variances could be granted, but there would be a broad guidance document in place.

Part two of this plan would be to concentrate our attention on Rank #1 sites. In the United States, this means the Great Plains. There are two simple reasons to emphasize this region. First, the United States' greatest wind resource is there. Second, the small-scale farmers in the area generally welcome the turbines. The message is this: When contentious sites breed contempt, avoid them, at least for now, even if the resource base is attractive and the load centers are nearby. Admit that the wind power alternative is uniquely visible and interferes with scenic vistas and cease trying to force-feed developments down the throats of a resistant public. This is not good for the future of wind power.

On the other hand, there are places where wind power development is welcome. Small farms of the Great Plains have been losing ground for decades to consolidation and the vagaries of weather; they need an economic boost to stay viable. The owners of these farms have put out the welcome mat for wind developers in places such as along Buffalo Ridge, on the border between northwest Iowa and Minnesota, and even farther west in places like Lamar, Colorado. As Chris Rundell, a local rancher, phrased it: "The wind farm has installed a new spirit of community in Lamar... it's intangible but very real." They are embracing a new acronym, PIMBY—Please In My Back Yard.

Seeing the wind development in the Great Plains in recent years is a continuation of history, if in a slightly different form: Where a century ago hundreds of thousands of farm windmills made the local agricultural life possible, wind power is again proving its worth to those who would live there. It is bringing needed cash into the local economy and slowing a multiyear trend of farm aban-

economy and slowing a multiyear trend of farm abandonment and consolidation. The same lands that early wind machines helped develop, new wind machines are helping preserve.[28]

NOTES

1. See http://www.state.hi.us/dbedt/ert/wwg/windy.html#molokai.

2. European Wind Energy Association, "Wind Power Expands 23% in Europe but Still Is Only a 3-Member State Story," press release, 3 February 2004, www.ewea.org/documents/0203_EU2003 figures&x005F;final6.pdf.

3. Pacific Northwest Laboratory, *An Assessment of the Available Windy Land Area and Wind Energy Potential in the Contiguous United States* (Washington, DC: U.S. Department of Energy (DOE), 1991).

4. American Wind Energy Association, "Wind Energy: An Untapped Resource," fact sheet, 13 January 2004, http://www.awea.org/pubs/factsheets/WindEnergyAnUntappedResource.pdf.

5. P. Gipe, "More Than First Thought? Wind Report Stirs Minor Tempest," *Renewable Energy World* 6, no. 5 (2003), available at http://www.jxj.com/magsandj/rew/2003_05/wind_report.html. See also C. L. Archer and M. Z. Jacobson, "The Spatial and Temporal Distributions of U.S. Winds and Wind Power at 80 M Derived from Measurements," *Journal of Geophysical Research* 108, no. D9 (2003): 4289.

6. North Dakota has the capacity to produce 138,000 MW; Texas, 136,000; Kansas, 122,000; South Dakota, 117,000 MW; and Montana, 116,000 MW.

7. American Wind Energy Association; see also the National Wind Technology Center Web site (http://www.nrel.gov/wind/).

8. Craig Cox, senior associate, Interwest Energy Alliance, personal communication with author, 27 April 2004.

9. Paul Gipe, executive director, Ontario Sustainable Energy Association, personal communication with author, 20 April 2004.

10. American Wind Energy Association, note 4 above.

11. American Wind Energy Association, note 4 above.

12. Lawrence Flowers, technical director, National Renewable Energy Laboratory, personal communication with author, 22 June 2004.

13. M. J. Pasqualetti and E. Butler, "Public Reaction to Wind Development in California," *International Journal of Ambient Energy* 8, no. 3 (1987): 83–90.

14. M. J. Pasqualetti, "Accommodating Wind Power in a Hostile Landscape," in M. J. Pasqualetti, P. Gipe, and R. Righter, eds., *Wind Power in View: Energy Landscapes in a Crowded World*, (San Diego, CA: Academic Press, 2002), 153–71; M. J. Pasqualetti, "Morality, Space, and the Power of Wind-Energy Landscapes," *The Geographical Review* 90, no. 3 (2001): 381–94,; and M. J. Pasqualetti, "Wind Energy Landscapes: Society and Technology in the California Desert," *Society and Natural Resources* 14, no. 8 (2001): 689–99.

15. L. Marx, *The Machine in the Garden* (Oxford, UK: Oxford University Press, 1964).

16. C. C. Reith and B. M. Thomson, eds., *Deserts As Dumps? The Disposal of Hazardous Materials in Arid Ecosystems* (Albuquerque: University of New Mexico Press, 1992).

17. Pasqualetti, "Wind Energy Landscapes: Society and Technology in the California Desert," note 14 above.

18. Total worldwide wind energy installations were 1,000 MW in 1985, 18,000 MW in 2000, and nearly 40,000 MW in 2003, growing at about 35 percent per annum. See Solarbuzz, *Fast Solar Energy Facts: Solar Energy Global*, http://www.solarbuzz.com/FastFactsIndustry.htm.

19. See http://www.saveoursound.org/windspin.html.

20. Today's support started with Public Utility Regulatory Policies Act (PURPA), the 1978 law that promoted alternative energy sources and energy efficiency by requiring utilities to buy power from independent companies that could produce power for less than what it would have cost for the utility to generate the power, the so-called "avoided cost." In the past 20 years, electricity feed-in laws have been popular in Denmark, Germany, Italy, France, Portugal, and Spain. Private generators, or producers, charge a feed-in tariff for the price-per-unit of electricity the suppliers or utility buy. The rate of the tariff is determined by the federal government. In other words, the government sets the price for electricity in the country. Because the producer is guaranteed a price for the electricity, if he or she meets certain criteria, feed-in laws help attract new generation capacity. During the past decade, Germany became the world leader in wind development. Much of this success is due to *Stromeinspeisungsgesetz*, literally meaning the "Law on Feeding Electricity from Renewable Sources into the Public Network." The original Electricity Feed Law set the price for renewable electricity sources at 90 percent the retail residential price. In 2001, the German feed law was modified to a simple, fixed price for each renewable technology. See http://www.geni.org/globalenergy/policy/renewableenergy/electricityfeed-inlaws/germany/index.shtml.

21. Union of Concerned Scientists, "How Wind Power Works," briefing, http://www.ucsusa.org/CoalvsWind/brief.wind.html.

22. Ibid.

23. The U.S. Production Tax Cut, by its emphasis on the actual generation and transmission of electricity, not just the construction of equipment, provides additional incentives for greater technical efficiency.

24. See http://www.interwestenergy.org.

25. Reuters went on to report that the "Wind industry backers say the gaps have created a roller coaster in U.S. wind production growth because companies become fearful of investing in the alternative energy source. They say the tax-break gaps hamper wind power growth in the United States which grew last year at a rate of only 10 percent compared to global growth of 28 percent." See "Wind Power Tax Credit Expires in December," *Reuters World Environment News*, 27 November 2003, http://www.planetark.com/dailynewsstory.cfm/newsid/22956/newsDate/27-Nov2003/story.htm.

26. *First Quarter Report: Wind Industry Trade Group Sees Little To No Growth In 2004, Following Near-Record Expansion In 2003*, http://www.awea.org/news/news0405121qt.html

27. For more information about green tags for wind power, see http://www.sustainablemarketing.com/wind.php?google.

28. For several economic summaries, see S. Clemmer, *The Economic Development Benefits of Wind Power*, presentation at Harvesting Clean Energy Conference, Boise, Idaho, 10 February 2003, available at http://www.eere.energy.gov/windpoweringamerica/pdfs/wpa/34600_wind_handbook.pdf.

Martin J. Pasqualetti *is a professor of geography at Arizona State University in Tempe. His research interests include renewable energy and the landscape impacts of energy development and use. He is a coeditor of and contributor to* Wind Power in View: Energy Landscapes in a Crowded World *(Academic Press, 2002). He also contributed articles on wind power to the* Encyclopedia of Energy *(Academic Press, 2004). His research has appeared in* The Geographical Review *and* Society and Natural Resources. *He thanks Paul Gipe and Robert Righter for reading the manuscript for this article and offering many helpful suggestions. Pasqualetti may be reached at (480) 965-7533 or via e-mail at pasqualetti@asu.edu.*

From *Environment*, Vol. 46, No. 7, September 2004, pp. 23–38. Reprinted by permission of the Helen Dwight Reid Educational Foundation. Published by Heldref Publications, 1319, Eighteenth St., NW, Washington, DC 20036-1802. Copyright © 2004. www.heldref.org

Hydrogen: Waiting For the Revolution

Everybody agrees it's the future fuel of choice. Why hasn't the future arrived?

By Bill Keenan

Here's how you'll live in the Hydrogen Age: Your car, powered by hydrogen fuel cells and electric motors, quietly drives along smog-free highways. At night, when you return your vehicle to the garage, you hook up its fuel cell to a worldwide distributed-energy network; the central power grid automatically purchases your battery's leftover energy, offsetting your overall energy costs.

In the garage, you also have a suitcase-sized electrolyzer, or other conversion device, plugged into the electrical system to pump a fresh batch of hydrogen into your car. (The fuel cell uses hydrogen to produce electricity, which powers the motor.) If you need a refill as you're driving along one of the nation's highways, you pull up to a clean, quiet hydrogen fueling station to top off in less time than it takes today to fill a car with gasoline.

The electricity in your home will also come from hydrogen, either via small local fuel-cell power plants or residential fuel cells in your basement. "Moreover," says Jeremy Rifkin, president of the Foundation on Economic Trends and author of *The Hydrogen Economy*, "sensors attached to every appliance or machine powered by electricity—refrigerators, air-conditioners, washing machines, security alarms—will provide up-to-the-minute information on energy prices, as well as on temperature, light, and other environmental conditions, so that factories, offices, homes, neighborhoods, and whole communities can continuously and automatically adjust their energy consumption to one another's needs and to the energy load flowing through the system."

The U.S. Department of Energy is only slightly less enthusiastic, maintaining in a report that in the hydrogen economy, "America will enjoy a secure,

clean, and prosperous energy sector that will continue for generations to come. It will be produced cleanly, with near-zero net carbon emissions, and it will be transported and used safely. [Hydrogen] will be the fuel of choice for American businesses and consumers."

The new energy regime will have economic and political ramifications as well. Oil companies and utility companies will merge and morph into "energy companies" with a focus on generating renewable energy and local power distribution, including purchasing power from residential customers. Distributed energy production will also result in a worldwide "democratization of energy," bringing low-cost power to underdeveloped areas.

Oil and Hydrogen Don't Mix

Driving the interest in a hydrogen-based energy system: threats to the economy, the environment, and national security. Oil production, by current estimates, will likely peak sometime between 2020 and 2040. At this point, the world's economies will have consumed half of the known oil reserves, with two-thirds of the remaining oil in the volatile Middle East. As a result, prices will rise dramatically, and global consumers will experience increasingly frequent shortages.

Global warming is another significant threat that a shift to hydrogen might ameliorate. The release of carbon dioxide into the atmosphere from the burning of fossil fuels such as coal, oil, and natural gas makes up about 85 percent of greenhouse-gas emissions in the United States. This increase has resulted in an unprecedented rate of global warming, according to most scientific experts. The thinning of the polar ice caps, the retreat of glaciers around the world, the spread of tropical diseases to more temperate climates, and the rising of global sea levels are all evidence of global warming. Says

---

The Mechanics of Hydrogen

While still untested on a large scale, the promise of a hydrogen economy is based on a number of undeniable realities. Hydrogen can be burned or converted into electricity in a way that creates virtually no pollution. It is also Earth's most abundant element, available everywhere in the world. While hydrogen is scarce naturally in pure form, it can be generated easily by reforming gasoline, methanol, natural gas, and other readily available resources. It can also be created by electrolysis, a process by which electricity is run through water to separate the oxygen and hydrogen molecules.

The fuel cell, which combines oxygen in the air with hydrogen to create electricity and water, is the vital link in the hydrogen vision. It closes the energy loop and allows electricity to be stored and transported via hydrogen and then reconverted back into electricity.

In an ideal future, renewable-energy sources such as wind, solar, or water power will be used to create hydrogen through electrolysis. The hydrogen can be converted again to electricity locally by means of a fuel cell to power a car, provide energy for a home, power a laptop, or operate any number of other products.

—B.K

Rifkin: "Weaning the world away from a fossil-fuel energy regime will limit carbon-dioxide emissions to only twice their pre-industrial levels and mitigate the effects of global warming on the Earth's already beleaguered biosphere."

Add to these threats the burden of growing world populations, an increasingly unstable political situation in the Middle East, and the likelihood of longer and more frequent blackouts and brown-outs resulting from an aging and vulnerable power grid in the United States, and the promise of a safe, pollution-free, and distributed power system based on hydrogen becomes increasingly attractive.

Pathways and Roadblocks

Does all this sound too good to be true? It is: The hydrogen economy faces serious obstacles. More than 90 percent of the hydrogen produced today comes from reformulated natural gas generated through a process that creates a significant amount of carbon dioxide. Energy for this process, or for electrolysis, a more expensive way of generating hydrogen would also come from power plants fueled by oil or natural gas. So in the near term, a shift to hydrogen will not greatly reduce the world's dependence on fossil fuels and, in fact, may well hasten the greenhouse effect and global warming by increasing carbon-dioxide emissions.

Consequently, a lot of discussion about the hydrogen economy revolves around the various "pathways," or means of producing hydrogen. Atakan Ozbek, director of energy research at ABI Research, a technology-research think tank, points out that while hydrogen can come from virtually any fuel, energy from oil and gas is currently cheaper and more efficient than energy from renewable resources such as wind, sun, or water. Then, too, in the event of an oil crisis and resultant electricity shortage, coal will likely be pressed into service, regardless of the environmental cost. Nuclear power plants can also provide electricity to create hydrogen, but nuclear energy's high cost—plus the still-hot controversy over waste disposal—make such a pathway less than certain.

"What we're trying to find out right now," Ozbek says, "is how to get hydrogen to the fuel cell in a way that is economically feasible and makes sense engineering-wise."

Environmental considerations are paramount: If coal is reintroduced in a large way into our "energy portfolio"—whether to produce hydrogen or as part of our existing energy plan to replace oil—carbon-dioxide emissions will rise significantly.

The Department of Energy roadmap anticipates this, and the DOE is funding research into the "sequestration" of carbon-dioxide gases created by coal processing and natural-gas reformation. This would involve capturing these gases at some point in the energy process and permanently storing them underground or in the ocean.

To many, this is unrealistic. Jon Ebacher, vice president of power-systems technology for GE Energy, won't say that sequestration is impossible, but his comments fall short of an endorsement. Even a fairly efficient coal plant, Ebacher says, produces millions of tons of carbon dioxide each year. "So if you're going to sequester carbon dioxide from all of the plants that use hydrocarbon fuels," he says, "that's a pretty massive undertaking."

Only a hydrogen economy based 100 percent on renewable power would result in zero emissions—the vision that has captured so many imaginations. And that vision remains decades away. In the meantime, Ebacher says, "natural gas can see us through a transition period until we get solar and other renewable-energy efficiencies up to a much higher level." That transition period, he suggests, might last twenty-five to fifty years.

Another potential roadblock: transport and storage of hydrogen. Less dense than other fuels, the gas must be compressed or liquefied to be stored or moved efficiently, adding to costs and inconvenience. While the existing natural-gas infrastructure would seem to offer a convenient pathway to hydrogen delivery, this can't be done without a major retrofit. Indeed, Ebacher says, almost all of the country's existing natural-gas pipeline would have to be modified to handle hydrogen.

Finally, fuel-cell researchers must make significant advances. The power produced by a fuel cell is significantly more expensive per unit than that produced by an internal-combustion engine. Fuel-cell vehicle development is also beset by problems and costs related to type of fuel, storage, and performance. A number of prototype and "concept car" fuel-cell vehicles have been produced and displayed at auto shows and fuel-cell conferences around the world—but at a development cost of about $250,000 or more per vehicle. GM estimates that it

spent between $1 million and $2 million to develop its Hy-wire fuel-cell concept car. A consumer version would cost far less, obviously, but likely would still take sticker shock to a whole new dimension.

Putting a Brake on Hydrogen Cars

Linking the hydrogen age to cars could be a critical policy mistake, according to Joseph Romm, former acting assistant secretary for the DOE's Office of Energy Efficiency and Renewable Energy and author of *The Hype About Hydrogen*. Despite car-company promises to have fuel-cell vehicles in dealer showrooms by 2010, if not sooner, Romm argues that the cost of fuel cells, problems with onboard storage of hydrogen in vehicles, and the issues related to creating a hydrogen delivery infrastructure are likely to push the market for hydrogen fuel-cell vehicles well into the future.

The focus on hydrogen as an immediate goal in the transportation sector amounts to confusing a means (hydrogen) with an end (greenhousegas reduction), Romm explains. This could have harmful consequences, since, he estimates, it will take thirty to fifty years for hydrogen vehicles to have a significant impact on greenhouse gases. A recent National Academy of Sciences study seconds this point, stating, "In the best-case scenario, the transition to a hydrogen economy would take many decades, and any reductions in oil imports and carbon-dioxide emissions are likely to be minor during the next twenty-five years."

"If the goal is to reduce greenhouse gases," Romm argues, "then there are technologies available right now that can have a more immediate effect"—hybrid vehicles, for instance. And diverting existing (and limited) natural-gas supplies to create hydrogen for vehicles "would make that fuel less available where its use could result in a more immediate reduction in greenhouse-gas emissions—in replacing existing oil and coal-burning electric-power plants in the nation's energy grid with cleaner natural-gas power plants."

=============================

As a fuel, hydrogen is "simply a better mousetrap."

In fact, some hydrogen-technology companies have back-burnered research

As GE Goes, So Goes The Nation?

Fuel cells will probably not be a viable market until a company like General Electric gets into the business in a big way, say critics of the hydrogen economy.

GE is indeed researching fuel cells, albeit cautiously, in keeping with its approach to most other energy markets. "We do have an investment in fuel cells," says Jon Ebacher, the company's vice president of power-systems technology." I don't know if it will ever get to the dimensions where it will work at the huge volumes that were once forecast, but I think it's quite viable in niche markets." Right now, he sees a possible market in "industrial facilities that have isolated power needs, where you have a maintenance crew that deals with heating, ventilating, and air-conditioning." But, he says, "there's a distance between where they are today and the huge potential in consumer markets that was forecast at one time."

GE has also invested substantially in researching a type of fuel-cell system that would employ a gas turbine and hydrogen system working as a combined cycle. Right now, Ebacher says, GE is considering creating a power plant based on this system by 2013. It could be sooner, depending on external factors, including political developments around both fuel and the environment, the price of fuel, and the types of fuel that are available. But there are still unresolved technical challenges that could push that back.

Natural-gas prices in particular are an important barometer. "During the California energy crisis, the price of gas spiked up to $7 per million BTUs," Ebacher says. Higher gas prices, he says, "could spark some other research efforts that may come in front of fuel cells—coal, for instance. It's possible to run a combined-cycle system on coal. You put a chemical plant beside a combined-cycle power plant to process the coal into a gaseous fuel to run the electrical plant." But right now, Ebacher says, "the capital cost of doing that doesn't cross the goal line. However, if the price of natural gas or its availability gets in a bad place, all of a sudden the capital cost of doing that might not look so bad.

"It all revolves around availability, economics, and the environment—where the pressures are, what are the levers. But if you talk about running out of hydrocarbon fuels, then you would have to say that hydrogen had better be in the cards."

—B. K

and development on transportation applications. "The horizons for fuel-cell vehicles keep getting pushed out further and further, and it's unlikely that somebody's going to license and commit to a uniform, standardized hydrogen technology for at least ten to fifteen years," says Stephen Tang, an industry consultant and former president and CEO of Eatontown, NJ.-based Millennium Cell, which makes a system called Hydrogen on Demand that supplies hydrogen to fuel cells.

To pay the bills in the meantime, Tang says, "Millennium Cell has targeted markets that it believes can tolerate the price of hydrogen and fuel cells, such as consumer electronic devices, standby power, and military portables. In all of those markets, you're competing with an incumbent technology that is rather expensive in its own right and also has some limitations in performance. In these markets, then, we can focus on hydrogen as a performance fuel and not focus so much on the environmental benefits or the energy-independence benefits—attributes that buyers have difficulty valuing. It's simply a better mousetrap: Hydrogen allows you to run your cell phone much longer, or your laptop much longer, without being a slave to the energy grid or inferior batteries."

Who Will Lead?

Despite the limitations, there is growing momentum for hydrogen vehicles.

Hybrid vehicles may be a "bridging technology" toward the hydrogen age, but it's one that "doesn't at all curb the nation's appetite for oil," says Chris Borroni-Bird, GM's director of design-technology fusion. Therefore, the automaker directs about a third of its R&D—over $1 billion thus far and involving more than six hundred people—toward fuel cells. The company insists that it will have a commercially viable fuel-cell vehicle available by the end of the decade.

In other business sectors, investment in hydrogen technology is slowly returning after the boom in hydrogen technology stocks in 1999-2000 and the subsequent bust that lasted until last year. "Behind a lot of the hype, there was tremendous capital inflow in the mid-1990s going into 2000," Tang says. Unfortunately, the number of commercial products—and the resulting revenue—in the industry have been "underwhelming" relative to investment dollars. That has made the investment community more cautious so far, but things are changing. "Right now there is a much more realistic view of the possibilities," Tang says. "The investor today is looking more toward interesting niche strategies and early market penetration rather than the hope of the mass market, the home run where fifty million cars are going to be sold with your product in it."

With the investment community poised and the technology issues coming together, says ABI Research's Ozbek, "Everything is feeding into this giant equation—you can consider it a giant chemical reaction—and once everything has been fed in and the equation solved, it's going to change the whole energy infrastructure." Federal support and direction will be especially important. While Ozbek considers President Bush's $1.7 billion State of the Union pledge for energy research a good start, he would like the government to provide such research incentives as Japan and the European Union have in recent years.

And though the president disappointed many hydrogen proponents by making no specific mention of hydrogen-energy R&D in his 2004 address, his proposed 2005 budget did increase funding for hydrogen research. The federal government, Ozbek argues, should provide enhanced tax credits for buyers of fuel-cell vehicles and fuel credits for energy companies and other investing in building a hydrogen infrastructure. Jeremy Rifkin agrees, urging the federal government to take the lead by establishing benchmarks—mandating tougher fuel-efficiency standards and requiring a greater use of renewable energy sources by power companies—as the European Union currently does.

Is GM's Hy-Wire The Car of the Future?

General Motors has had a reputation for being rather conservative when it comes to both new technological developments and vehicle design, but it seems to have leapt ahead of other carmakers with its concept car, the Hy-wire.

The idea, says Chris Borroni-Bird, director of design-technology fusion for GM, is that "if you design a vehicle around the fuel cell and hydrogen tanks, you might be able to create a better vehicle than if you just put those same systems in a car designed for an internal-combustion engine."

The Hy-wire design puts the fuel cell and hydrogen storage tanks into a skateboard-like chassis that allows for greater flexibility and interchangeability of body types. Customized car bodies are then effectively "docked into" the uniform chassis.

And because the fuel cell can provide much greater electrical output than today's batteries, GM's designers have replaced mechanical and hydraulic systems for steering and braking with an electronically controlled one. "This system provides more design freedom, because those electrical wires can be routed in numerous ways, replacing a fixed steering column," Borroni-Bird says.

The Hy-wire prototype has no gas engine, no brake pedals, and no instrument panel. The fuel cell enables you to operate everything by wire. The electronic controls are included in a compact handgrip console that extends from the floor from between the front seats of the vehicle. Drivers can steer, brake, or accelerate with the controls built into the handgrips.

Because GM puts the hydrogen directly on board the vehicle, there is no need for the car to convert fossil fuels or other renewable sources into hydrogen. As a result, it can claim to offer a zero-emission vehicle and market the car to be compatible with a network of hydrogen fueling stations.

To that end, GM "applauds any hydrogen infrastructure projects, anywhere in the world," says Tim Vail, GM's director of business development for fuel-cell activities. Yet it will take a lot of applause to get the government to invest the estimated $11 billion to get a sufficient mass of hydrogen refueling stations to support 1 million vehicles, in proximity to 70 percent of the nation's population. "But," says the optimistic Vail, "$11 billion is nothing compared to past infrastructure projects such as the highways or the railroads. So it's not that big an issue to overcome. You just have to have the will to do it."

—B. K.

California's Hydrogen Highway

One state isn't waiting for action from companies or the federal government. In California, the new Schwarzenegger administration has committed to an energy plan that aims to create a "hydrogen highway" in the state by 2010. The ambitious plan proposes the construction of hydrogen fueling stations every twenty miles along the state's twenty-one major interstate highways. By taking this step to break the chicken-and-egg dilemma (which comes first, the vehicle or the fueling infrastructure?) and by continuing to impose strict mandates on automakers for fuel efficiencies, California could jumpstart the hydrogen economy.

"The pieces are all on the table," says Terry Tamminen, secretary of California's Environmental Protection Agency, "and there have been demonstration projects, but they have not been pulled together into any kind of unified vision, something that average people can use and where we can more fully commercialize the technology. So we're taking a lot of this work that's already been done, bringing it together, adding some timetables and leadership, and then of course asking for some federal money to help

with the pieces that aren't paid for by private industry or other investments."

California could jumpstart the hydrogen economy.

California already has several hydrogen fueling stations, serving research projects and some municipal fleets, and about a dozen more are in the works. For instance, SunLine Transit Agency, a local public-transit company, now operates a hydrogen fueling station that it uses to test its hydrogen-powered buses. And AC Transit, which provides public transportation in the San Francisco Bay area, expects to have three fuel-cell-powered buses later this year.

The state's goal is to provide an infrastructure of fueling stations to support a consumer market for fuel-cell vehicles. "If we can deliver such a network by a certain date," Tamminen explains, "we can then ask car companies to deliver on their promises to start delivering cars to showrooms."

One of the things driving California's plan is a California Energy Comission report that, Tamminen says, "includes credible evidence that in three to five years we are going to have serious shortages of refined fuels in

the state. Not because there's not enough petroleum under the sands of Iraq but, rather, because we don't have enough refinery capacity in the state—or in the country—to keep up with the demand created by longer commutes, poorer fuel economy, and a growing population. The report predicts a likelihood of $3 to $5 per gallon gasoline prices and periodic shortages.

"During the oil embargo of the mid-1970s, we had twenty-four thousand retail gasoline outlets in the state, compared to ten thousand today. If there are shortages, not only will there be gas lines—they will be twice as long."

Consequently, it's not a question of if but when we move toward a hydrogen economy. Even Romm, who is dubious about short-term prospects for hydrogen, concludes: "The longer we wait to act, and the more inefficient, carbon-emitting infrastructure that we lock into place, the more expensive and the more onerous will be the burden on all segments of society when we finally do act."

BILL KEENAN *is a freelance business writer and former editor of* Selling *magazine.*

UNIT 4

Biosphere: Endangered Species

Unit Selections

Key Points to Consider

- Why is the spread of invasive species through the global trading network so difficult to control and what kinds of damage do invasive species produce? What suggestions would you make to remedy the problem?

Student Website

www.mhcls.com/online

Internet References

Further information regarding these websites may be found in this book's preface or online.

Endangered Species
   http://www.endangeredspecie.com/
Friends of the Earth
   http://www.foe.co.uk/index.html
Natural Resources Defense Council
   http://nrdc.org
Smithsonian Institution Web Site
   http://www.si.edu
World Wildlife Federation (WWF)
   http://www.wwf.org

Tragically, the modern conservation movement began too late to save many species of plants and animals from extinction. In fact, even after concern for the biosphere developed among resource managers, their effectiveness in halting the decline of herds and flocks, packs and schools, or groves and grasslands has been limited by the ruthlessness and efficiency of the competition. Wild plants and animals compete directly with human beings and their domesticated livestock and crop plants for living space and for other resources such as sunlight, air, water, and soil. As the historical record of this competition in North America and other areas attests, since the seventeenth century human settlement has been responsible—either directly or indirectly—for the demise of many plant and wildlife species. It should be noted that extinction is a natural process—part of the evolutionary cycle—and not always created by human activity; but human actions have the capacity to accelerate a natural process that might otherwise take a millennia.

In "Strangers in Our Midst: The Problem of Invasive Alien Species," scientist Jeffrey McNeely of the World Conservation Union notes that "invasive" or non-native species that become established in environments and spread rapidly because they lack natural enemies are not only very damaging to human economic and other interests but loom large as one of the world's most serious biological threats. While the migration of people, plants, and animals has been a feature of human history for millennia, as globalization and an increasing emphasis on international trade has dominated the late 20th and early 21st centuries, natural barriers to the ready movement of plants and animals without human aid have disappeared. But no nation on earth has, as yet, developed an effective coordinated strategy to deal with this more pervasive problem.

In "Markets for Biodiversity Services: Potential Roles and Challenges," authors Michael Jenkins, Sara Scherr, and Mira Inbar (all with non-profit NGOs) argue that while historically it may have been the role of government to deal with such problems as invasive species and the loss of biodiversity, an alternative based in the market exists. Public sector financing for protection of the biosphere is facing increasingly severe challenges, partic-

ularly in countries where the prevailing political philosophies may deny that problems exist. In such situations, the authors contend, the lower cost approach to protecting and preserving biodiversity may be to pay landowners to manage their lands in ways that will preserve natural systems and conserve native species. At least some of this payment could come from conservation organizations such as the Nature Conservancy, but other logical sources include private corporations, research institutes, and private individuals. What is essential is to find market-based mechanisms such as eco-tourism that will convince resource owners and managers that good stewardship creates economic value.

Strangers in Our Midst:
The Problem of Invasive Alien Species

By Jeffrey A. McNeely

Invasive alien species—non-native species that become established in a new environment then proliferate and spread in ways that damage human interests—are now recognized as one of the greatest biological threats to our planet's environmental and economic well-being.[1]

Most nations are already grappling with complex and costly invasive-species problems: Zebra mussels (*Dreissena polymorpha*) from the Caspian and Black Sea region affect fisheries, mollusk diversity, and electric-power generation in Canada and the United States; water hyacinth (*Eichornia crassipes*) from the Amazon chokes African and Asian waterways; rats originally carried by the first Polynesians exterminate native birds on Pacific islands; and deadly new disease organisms (such as the viruses causing SARS, HIV/AIDS, and West Nile fever) attack human, animal, and plant populations in temperate and tropical countries. For all animal extinctions where the cause is known, invasive alien species are the leading culprits, contributing to the demise of 39 percent of species that have become extinct since 1600.[2] The 2000 IUCN Red List of Threatened Species reported that invasive alien species harmed 30 percent of threatened birds and 15 percent of threatened plants.[3] Addressing the problem of these invasive alien species is urgent because the threat is growing daily and the economic and environmental impacts are severe.

A key question is whether the global reach of modern human society can be matched by an appropriate sense of responsibility. One critical element of this question is the definition of "native," a concept with challenging spatial and temporal dimensions. While every species is native to a particular geographic area, this is just a snapshot in time, because species are constantly expanding and contracting their ranges, sometimes with human help. For example, Britain has nearly 40 more species of birds today than were recorded 200 years ago. About a third of these are deliberate introductions, such as the Little Owl (*Athene noctua*), while the others are natural colonizations that may be taking advantage of climate change.[4]

An invasive alien species is not a "bad" species but rather one "behaving badly" in a particular context.

According to one view, local biological "enrichment" by non-native species always harms native species at some level, so any introduction should be regarded, at least in principle, as undesirable. An opposing view is that because species are constantly expanding or contracting their range, new species—especially those that are beneficial to people, such as crops, ornamental plants, and pets—should be welcomed as "increasing biodiversity" unless they are clearly harmful. According to this perspective, in the case of British birds noted above, only those introduced by people and that are causing ecological or economic damage, such as pigeons, are considered to be invasive.

All continental areas have suffered from invasions of alien species, losing biological diversity as a result, but the problem is especially acute on islands in general and for small islands in particular. The physical isolation of islands over millions of years has favored the evolution of unique species and ecosystems, so islands often have a high proportion of endemic species. The evolutionary processes associated with isolation have also meant that island species are especially vulnerable to predators, pathogens, and parasites from other areas. More than 90 percent of the 115 birds known to have become extinct over the past 400 years were endemic to islands.[5] Most of these evolved in the absence of mammalian predators, so the arrival of rats and cats carried by people has had a devastating impact.

Island plants are also affected. For example, the tree *Miconia calvescens* replaced the forest canopy on more than 70 percent of the island of Tahiti over a 50-year time span, starting with a few trees in two botanical gardens. Some 40–50 of the 107 plant species endemic to the island of Tahiti are believed to be on the verge of extinction primarily due to this invasion.[6] Introduced animals also can affect plants. For example, goats introduced on St. Clemente Island, California, have caused the extinction of eight endemic species of plants and have endangered eight others.[7]

An invasive alien species is not a "bad" species but rather one "behaving badly" in a particular context, usually due to inappropriate human agency or intervention. A species may be so threatened in its natural range that it is given legal protection, yet it may generate massive ecological and other damage elsewhere.

The degradation of natural habitats, ecosystems, and agricultural lands (through loss of vegetation and soil and pollution of land and waterways) that has occurred throughout the world has made it easier for non-native species to become invasive, opening up new possibilities for them. For all of these reasons, and others that will become apparent below, the issue of invasive alien species is receiving growing international attention.

The Vectors: How Species Move Around the World

The natural barriers of oceans, mountains, rivers, and deserts have provided the isolation that has enabled unique species and ecosystems to evolve. But in just a few hundred years, these barriers have been overcome by technological changes that helped people move species vast distances to new habitats, where some of them became invasive. The growth in the volume of international trade, from US$192 billion in 1960 to almost $6 trillion in 2003,[8] provides more opportunities than ever for species to be spread either accidentally or deliberately.

Some movement seems accidental, or at least incidental, in that transporting the species was not the purpose of the transporter. For example, ballast water is now regarded as the most important vector for transoceanic movements of shallow-water coastal organisms, dispersing fish, crabs, worms, mollusks, and microorganisms from one ocean to another. Enclosed water bodies like San Francisco Bay are especially vulnerable. The bay already has at least 234 invasive alien species, causing significant economic damage. California has one of the toughest ballast water laws in the nation, requiring ships from foreign ports to exchange their ballast water 200 miles from the California coastline, but enforcement remains spotty at best.

Ballast water may also be important in the epidemiology of waterborne diseases affecting plants and animals. One study measured the concentration of the bacteria *Vibrio cholerae*—which cause human epidemic cholera—in the ballast water of vessels arriving to the Chesapeake Bay

from foreign ports, finding the bacteria in plankton samples from all ships.[9]

Other invasives are hitchhikers on global trade. For example, the Asian long-horned beetle (*Anoplophora glabripennis*) is one of the newest and most harmful invasive species in the United States. Originating in northeastern Asia, it finds its way to the United States through packing crates made of low-quality timber (that which is too infested for other uses). The number of insects found in materials imported from China increased from 1 percent of all interceptions in 1987 to 20 percent in 1996.[10] Outbreaks were reported in and around Chicago as early as 1992, in Brooklyn in August 1996, and in California in 1997. The beetle finds a congenial home among native maples, elders, elms, horse chestnuts, and others. The U.S. Department of Agriculture predicted that if the beetle becomes established, it could denude Main Street, USA, of shade trees, affect lumber and maple sugar production, threaten tourism in infested areas, and reduce biological diversity in forests.[11]

Another dangerous trade-related species for North America is the Asian gypsy moth (*Lymantria dispar*), which was first reported in the United States in 1991, entering as egg masses attached to ships or cargo from eastern Siberia. The caterpillars of this species are known to feed on more than 600 species of trees, and as moths, the females can disperse themselves over long distances. Scientists fear that this species could cause vastly more damage than the European gypsy moth, which already defoliates 1.5 million hectares of forest per year in North America.

With almost 700 million people crossing international borders as tourists each year, the opportunities for them to carry potential invasive species, either knowingly or unknowingly, are profound and increasing. Many tourists return with living plants that may become invasive, or carry exotic fruits that may be infested with invasive insects that can plague agriculture back home. Travelers may also carry diseases between countries, as apparently happened with the SARS virus. Tourism is considered an especially efficient pathway for invasive alien species

on subAntarctic islands such as South Georgia. Visitors to the island reached 15,000 in 1999. Part of the problem is that many tourists are visiting similar islands on the same trip, increasing the chances of a seed, fruit, or insect being carried, more than would be expected from a single landing of a few people who spend an extended time on one island.[12]

Many species are introduced on purpose but have unintended consequences. One example of purposeful introduction gone wrong is the extensive stocking program that introduced African tilapia *Oreochromis* into Lake Nicaragua in the 1980s, resulting in the decline of native populations of fish and the imminent collapse of one of the world's most distinctive freshwater ecosystems. The alteration of Lake Nicaragua's ecosystem is likely to have effects on the planktonic community and primary productivity of the entire lake—Central America's largest—destroying native fish populations and likely leading to unanticipated consequences.[13]

Sport fishers have also had an influence, importing their favorite game fish into new river systems, where they can have significant negative impacts on native species. For example, the northern pike (*Esox lucius*) has invaded rivers in Alaska and is replacing native species of salmon. While the northern pike occurs naturally in some parts of Alaska, it was introduced to the salmon-rich south-central area in the 1950s, probably by a fisherman who brought it to Bulchitna Lake. Flooding in the 1980s subsequently spread the pike into the streams of the Susitna and Matanuska river basins. Pike have now occupied at least a dozen lakes and four rivers in some of the richest salmon and trout habitat in the Pacific Northwest. Rainbow trout are an even greater threat. Originating in western North America, they have been introduced into 80 new countries, often with devastating impacts on native fish.

Pets are also a problem. Domestic cats can plunder ecosystems that they did not previously inhabit. On Marion Island in the sub-Antarctic Indian Ocean, cats were estimated to kill about 450,000 seabirds annually.[14] Exotic pets may escape—or be released when they have outlived their novelty—and become established in their new home. Stories of crocodiles in the

Manhattan's sewer system are probably fanciful, but many former pets are becoming established in the wild. For example, Monk parakeets (*Myiopsitta monachus*), descended from former pets that were released possibly in the 1960s, have invaded some 76 localities in 15 U.S. states.[15] Native to southern South America, they are the only parrots that build their own nests, some of which support several hundred individuals and have separate families living in different chambers. Some believe that they soon will become widespread throughout the lower 48 states, posing a significant threat to at least some agricultural lands by feeding on ripening crops. And Burmese pythons (*Python molurus*) have become established in Everglades National Park, where they reach a very large size and prey on many native species, even alligators.

Pet stores often advertise invasive species that are legally controlled. For example, the July 2000 issue of the magazine *Tropical Fish Hobbyist* recommended several species of the genus *Salvinia* as aquarium plants, even though they are considered noxious weeds in the United States and prohibited by Australian quarantine laws.

The globalization of trade and the power of the Internet offer new challenges, as sales of seeds and other organisms by mail order or over the Internet pose new and very serious risks to the ecological security of all nations. Controls on harvest and export of species are required as part of a more responsible attitude of governments toward the potential of spreading genetic pollution through invasive species. Further, all receiving countries want to ensure that they are able to control what is being imported. Virtually all countries in the world have serious problems in this regard, an issue that some countries are calling "biosecurity."

The Science of Understanding Invasions

Biodiversity is dynamic, and the movement of species around the world is a continuing process that is accelerating through expanding global trade. By trying to identify which species are especially likely to become invasive, and

hence harmful to people, ecologists are improving the quality of invasion biology as a predictive science so that people can continue to benefit from global biodiversity without paying the costs resulting from species that later become harmful.

Previous examples indicate the characteristics that can make a species invasive. For instance, coastal ecosystems are frequently invaded by microorganisms from ballast water for three main reasons. First, concentrations of bacteria and viruses exceed those reported for other taxonomic groups in ballast water by 6 to 8 orders of magnitude, and the probability of successful invasion increases with inoculation concentration. Second, the biology of many microorganisms combines a high capacity for increase, asexual reproduction, and the ability to form dormant resting stages. Such flexibility in life history can broaden the opportunity for successful colonization, allowing rapid population growth when suitable environmental conditions occur. And third, many microorganisms can tolerate a broad range of environmental conditions, such as salinity or temperature, so many sites may be suitable for colonization.[16] Insects are a major problem because they can lay dormant or travel as egg masses and are difficult to detect. The African tilapia introduced to Lake Nicaragua adapted well, because they are able to grow rapidly; feed on a wide range of plants, fish, and other organisms; and form large schools that can migrate long distances. Further, they are maternal mouth brooders, so a single female can colonize a new environment by carrying her young in her mouth.[17] Rapid growth, generalized diet, ability to move large distances, and prolific breeding are all characteristics of successful invaders.

It is not always simple, however, to distinguish a beneficial non-native species from one at significant risk of becoming invasive. A non-native species that is useful in one part of a landscape may invade other parts of the landscape where its presence is undesirable, and some species may behave well for decades before suddenly erupting into invasive status. The Nile Perch (*Lates niloticus*), for example, was introduced

to Lake Victoria in the 1950s but did not become a problem until the 1980s, when it was a key factor in the extinction of as many as half of the lake's 500 species of endemic fish, attractive prey for the perch.[18] That said, ecologists over the past several decades have agreed on some broad principles for guiding risk assessment. First, the probability of a successful invasion increases with the initial population size and with the number of attempts at introduction. While it is possible for a species to invade with a single gravid female or fertile spore, the odds of doing so are very low. Second, among plants, the longer a non-native plant has been recorded in a country and the greater the number of seeds or other propagules that it produces, the more likely it will become invasive. Third, species that are successful invaders in one situation are likely to be successful in other situations; rats, water hyacinth, microorganisms, and many others fall into this category. Fourth, intentionally introduced species may be more likely to become established than are unintentionally introduced species, at least partly because the vast majority of these have been selected for their ability to survive in the environment where they are introduced. Fifth, plant invaders of croplands and other highly disturbed areas are concentrated in herbaceous families with rapid growth and a wide range of environmental tolerances, while invaders of undisturbed natural areas are usually from woody families, especially nitrogen-fixing species that can live in nitrogen-poor soils.[19] And sixth, fire, like disturbance in general, increases invasion by introduced species. So ecosystems that are naturally prone to fire, such as the fynbos of South Africa, coastal chaparral in California, and maquis in the Mediterranean,[20] can be heavily invaded if fire-liberated seeds of invasive species are available. (These are all shrub communities adapted to cool, wet winters and hot, dry summers, where fire is a regular phenomenon. They are also rich in species: Fynbos have about 8,500 species that include many endemic *Proteaceae*; chaparral have about 5,000 species; and maquis have 25,000—of which about 60 percent are endemic to the Mediterranean region.[21])

INVASIVE ALIEN SPECIES AND PROTECTED AREAS

Protected areas are widely perceived as being devoted to conserving natural ecosystems. Ironically, protected areas are in fact heavily damaged by invasive alien species, and many protected-area managers consider this their biggest problem. Some examples:

- Galapagos National Park, a World Heritage site, is being affected by numerous invasive alien species, including pigs, goats, feral cats, fire ants, and mosquitoes.

- Kruger National Park, South Africa's largest, has recorded 363 alien plant species, including water weeds that pose a serious threat to the park's rivers.

- In the Wadden Sea, a biosphere reserve and Ramsar site protected by the Netherlands, Germany, and Denmark, the Pacific oyster has invaded, having escaped captive management. It is disrupting tourism because of its sharp shells. It has also carried with it numerous other invasive alien species.

- The Wet Tropic World Heritage Area of North Queensland, Australia, is infested by numerous invasive alien species, of which the worst is the pond apple from Florida, which has invaded creeks and riverbanks, wetlands, melaleuca swamps, and mangrove communities. Feral pigs, another invasive species, help to spread the species. The pond apple is now rare in its native range in the Florida Everglades.

- Everglades National Park in Florida, another World Heritage site, is threatened by the invasion of melaleuca from Queensland, demonstrating that species that may behave well in their natural habitat can be a serious problem when they invade somewhere else.

- Tongariro National Park, New Zealand, is also a World Heritage site, but a third of its territory has been infested by heather, a European plant deliberately introduced into New Zealand by an early park warden in 1912 in an attempt to reproduce the moors of Scotland.

These are just a few examples among many that could be cited that demonstrate that even the most strictly protected areas can be extremely vulnerable to invasion by non-native species.

Other ecological factors that may favor nonindigenous species include a lack of controlling natural enemies, the ability of an alien parasite to switch to a new host, an ability to be an effective predator in the new ecosystem, the availability of artificial or disturbed habitats that provide an ecosystem the aliens can easily invade, and high adaptability to novel conditions.[22]

It is sometimes argued that systems with great species diversity are more resistant to new species invading. However, a study in a California riparian system found that the most diverse natural assemblages are in fact the most invaded by non-native plants, and protected areas worldwide are heavily invaded by non-native plants and animals.[23] Dalmatian toadflax (*Linaria dalmatica*) is invading relatively undisturbed shrub-steppe habitat in the Pacific Northwest, wetland nightshade (*Solanum tampicense*) is invading cypress wetlands in central and south Florida, and garlic mustard (*Allilaria officinalis*) is often found in relatively undisturbed systems in the northern parts of North America.

This work helps resolve the controversy over the relationship between biodiversity and invasions, suggesting that the scale of investigation its a critical factor. Theory suggests that non-native species should have a more difficult time invading a diverse ecosystem, because the web of species interactions should be more efficient in using resources such as nutrients, light, and water than would fewer species, leaving fewer resources available for the nonnative species. But even in well-protected landscapes such as national parks, invaders often seem to be more successful in diverse ecosystems. Even though diversity does matter in fending off invasives, its effects are negated by other factors at larger scales. The most diverse ecosystems might be at the greatest risk of invasion, while losses of species, if they affect community-scale diversity, may erode invasion resistance.[24]

The Economic Impacts of Invasion

One reason invasive alien species are attracting more attention is that they are having substantial negative impacts on numerous economic sectors, even beyond the obvious impacts on agriculture (weeds), forestry (pests), and health (diseases or disease vectors). The probability that any one introduced species will become invasive may be low, but the damage costs and costs of control of the species that do become invasive can be extremely high (such as the recent invasion of eastern Canada by the European brown spruce longhorn beetle (*Tetropium fuscum*), which threatens the Canadian timber industry).

Estimates of the economic costs of invasive alien species include considerable uncertainty, but the costs are profound—and growing (see Table 1).

Most of these examples come from the industrialized world, but developing countries are experiencing similar, and perhaps proportionally greater, damage. Invasive alien insect pests—such as the white cassava mealybug (*Phenacoccus herreni*) and larger grain borer (*Prostephanus truncates*) in Africa—pose direct threats to food security. Alien weeds constrain efforts to restore degraded land, regenerate forests, and improve utilization of water for irrigation and fisheries. Water hyacinth and other alien water weeds that choke waterways currently cost developing countries in Africa and Asia more than US$100 million annually. Invasive alien species pose a threat to more than $13 billion of current and planned World Bank funding to projects in the irrigation, drainage, water supply, sanitation, and power sectors.[25] And a study of three developing nations (South Africa, India, and Brazil) found annual losses to introduced pests of $138 billion per year.[26]

Table 1. Indicative costs of some invasive alien species (in U.S. dollars)

Species	Economic variable	Economic impact
Introduced disease organisms	Annual cost to human, plant, and animal health in the United States	$41 billion per year[a]
A sample of alien species of plants and animals	Economic costs of damage in the United States	$137 billion per year[b]
Salt cedar	Value of ecosystem services lost in western United States	$7–16 billion over 55 years[c]
Knapweed and leafy spurge	Impact on economy in three U.S. states	Direct costs of $40.5 million per year; indirect costs of $89 million[d]
Zebra mussel	Damages to U.S. industry	Damage of more than $2.5 billion to the Great Lakes fishery between 1998–2000;[e] $5 billion to U.S. industry by 2000[f]
Most serious invasive alien plant species	Costs 1983–1992 of herbicide control in England	$344 million per year for 12 species[g]
Six weed species	Costs in Australia agroecosystems	$105 million per year[h]
Pinus, Hakeas, and Acacia	Costs on South African floral kingdom to restore to pristine state	$2 billion total for impacts felt over several decades[i]
Water hyacinth	Costs in seven African countries	$20–50 million per year[j]
Rabbits	Costs in Australia	$373 million per year (agricultural losses)[k]
Varroa mite	Economic cost to beekeeping in New Zealand	An estimated $267–602 million over the next 35 years[l]

[a] P. Daszak, A. Cunningham, and A. D. Hyatt, "Emerging Infectious Diseases of Wildlife: Threats to Biodiversity and Human Health," *Science*, 21 January 2000, 443–49.

[b] D. Pimentel, L. Lach, R. Zuniga, and D. Morrison, "Environmental and Economic Costs of Non-indigenous Species in the United States," *BioScience* 50 (2000): 53–65.

[c] E. Zavaleta, "Valuing Ecosystem Services Lost to Tamarix Invasion in the United States," in H. A. Mooney and R. J. Hobbs, eds., *Invasive Species in a Changing World* (Washington DC: Island Press, 2000).

[d] D. A. Bangsund, F. L. Leistritz, and J. A. Leitch, "Assessing Economic Impacts of Biological Control of Weeds: The Case of Leafy Spurge in the Northern Great Plains of the United States," *Journal of Environmental Management* 56 (1999): 35–43; and D. A. Bangsund, S. A. Hirsch, and J. A. Leitch, *The Impact of Knapweed on Montana's Economy* (Fargo, ND: Department of Agricultural Economics, North Dakota State University, 1996).

[e] P. C. Focazio, "Coordinated Issue Area: Aquatic Nuisance, Non-Indigenous, and Invasive Species," *Coastlines* 30, no.1 (2001): 4–5.

[f] "Coatings to Repel Zebra Mussels," U.S. Army Construction Engineering Research Laboratory fact sheet, http://www.cecer.army.mil/facts/sheets/FL10.html.

[g] M. Williamson, "Measuring the Impact of Plant Invaders in Britain," in S. Starfinger, K. Edwards, I. Kowarik, and M. Williamson, eds., *Plant Invasions: Ecological Mechanisms and Human Responses* (Leiden, Netherlands: Backhuys, 2000), 57–70.

[h] A. Watkinson, R. Freckleton, and P. Dowling, "Weed Invasions of Australian Farming Systems: From Ecology to Economics," in C. Perrings, M. Williamson, and S. Dalmazzone, eds., *The Economics of Biological Invasions* (Cheltenham, UK: Edward Elgar, 2000), 94–116.

[i] J. Turpie and B. Heydenrych, "Economic Consequences of Alien Infestation of the Cape Floral Kingdom's Fynbos Vegetation," in Perrings, Williamson, and Dalmazzone, ibid., pages 152–82.

[j] S. Joffe and S. Cook, *Management of the Water Hyacinth and Other Aquatic Weeds: Issues for the World Bank* (Cambridge, UK: Commonwealth Agriculture Bureau International (CABI) Bioscience, 1997).

[k] P. White and G. Newton-Cross, "An Introduced Disease in an Invasive Host: The Ecology and Economics of Rabbit Carcivirus Disease (RCD) in Rabbits in Australia," in Perrings, Williamson, and Dalmazzone, note k above, pages 117–37.

[l] R. Wittenberg and M. J. W. Cock, eds., *Invasive Alien Species: A Tool Kit of Best Prevention and Management Practices* (Wallingford, UK: Global Invasive Species Programme and CABI, 2001).

SOURCE: J. A. McNeely.

In addition to the direct costs of managing invasives, the economic costs also include their indirect environmental consequences and other nonmarket values. For example, invasives may cause changes in ecological services by disturbing the operation of the hydrological cycle, including flood control and water supply, waste assimilation, recycling of nutrients, conservation and regeneration of soils, pollination of crops, and seed dispersal. Such services have current-use value and option value (the potential value of such services in the future). In the South African fynbos, for example, the establishment of invasive tree species—which use more water than do native species—has decreased water supplies for nearby communities and increased fire hazards, justifying government expenditures equivalent to US$40 million per year for both manual and chemical control.[27]

> **Customs and quarantine practices, developed in an earlier time, are inadequate safeguards against the rising tide of species that threaten native biodiversity.**

Many people in today's globalized economy are driven especially by economic motivations. Those who are importing non-native species are usually doing so with a profit motive and often seek to avoid paying for possible associated negative impacts if those species become invasive. The fact that these negative impacts might take several decades to appear make it all the easier for the negative economic impacts to be ignored. Similarly, those who are ultimately responsible for such "accidental" introductions (for example, through infestation of packing materials or organisms carried in ballast water) seek to avoid paying the economic costs that would be required to prevent these "accidental," but predictable, invasions. In both cases, the potential costs are externalized to the larger society, and to future generations.

Responses

Customs and quarantine practices, developed in an earlier time to guard against diseases and pests of economic importance, are inadequate safeguards against the rising tide of species that threaten native biodiversity. Globally, about 165 million 6-meter-long, sealed containers are being shipped around the world at any given time. This number is far larger than custom officers can reasonably be expected to examine in detail. In the United States, some 1,300 quarantine officers are responsible for inspecting 410,000 planes and more than 50,000 ships, with each ship carrying hundreds of containers. While they intercept alien species nearly 50,000 times a year, it is highly likely that at least tens of thousands more enter the country uninspected each year. In Europe, inspection at the port of entry is also desperately overextended, and once a container enters the European Union, no further border inspections are done. This is a recipe for disaster.

Instead, a different set of strategies is now needed to deal with invasive species. These include prevention (certainly the most preferable), early eradication, special containment, or integrated management (often based on biological control). Mechanical, biological, and chemical means are available for controlling invasive species of plants and animals once they have arrived. Early warning, quarantine, and various other health measures are involved to halt the spread of pathogens.[28]

The international community has responded to the problem of invasive alien species through more than 40 conventions or programs, and many more are awaiting finalization or ratification.[29] The most comprehensive is the 1992 Convention on Biological Diversity, which calls on its 188 parties to "prevent the introduction of, control, or eradicate those alien species which threaten ecosystems, habitats, or species" (Article 8h).[30] A much older instrument, one that is virtually universally applied, is the 1952 International Plant Protection Convention, which applies primarily to plant pests, based on a system of phytosanitary certificates. Regional agreements further strengthen this convention. Other instruments deal with invasive alien species in specific regions (such as Antarctica), sectors (such as fishing in the Danube River), or vectors (such as invasive species in ballast water, through the International Maritime Organization). The fact that the problem continues to worsen in-

dicates that the international response to date has been inadequate.

On the national level, some legal measures can offer very straightforward methods of preventing or managing invasions. For example, to deal with the problem of Asian beetle invasions, the United States now requires that all solid-wood packing material from China must be certified free of bark (under which insects may lurk) and heat-treated, fumigated, or treated with preservatives. China might reasonably issue a reciprocal regulation, as North American beetles are a hazard there.

The nursery industry is by far the largest intentional importer of new plant taxa. Issuing permits for imported species is a good way for the agencies responsible for managing such invasions to keep track of what is being traded and moved around the country. Some people believe that it is impossible to issue a regulation containing a list of permitted and prohibited species, at least partly because the ornamental horticulture industry is always seeking new species. But the Florida Nurserymen and Growers Association recently identified 24 marketed species on a black list drawn up by Florida's Exotic Pest Plant Council and decided to discourage trade in 11 of the species (the least promising sellers in any case).[31]

Sometimes nature itself can fight back against invasive alien species, at least when they reach plague proportions. For example, the zebra mussels that have invaded the North American Great Lakes with disastrous effects are now declining because a native sponge (*Eunapius fragilis*) is growing on the mussels, preventing them from opening their shells to feed or to breathe. The sponge has become abundant in some areas, while the zebra mussel population has fallen by up to 40 percent, although it is not yet clear whether the sponges will be effective in controlling the invasive mussels in the long term.[32]

Biological control—the intentional use of natural enemies to control an invasive species—is an important tool for managers. Some early efforts at biological control agents had disastrous effects, such as South American cane toads (*Bufo marinus*) in Australia, Indian com-

mon mynahs (*Acridotheres tristis*) in Hawaii, and Asian mongooses (*Herpestes javanicus*) in the Caribbean. Not only did these species not deal with the problem species upon which they were expected to prey, but they ended up causing havoc to native species and ecosystems. On the other hand, biological control programs are now much more carefully considered and in many cases are the most efficient, most effective, cheapest, and least damaging to the environment of any of the options for dealing with invasives that have already arrived.[33] Examples include the use of a weevil (*Cyrtobagous salviniae*) to control salvinia fern (*Salvinia molesta*), another weevil (*Neohydronomus affinis*) to control water lettuce (*Pistia stratiotes*), and a predatory beetle (*Hyperaspis pantherina*) to control orthezia scale (*Orthezia insignis*) that threatened the endemic national tree of Saint Helena (*Commidendrum robustum*).[34]

Those seeking to use viruses or other disease organisms to control an invasive species need to understand ecological links. When millions of rabbits died after the intentional introduction of the myxomatosis virus in the United Kingdom, for example, populations of their predators, including stoats, buzzards, and owls, declined sharply. The impact affected other species indirectly, leading to local extinction of the endangered large blue butterfly (*Maculina arion*) because of reduced grazing by rabbits on heathlands, which removed the habitat for an ant species that assists developing butterfly larvae.[35] But the use of the myxoma virus in conjunction with 1080 poison on the Phillip Island in the South Pacific successfully eradicated invasive rabbits, allowing the recovery of the island's vegetation (including the endemic *Hibiscus insularis*).[36]

At small scales of less than one hectare, it appears possible with current technology to eradicate invasive species of plants through use of herbicides, fire, physical removal, or a combination of these, but the costs of eradication rise quickly as the area covered increases. With the right approach and technology, invasive alien mammals can be eradicated from islands of thousands of hectares in size. Rat eradication from islands of larger than 2,000 hectares has been successful, and large mammals have been removed from much bigger ones than that, primarily by hunting and trapping.

Environmentally sensitive eradication also requires the restoration of the community or ecosystem following the removal of the invasive. For example, the eradication of Norway rats from Mokoia Island in New Zealand was followed by greatly increased densities of mice, also alien species. Similarly, the removal of Pacific rats (*Rattus exulans*) from Motupao Island, New Zealand, to protect a native snail led to increases of an exotic snail to the detriment of the natives. And on Motunau Island, New Zealand, the exotic box-thorn (*Lycium ferocissimum*) increased after the control of rabbits. On Santa Cruz Island, off the west coast of California, removing goats led to dramatic increases in the abundance of fennel (*Foeniculum vulgare*) and other alien species of weeds. Thus reversing the changes to native communities caused by non-native species will often require a sophisticated understanding of ecological relationships. It is now well recognized that eradication programs are only the first step in a long process of restoration.[37] Sometimes native species become dependent on invasive ones, causing dilemmas for managers. For example, giant kangaroo rats (*Dipodomys ingens*) in the American West continually modify their burrow precincts by digging tunnels, clipping plants, and other activities. This chronic disturbance to soil and vegetation sometimes promotes the establishment of invasive species of plants that were originally imported as ornamentals from the Mediterranean so that they constitute a very large proportion of the vegetation on giant kangaroo rat territories. They have significantly larger seeds than do native species so are favored by the grain-eating kangaroo rats.[38] Because the kangaroo rats depend on non-native plant species for food and the non-native plant species depend upon the kangaroo rats to disturb their habitat continually, the relationship is mutualistic. This strong relationship may also inhibit population growth of native grassland plants that occupy disturbed habitats but have difficulty competing with nonnative weeds for resources. This mutualism presents an intractable conservation management dilemma, suggesting that it may be impossible to restore valley grasslands occupied by endangered kangaroo rats to conditions where native species dominate.

High-tech management measures are also being tried. For example, Australian scientists are planning to insert a gene known as "daughterless" into invasive male carp (*Cyprinus carpio*) in the Murray-Darling River, the country's longest, thereby ensuring that their offspring are male. The objective is to release them into the wild, sending wild carp populations into a decline and making room for the native species that are being threatened by the invasive European carp.[39] Using genetic modification can help eradicate an invasive alien species, but if the detrimental gene is released into nature and starts to flourish, many other species could be negatively affected. Thus the precautionary approach needs to be applied to control techniques as well as to introductions.

The problems of invasive alien species are so serious that actions must be taken even before we can be "certain" of all of their effects. However, mechanical removal, biocontrol, chemical control, shooting, or any other approach to controlling alien invasive species needs to be carefully considered prior to use to ensure that the implications have been fully and carefully considered, including impacts on human health, other species, and so forth. A public information program is also needed to ensure that the proposed measures are likely to be effective as well as socially and politically acceptable. Many animal-rights groups oppose the killing of any species of wildlife, for instance, even if they are causing harm to native species of plants and animals. The recent controversy surrounding the population of mute swans in the Chesapeake Bay is a good example.[40]

Conclusions

Ecosystems have been significantly influenced by people in virtually all parts of the world; some have even called these "engineered ecologies." Thus, a much more conscious and better-informed management of ecosystems—one that deals with non-native species—is critical.

In just a few hundred years, major global forces have rendered natural barriers ineffective, allowing non-native species to travel vast distances to new habitats and become invasive alien species. The globalization and growth in the volume of trade and tourism, coupled with the emphasis on free trade, provide more opportunities than ever for species to be spread accidentally or deliberately. This inadvertent ending of millions of years of biological isolation has created major ongoing environmental problems that affect developed and developing countries, with profound economic and ecological implications.

Because of the potential for economic and ecological damage when an alien species becomes invasive, every alien species needs to be treated for management purposes as if it is potentially invasive, unless and until convincing evidence indicates that it is harmless in the new range. This view calls for urgent action by a wide range of governmental, intergovernmental, private sector, and civil institutions.

A comprehensive solution for dealing with invasive alien species has been developed by the Global Invasive Species Programme.[41] It includes 10 key elements:

- *An effective national capacity to deal with invasive alien species.* Building national capacity could include designing and establishing a "rapid response mechanism" to detect and respond immediately to the presence of potentially invasive species as soon as they appear, with sufficient funding and regulatory support; as well as implementing appropriate training and education programs to enhance individual capacity, including customs officials, field staff, managers, and policymakers. It could also include developing institutions at national or regional levels that bring together biodiversity specialists with agricultural quarantine specialists. Building basic border control and quarantine capacity and ensuring that agricultural quarantine, customs, and food inspection officers are aware of the elements of the Biosafety Protocol are other ways to deal with invasive alien species on a national level.

- *Fundamental and applied research at local, national, and global levels.* Research is required on taxonomy, invasion pathways, management measures, and effective monitoring. Further understanding on how and why species become established can lead to improved prediction on which species have the potential to become invasive; improved understanding of lag times between first introduction and establishment of invasive alien species; and better methods for excluding or removing alien species from traded goods, packaging material, ballast water, personal luggage, and other methods of transport.

> The problems of invasive alien species are so serious that actions must be taken even before we can be "certain" of all their effects.

- *Effective technical communications.* An accessible knowledge base, a planned system for review of proposed introductions, and an informed public are needed within countries and between countries. Already, numerous major sources of information on invasive species are accessible electronically and more could also be developed and promoted, along with other forms of media.
- *Appropriate economic policies.* While prevention, eradication, control, mitigation, and adaptation all yield economic benefits, they are likely to be undersupplied, because it is difficult for policymakers to identify specific beneficiaries who should pay for the benefits received. New or adapted economic instruments can help ensure that the costs of addressing invasive alien species are better reflected in market prices. Economic principles relevant to national strategies include

ensuring that those responsible for the introduction of economically harmful invasive species are liable for the costs they impose; ensuring that use rights to natural or environmental resources include an obligation to prevent the spread of potential invasive alien species; and requiring importers of such potential species to have liability insurance to cover the unanticipated costs of introductions.

- *Effective national, regional, and international legal and institutional frameworks.* Coordination and cooperation between the relevant institutions are necessary to address possible gaps, weaknesses, and inconsistencies and to promote greater mutual support among the many international instruments dealing with invasive alien species.
- *A system of environmental risk analysis.* Such a system could be based on existing environmental impact assessment procedures that have been developed in many countries. Risk analysis measures should be used to identify and evaluate the relevant risks of a proposed activity regarding alien species and determine the appropriate measures that should be adopted to manage the risks. This would also include developing criteria to measure and classify impacts of alien species on natural ecosystems, including detailed protocols for assessing the likelihood of invasion in specific habitats or ecosystems.
- *Public awareness and engagement.* If management of invasive species is to be successful, the general public must be involved. A vigorous public awareness program would involve the key stakeholders who are actively engaged in issues relevant to invasive alien species, including botanic gardens, nurseries, agricultural suppliers, and others. The public can also be involved as volunteers in eradication programs of certain non-native species, such as woody invasives of national parks for suggested actions that individuals can take.)
- *National strategies and plans.* The many elements of controlling inva-

sive alien species need to be well coordinated, ensuring that they are not simply passed on to the Ministry of Environment or a natural resource management department. A national strategy should promote cooperation among the many sectors whose activities have the greatest potential to introduce them, including military, forestry, agriculture, aquaculture, transport, tourism, health, and water-supply sectors. The government agencies with responsibility for human health, animal health, plant health, and other relevant fields need to ensure that they are all working toward the same broad objective of sustainable development in accordance to national and international legislation. Such national strategies and plans can also encourage collaboration between different scientific disciplines and approaches that can seek new approaches to dealing with problems caused by invasive alien species.

- *Invasive alien species issues built into global change initiatives.* Global change issues relevant to invasives begin with climate change but also include changes in nitrogen cycles, economic development, land use, and other fundamental changes that might enhance the possibilities of these species becoming established. Further, responses to global change issues, such as sequestering carbon, generating biomass energy, and recovering degraded lands, should be designed in ways that use native species and do not increase the risk of the spread of non-native invasives.
- *Promotion of international cooperation.* The problem of invasive alien species is fundamentally international, so international cooperation is essential to develop the necessary range of approaches, strategies, models, tools, and potential partners to ensure that the problems of such species are effectively addressed. Elements that would foster better international cooperation could include developing an international vocabulary, widely agreed upon and adopted; cross-sector collaboration

WHAT CAN AN INDIVIDUAL DO?

While the problem of invasive alien species seems daunting, an individual can make an important contribution to the problem, and if thousands of individuals work toward reducing the spread of invasive aliens, real progress can be made. Here are some steps that can be taken:

- Become informed about the issue.
- Grow native plants, keep native pets, and avoid releasing non-natives into the wild.

- Avoid carrying any living materials when traveling.
- Never release plants, fish, or other animals into a body of water unless they came out of that body of water.
- Clean boats before moving them from one body of water to another, and avoid using non-native species as bait.
- Support the work of organizations that are addressing the problem of invasive alien species.

among international organizations involved in agriculture, trade, tourism, health, and transport; and improved linkages among the international institutions dealing with phytosanitary, biosafety, and biodiversity issues and supporting these by strong linkages to coordinated national programs.

Because the diverse ecosystems of our planet have become connected through numerous trade routes, the problems caused by invasive alien species are certain to continue. As with maintaining and enhancing health, education, and security, perpetual investments will be required to manage the challenge they present. These 10 elements will ensure that the clear and present danger of invasive species is addressed in ways that build the capacity to address any future problems arising from expanding international trade.

NOTES

1. H. A. Mooney, J. A. McNeely, L. E. Neville, P. J. Schei, and J. K. Waage, eds., *Invasive Alien Species: Searching for Solutions* (Washington, DC: Island Press, 2004).

2. B. Groombridge, ed., *Global Biodiversity: Status of the Earth's Living Resources* (Cambridge, UK: World Conservation Monitoring Centre, 1992).

3. C. Hilton-Taylor, IUCN *Red List of Threatened Species* (Gland, Switzerland: IUCN–The World Conservation Union (IUCN). 2000).

4. R. May, "British Birds by Number," *Nature*, 6 April 2000, 559–60.

5. Ibid., and Groombridge, note 2 above.

6. J-Y. Meyer, "Tahiti's Native Flora Endangered by the Invasion of *Miconia calvesens*," *Journal of Geography* 23 (1997): 775–81.

7. D. Pimentel, L. Lach, R. Zuniga, and D. Morrison, "Environmental and Economic Costs of Nonindigenous Species in the United States," *BioScience* 50 (2000): 53–65.

8. World Trade Organization (WTO), *International Trade Statistics 2003* (Geneva: WTO, 2004).

9. G. M. Ruiz et al., "Global Spread of Microorganisms by Ships," *Nature*, 2 November 2000, 49–50.

10. J. E. Pasek, "Assessing Risk of Foreign Pest Introduction via the Solid Wood Packing Material Pathway," presentation made at the North American Plant Protection Organization Symposium on Pet Risk Analysis, Puerto Vallerta, Mexico, 18–21 March 2002.

11. U.S. Department of Agriculture, *Agricultural Research Service Research to Combat Invasive Species*, **www.invasivespecies.gov/ toolkit/arsisresearch.doc** (accessed 27 April 2004).

12. S. L. Chown and K. J. Gaston, "Island-Hopping Invaders Hitch a Ride with Tourists in South Georgia, *Nature*, 7 December 2000, 637.

13. K. R. McKaye et al., "African Tilapia in Lake Nicaragua," *BioScience* 45 (1995): 406–11.

14. L. Winter, "Cats Indoors!" *Earth Island Journal*, Summer 1999, 25–26.

15. G. Zorpette, "Parrots and Plunder," *Scientific American*, July 1997, 15–17.

16. Ruiz et al., note 9 above.

17. McKaye et al., note 13 above.

18. A. J. Ribbink, "African Lakes and Their Fishes: Conservation Scenarios and Suggestions," *Environmental Biology of Fishes* 19 (1987): 3–26; T. Goldschmit, F. Witte, and J. Wanink, "Cascading Effects of the Introduced Nile Perch on the Detritivorousphytoplanktivorous Species in the Sublittoral Areas of Lake Victoria," *Conservation Biology* 7 (1993): 686–700; and R. Ogutu-Ohwayo, "Nile Perch in Lake Victoria: The Balance between Benefits and Negative Impacts of Aliens," in O. T. Sandlund, P. J. Schei, and A. Viken, eds., *Invasive Species and Biodiversity Management* (Dordrecht, Netherlands: Kluwer Academic Publishers, 1999), 47–64.

19. M. L. McKinney and J. L. Lockwood, "Biotic Homogenization: A Few Winners Replacing Many Losers in the Next Mass Extinction," *Tree* 14 (1999): 450–53.

20. C. M. D'Antonio, T. L. Dudley, and M. Mack, "Disturbance and Biological Invasions: Direct Effects and Feedbacks," in L. Locker, ed., *Ecosystems of Disturbed Ground* (Amsterdam: Elziveer, 1999).

21. B. Groombridge and M. D. Jenkins, *World Atlas of Biodiversity: Earth's Living Resources in the 21st Century* (Berkeley, CA: University of California Press, 2002)

22. Pimentel, Lach, Zuniga, and Morrison, note 7 above.

23. N. L. Larson, P. J. Anderson, and W. Newton, "Alien Plant Invasion in Mixed-Grass Prairie: Effects of Vegetation Type and Anthropogenic Disturbance," *Ecological Applications* 11 (2001): 128–41; and J. M. Levine, "Species Diversity and Biological Invasions: Relating Local Process to

Community Pattern," *Science*, 5 May 2000, 852–54.

24. Ibid.

25. S. Noemdoe, "Putting People First in an Invasive Alien Clearing Programme: Working for Water Programme," in J. A. McNeely, ed., *The Great Reshuffling: Human Dimensions of Invasive Alien Species*, (Gland, Switzerland: IUCN, 2001), 121–26

26. D. Pimentel et al., "Economic and Environmental Threats of Alien Plant, Animal, and Microbe Invasions," *Agriculture, Ecosystems and Environment* 84 (2001): 1–20.

27. Ibid.

28. J. Kaiser, "Stemming the Tide of Invading Species," *Science*, 17 September 2000, 1836–841.

29. C. Shine, N. Williams, and L. Gündling, *A Guide to Designing Legal and Institutional Frameworks on Alien Invasive Species* (Bonn, Germany: IUCN, 2000).

30. L. Glowka, F. Burhenne-Guilmin, and H. Synge, *A Guide to the Convention on Biological Diversity* (Gland, Switzerland: IUCN, 1994). See also K. Raustiala and D. G. Victor, "Biodiversity Since Rio: The Future of the Convention on Biological Diversity," *Environment*, May 1996, 16–20, 37–45.

31. Kaiser, note 28 above.

32. A. Ricciardi, F. Sneider, D. Kelch, and H. Reiswig, "Lethal and Sub-Lethal Effects of Sponge Overgrowth on Introduced Dreissenide Mussels in the Great Lakes—St. Lawrence River System," *Canadian Journal of Fisheries and Aquatic Sciences* 52: 2695–703.

33. M. S. Hoddle, "Restoring Balance: Using Exotic Species to Control Invasive Exotic Species," *Conservation Biology* 18 (2004): 38–49; S. M. Louda and P. Stiling, "The Double-Edged Sword of Biological Control in Conservation and Restoration," *Conservation Biology* 18 (2004): 50–53; and R. Wittenberg and M. J. W. Cock, eds., *Invasive Alien Species: A Tool Kit of Best Prevention and Management Practices* (Wallingford, UK: Global Invasive Species Programme and Commonwealth Agricultural Bureau International, 2001).

34. Wittenberg and Cock, ibid.

35. P. Daszak, A. Cunningham, and A. D. Hyatt. "Emerging Infectious Diseases of Wildlife: Threats to Biodiversity and Human Health," *Science*, 21 January 2000, 443–49.

36. P. Coyne, "Rabbit Eradication on Phillip Island," in Wittenberg and Cock, note 33 above, page 176.

37. R. C. Klinger, P. Schuyler, and J. D. Sterner, "The Response of Herbaceous Vegetation and Endemic Plant Species to the Removal of Feral Sheep from Santa Cruz Island, California," in C. R. Veitch and M. N. Klout, eds., *Turning the Tide: the Eradication of Invasive Species* (Gland, Switzerland, and Cambridge, UK: IUCN, 2002), 14 1–54.

38. P. Schiffman, "Promotion of Exotic Weed Establishment by Endangered Giant Kangaroo Rats (*Dipodomys ingens*) in a California Grassland," *Biodiversity and Conservation* 3 (1994): 524–37.

39. R. Nowak, "Gene Warfare: One Small Tweak and a Whole Species Will Be Wiped Out," *New Scientist*, 11 May 2002, 6.

40. See B. Engle, "No Swansong in the Chesapeake Bay," *Environment*, December 2003, 7.

41. The Global Invasive Species Programme (GISP) was established in 1997 as a consortium of the Scientific Committee on Problems of the Environment (SCOPE), CABI, and IUCN, in partnership with the United Nation Environment Programme and with funding from the Global Environment Facility (GEF). See J. A. McNeely, H. A. Mooney, L. E. Neville, P. J. Schei, and J. K. Waage, eds., *Global Strategy on Invasive Alien Species* (Gland, Switzerland: IUCN, 2001).

Jeffrey A. McNeely is chief scientist at IUCN-The World Conservation Union in Gland, Switzerland. His research focuses on a broad range of topics relating to conservation and sustainable use of biodiversity, with a particular focus in recent years on the relationship between agriculture and wild biodiversity, the relationship between biodiversity and human health, and the impacts of war on biodiversity. McNeely has written or edited more than 30 books, from *Mammals of Thailand* (Association for the Conservation of Wildlife, 1975), his first, to *Ecoagriculture: Strategies to Feed the World and Save Wild Biodiversity* (Island Press, 2003). He has also published extensively on biodiversity, protected areas, and cultural aspects of conservation. He may be reached at jam@iucn.org.

From *Environment*, July/August 2004, pp. 17-31. Reprinted by permission of the Helen Dwight Reid Educational Foundation. Published by Heldref Publications, 1319, Eighteenth St., NW, Washington, DC 20036-1802. Copyright © 2004.

Markets for Biodiversity Services

POTENTIAL ROLES AND CHALLENGES

By Michael Jenkins, Sara J. Scherr, and Mira Inbar

Historically, it has been the responsibility of governments to ensure biodiversity protection and provision of ecosystem services. The main instruments to achieve such objectives have been

- direct resource ownership and management by government agencies;
- public regulation of private resource use;
- technical assistance programs to encourage improved private management; and
- targeted taxes and subsidies to modify private incentives.

But in recent decades, several factors have stimulated those concerned with biodiversity conservation services to begin exploring new market-based instruments. The model of public finance for forest and biodiversity conservation is facing a crisis as the main sources of finance have stagnated, despite the recognition that much larger areas require protection. At the same time, increasing recognition of the roles that ecosystem services play in poverty reduction and rural development is highlighting the importance of conservation in

the 90 percent of land outside protected areas. It is thus urgent to find new means to finance the provision of ecosystem services, yet under current conditions private actors lack financial incentives to do so.

Crisis in Biodiversity Conservation Finance

Financing and management of natural protected areas has historically been perceived as the responsibility of the public sector. According to the United Nations Environment Programme, there are presently 102,102 protected areas worldwide, covering an area of 18.8 million square kilometers. Seventeen million square kilometers of these areas—11.5 percent of the Earth's terrestrial surface—are forests. Two-thirds of these have been assigned to one of the six protected-area management categories designated by the World Conservation Union (IUCN).

However, over the last few decades, severe cutbacks in the availability of public resources have undermined the effectiveness of such strategies. Protected ar-

eas in the tropics are increasingly dependent on international public or private donors for financing. Yet budgets for government protection and management of forest ecosystem services are declining, as are international sources from overseas development assistance. Land acquisition for protected areas and compensation for lost resource-based livelihoods are often prohibitively expensive. For example, it has been estimated that $1.3 billion would be required to fully compensate inhabitants in just nine central African parks.[1] The donation-driven model is often unsustainable, both economically and environmentally. Sovereignty is also an issue: About 30 percent of private forest concessions in Latin America and the Caribbean and 23 percent in Africa are already foreign owned. At the same time, public responsibility for nature protection is shifting with processes of devolution and decentralization, and new sources of financing for local governments to take on biodiversity and ecosystem service protection have not been forthcoming.

Table 1. Estimated financial flows for forest conservation (in millions, U.S. dollars)

Sources of finance	SFM (early 1990s)	SFM (early 2000)	PAS (early 1990s)	PAS (early 2000)
Official development assistance	$2,000–$2,200	$1,000–$1,200	$700–$770	$350–$420
Public expenditure	NA	$1,600	NA	$598
Philanthropy[a]	$85.6	$150	NA	NA
Communities[b]	$365–$730	$1,300–$2,600	NA	NA
Private companies	NA	NA	NA	NA

[a]Underestimates self-financing and in-kind nongovernmental organization contributions.
[b]Self-financing and in-kind contributions from indigenous and other local communities.

NOTE: In 1990, there were an estimated 100 million hectares of community-managed forests worldwide. SFM is "sustainable forest management." PAS stands for "protected area system."

SOURCE: A. Molnar, S. J. Scherr, and A. Khare, *Current Status and Future Potential of Markets for Ecosystem Services of Tropical Forests: An Overview* (Washington, DC: Forest Trends, 2004).

Moreover, scientific studies increasingly indicate that biodiversity cannot be conserved by a small number of strictly protected areas.[2] Conservation must be conceived in a landscape or ecosystem strategy that links protected areas within a broader matrix of land uses that are compatible with and support biodiversity conservation in situ. To achieve such outcomes, it will be essential to engage private actors in conservation finance on a large scale. Yet the markets for products from natural areas and forests face at least three serious challenges: declining commodity prices for traditionally important products, such as timber; competition from illegal sources; and poorly functioning, overregulated markets. Thus, private forest owners and landowners need to find new revenue streams to justify retaining forests on the landscapes and to manage them well in the context of declining commodity prices and competition in natural forests from illegal sources of timber.

Rural Development, Poverty Reduction, and Biodiversity

The vast majority of biodiversity resources in the world are found in populated landscapes, and it can be argued that the biodiversity that underpins ecosystem services critical to human health and livelihoods should have high priority in conservation efforts. An estimated 240 million rural people live in the world's high-canopy forest landscapes. In Latin America, for example, 80 percent of all forests are located in areas of medium to high human popula-

tion density.[3] Population growth in the world's remaining "tropical wilderness areas" is twice the global average. More than a billion people live in the 25 biodiversity "hotspots" identified by Conservation International; in 16 of these hotspots, population growth is higher than the world average.[4] While species richness is lower in drylands and other ecosystems not represented among the "hot spots," the species that play functional ecosystem roles are all the more important and difficult to replace.

Poor rural communities are especially dependent upon natural biodiversity. Low-income rural people rely heavily on the direct consumption of wild foods, medicines, and fuels, especially for meeting micronutrient and protein needs, and during "hungry" periods. An estimated 350 million poor people rely on forests as safety nets or for supplemental income. Farmers earn as much as 10 to 25 percent of household income from nontimber forest products. Bushmeat is the main source of animal protein in West Africa. The poor often harvest, process, and sell wild plants and animals to buy food. Sixty million poor people depend on herding in semiarid rangelands that they share with large mammals and other wildlife. Thirty million low-income people earn their livelihoods primarily as fishers, twice the number of 30 years ago. The depletion of fisheries has serious impacts on food security. Wild plants are used in farming systems for fodder, fertilizer, packaging, fencing, and genetic materials. Farmers rely on soil microorganisms to maintain soil fertility and structure for crop production, and they also rely on wild species in natural ecological communities for crop pollination and

pest and predator control. Wild relatives of domesticated crop species provide the genetic diversity used in crop improvement. The rural poor rely directly on ecosystem services for clean and reliable local water supplies. Ecosystem degradation results in less water for people, crops, and livestock; lower crop, livestock, and tree yields; and higher risks of natural disasters.

> **More than a billion people live in the 25 biodiversity "hotspots" identified by Conservation International; in 16 of these hotspots, population growth is higher than the world average.**

Three-quarters of the world's people living on less than $1 per day are rural. Strategies to meet the United Nations Millennium Development Goals in rural areas—to reduce hunger and poverty and to conserve biodiversity—must find ways to do so in the same landscapes. Crop and planted pasture production—mostly in low-productivity systems—dominate at least half the world's temperate, subtropical, and tropical forest areas; a far larger area is used for grazing livestock.[5] Food insecurity threatens biodiversity when it leads to overexploitation of wild plants and animals. Low farm productivity leads to depletion of soil and water resources and increases the pressure to clear additional land that serves as wildlife habitat. Some 40 percent of cropland in developing countries is degraded. Of more than 17,000 major protected areas, 45 percent (accounting for one-fifth of total protected areas) are heavily used for agriculture, while many of the rest are islands in a sea of farms, pastures, and production forests that are managed in ways incompatible for long-term species and ecosystem survival.[6]

Despite this high level of dependence by the poor on biodiversity, the dominant model of conservation seeks to exclude people from natural habitats. In India, for example, 30 million people are targeted for resettlement from protected areas.[7] From the perspective of poverty reduction and rural development, it is thus urgent to identify alternative conservation systems that respect the rights of forest dwellers and owners and address conservation objectives in the 90 percent of forests outside public protected areas. Markets for ecosystem services potentially offer a more efficient and lower-cost approach to forest conservation.[8]

Need for Financial Incentives to Provide Ecosystem Services

There is growing recognition that regulatory and protected area approaches, while critical, are insufficient to adequately conserve biodiversity. A fundamental problem is financial, especially for resources that lie outside protected areas. For these to be conserved, they need to be more valuable than the alternative uses of the land. And for such resources to be well managed, good stewardship needs to be more profitable than bad stewardship. The failure of forest owners and producers to capture financial benefits from conserving ecosystem benefits leads to overexploitation of forest resources and undersupply of ecosystem services.

This reality is hard for many people to accept, because most ecosystem services are considered "public goods." The "polluter pays" principle has argued that the right of the public to these services trumps the private rights of the landowner or manager. Yet good management has a cost. While the individual who manages his or her resources to protect biodiversity produces public benefits, the costs incurred are private. Under current institutions, those who benefit from these services have no incentive to compensate suppliers for these services. In most of the world, forest ecosystem services are not traded and have

no "price." Thus, where the opportunity costs of forest land for agricultural enterprises, infrastructure, and human settlements are higher than the use or income value of timber and nontimber forest products (NTFPs), habitats will be cleared and wild species will be allowed to disappear. Because they receive little or no direct benefit from them, resource owners and producers ignore the real economic and noneconomic values of ecosystem services in making decisions about land use and management.

> A lower-cost approach to securing conservation is to pay only for the biodiversity services themselves, by paying landowners to manage their assets so as to achieve biodiversity or species conservation.

Mechanisms are needed by which resource owners are rewarded for their role as stewards in providing biodiversity and ecosystem services. Anticipation of such income flows would enhance the value of natural assets and thus encourage their conservation. Compared to previous approaches to forest conservation, market-based mechanisms promise increased efficiency and effectiveness, at least in some situations. Experience with market-based instruments in other sectors has shown that such mechanisms, if carefully designed and implemented, can achieve environmental goals at significantly less cost than conventional "command-and-control" approaches, while creating positive incentives for continual innovation and improvement. Markets for ecosystem services could potentially contribute to rural development and poverty reduction by providing financial benefits from the sale of ecosystem services, improving human capital through associated training and education, and strengthening social capital through investment in local cooperative institutions.

New Market Solutions to Conserve Biodiversity

The market for biodiversity protection can be characterized as a nascent market. Many approaches are emerging to finan-

cially remunerate the owners and managers of land and resources for their good stewardship of biodiversity (see Table 2). Market mechanisms to pay for other ecosystem services—watershed services, carbon sequestration or storage, landscape beauty, and salinity control, for example—can be designed to conserve biodiversity as well. However, in general, biodiversity services are the most demanding to protect because of the need to conserve many different elements essential for diverse, interdependent species to thrive. Figure 1 illustrates potential market solutions and some of the complexities involved.

Land Markets for High-Biodiversity-Value Habitat

National governments (in the form of public parks and protected areas), NGO conservation organizations (for example, The Nature Conservancy), and individual conservationists have long paid for the purchase of high-biodiversity-value forest habitats. Direct acquisition can be expensive, as underlying land and use values are also included. Local sovereignty concerns arise when buyers are from outside the country—or even the local area—or where extending the area of noncommercial real estate reduces the local tax base. New commercial approaches are being developed to encourage the establishment of privately owned conservation areas, such as conservation communities (the purchase of a plot of land by a group of people mainly for recreation or conservation purposes), ecotourism-based land protection projects, and ecologically sound real estate projects being organized in Chile.[9] These build on growing consumer demand for housing and vacation in biodiverse environments.

Payments for Use or Management

A lower-cost approach to securing conservation is to pay only for the biodiversity services themselves, by paying landowners to manage their assets so as to achieve biodiversity or species conservation. It is likely that the largest-scale payments for land-use or management agreements belong to one of two categories. One encompasses government agroenvironmental payments made to farmers in North America and Europe for reforesting conservation

Table 2: Types of payments for biodiversity protection

Purchase of high-value habitat

Type	Mechanism
Private land acquisition	Purchase by private buyers or nongovernmental organizations explicitly for biodiversity conservation
Public land acquisition	Purchase by government agency explicitly for biodiversity conservation

Payment for access to species or habitat

Type	Mechanism
Bioprospecting rights	Rights to collect, test, and use genetic material from a designated area
Research permits	Right to collect specimens, take measurements in area
Hunting, fishing, or gathering permits for wild species	Right to hunt, fish, and gather
Ecotourism use	Rights to enter area, observe wildlife, camp, or hike

Payment for biodiversity-conserving management

Type	Mechanism
Conservation easements	Owner paid to use and manage defined piece of land only for conservation purposes; restrictions are usually in perpetuity and transferable upon sale of the land
Conservation land lease	Owner paid to use and manage defined piece of land for conservation purposes for defined period of time
Conservation concession	Public forest agency is paid to maintain a defined area under conservation uses only; comparable to a forest logging concession
Community concession in public protected areas	Individuals or communities are allocated use rights to a defined area of forest or grassland in return for commitment to protect the area from practices that harm biodiversity
Management contracts for habitat or species conservation on private farms, forests, or grazing lands	Contract that details biodiversity management activities and payments linked to the achievement of specified objectives

Tradable rights under cap-and-trade regulations

Type	Mechanism
Tradable wetland mitigation credits	Credits from wetland conservation or restoration that can be used to offset obligations of developers to maintain a minimum area of natural wetlands in a defined region
Tradable development rights	Rights allocated to develop only a limited total area of natural habitat within a defined region
Tradable biodiversity credits	Credits representing areas of biodiversity protection or enhancement that can be purchased by developers to ensure they meet a minimum standard of biodiversity protection

Support biodiversity-conserving businesses

Type	Mechanism
Biodiversity-friendly businesses	Business shares in enterprises that manage for biodiversity conservation
Biodiversity-friendly products	Eco-labeling

SOURCE: S. J. Scherr, A. White, and A. Khare, *Current Status and Future Potential of Markets for Ecosystem Services in Tropical Forests: An Overview* (Washington, DC: Forest Trends, 2003).

easements. The other category describes management contracts aiming to conserve aquatic and terrestrial wildlife habitat. In Switzerland, "ecological compensation areas," which use farming systems compatible with biodiversity conservation, have expanded to include more than 8 percent of total agricultural land. In the tropics, diverse approaches include nationwide public payments in Costa Rica for forest conservation and in Mexico for forested watershed protection.

Conservation agencies are organizing direct payments systems, such as conservation concessions being negotiated by Conservation International, and forest conservation easements negotiated by the *Cordão de Mata* ("linked forest") project with dairy farmers in Brazil's Atlantic Forest. The dairy farmers in the latter example receive, in exchange, technical assistance and investment resources to raise crop and livestock productivity. Some countries that use land taxes are using tax policies in innovative

ways to encourage the expansion of private and public protected areas.

Payment for Private Access to Species or Habitat

Private sector demand for biodiversity has tended to take the form of payments for access to particular species or habitats that function as "private goods" but in practice serve to cover some or all of the costs of providing broader ecosystem services. Pharmaceutical compa-

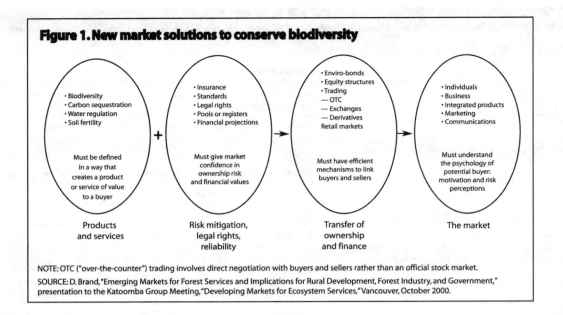

Figure 1. New market solutions to conserve biodiversity

- Biodiversity
- Carbon sequestration
- Water regulation
- Soil fertility

Must be defined in a way that creates a product or service of value to a buyer

Products and services

+

- Insurance
- Standards
- Legal rights
- Pools or registers
- Financial projections

Must give market confidence in ownership risk and financial values

Risk mitigation, legal rights, reliability

→

- Enviro-bonds
- Equity structures
- Trading
  — OTC
  — Exchanges
  — Derivatives
- Retail markets

Must have efficient mechanisms to link buyers and sellers

Transfer of ownership and finance

- Individuals
- Business
- Integrated products
- Marketing
- Communications

Must understand the psychology of potential buyer: motivation and risk perceptions

The market

NOTE: OTC ("over-the-counter") trading involves direct negotiation with buyers and sellers rather than an official stock market.

SOURCE: D. Brand, "Emerging Markets for Forest Services and Implications for Rural Development, Forest Industry, and Government," presentation to the Katoomba Group Meeting, "Developing Markets for Ecosystem Services," Vancouver, October 2000.

nies have contracted for bioprospecting rights in tropical forests. Ecotourism companies have paid forest owners for the right to bring tourists into their lands to observe wildlife, while private individuals are willing to pay forest owners for the right to hunt, fish, or gather nontimber forest products.

Tradable Rights and Credits within a Regulatory Framework

Multifactor markets for ecosystem services have been successfully established, notably for sulfur dioxide emissions, farm nutrient pollutants, and carbon emissions. These create rights or obligations within a broad regulatory framework and allow those with obligations to "buy" compliance from other landowners or users. Developing such markets for biodiversity is more complicated, because specific site conditions matter so much. The United States has operated a wetlands mitigation program since the early 1980s in which developers seeking to destroy a wetland must offset that by buying wetland banks conserved or developed elsewhere. A similar approach is used for "conservation banking," described in the box.

A variant of this approach is being designed for conserving forest biodiversity in Brazil by permitting flexible enforce-

ment of that country's "50 percent rule," which requires landholders in Amazon forest areas to maintain half of their land in forest. This rule is also applied in other regions in Brazil, where lesser proportional areas are set aside for forest use. Careful designation of comparable sites is required.

Another approach, biodiversity credits, is under development in Australia. In this system, legislation creates new property rights for private landholders who conserve biodiversity values on their land. These landholders can then sell resulting "credits" to a common pool. The law also creates obligations for land developers and others to purchase those credits. The approach requires that the "value" of the biodiversity unit can be translated into a dollar value.

Biodiversity-Conserving Businesses

Conservation values are beginning to inform consumer and investor decisions. Eco-labeling schemes are being developed that advertise or certify that products were produced in ways consistent with biodiversity conservation. The global trade in certified organic agriculture was worth $21 billion worldwide in 2000.[10] International organic standards are expanding to landscape-scale biodiversity impacts. The Rainforest Alliance and the Sustainable

Agriculture Network certify coffee, bananas, oranges, and other products grown in and around high-biodiversity-value areas. The Sustainable Agriculture Initiative is a coalition of multinational commercial food producers (Nestle, Dannon, Unilever, and others) who are seeking to ensure that all of the products they purchase along the supply chain come from producers who are protecting biodiversity. In 2002, more than 100 million hectares of forest were certified (a fourfold increase over 1996), although only 8 percent of the total certified area is in developing countries, and most of that is in temperate forests.

Current Market Demand

Available information suggests that biodiversity protection services are presently the largest market for ecosystem services. A team from McKinsey & Company, the World Resources Institute, and The Nature Conservancy estimated the annual international finance for the conservation market (conservation defined as protecting land from development) at $2 billion, with the forest component a large share of that.[11] Buyers are predominantly development banks and foundations in the United States and Europe.

A study by the International Institute for Environment and Development (IIED) of 72 cases of markets for forest biodiversity protection services in 33 countries found that the main buyers of biodiversity services (in declining order of prevalence) were private corporations, international NGOs and research institutes, donors, governments, and private individuals.[12] Communities, public agencies, and private individuals predominate as sellers. Most of these cases took place in Latin America and in Asia and the Pacific. Only four cases were found in Europe and Russia and one was found in the United States.

Three-quarters of the cases in the IIED study were international markets, and the rest were distributed among regional, national, and local buyers. International actors—as well as many on the national level—who demand biodiversity protection services tend to focus on the most biodiverse habitats (in terms of species

richness) or those perceived to be under the greatest threat globally (for example, places like the Amazon, where there are a high number of endemic species and where habitat area has greatly declined). Most of the private corporations were interested in eco-labeling schemes for crops or timber, investment in biodiversity-friendly companies, horticultural companies concerned with ecosystem services, or pharmaceutical bioprospecting. Such private payments are usually site-specific. Local actors more commonly focus on protecting species or habitats of particular economic, subsistence, or cultural value.

Projected Growth in Market Demand

The fastest-growing component of future market demand for biodiversity services is likely to be in eco-labeling of crop, livestock, timber, and fish products for export and for urban consumers. In 1999, the value of the organic foods market was US$14.2 billion. Its value is growing at 20–30 percent a year in the industrialized world, as the international organic movement is strengthening standards for biodiversity conservation.[13] Pressures continue to increase on major international trading and food processing companies to source from suppliers who are not degrading ecosystem services. Donor and international NGO conservation will continue to expand as NGOs begin to establish entire research departments aimed at developing new market-based instruments. Voluntary biodiversity offsets are also a promising source of future demand, as many large companies are seeking ways to maintain their "license to operate" in environmentally sensitive areas, and offsets are of increasing interest to them.

The costs of and political resistance to land acquisition are rising. Construction of biological corridors in and around production areas is an increasingly important conservation objective. At the same time, however, many of the most important sites for biodiversity conservation are in more densely populated areas with high opportunity costs for land. Thus we are likely to see a major shift from land acquisition to various types of

CONSERVATION BANKING IN THE UNITED STATES

Amendments to the United States Endangered Species Act in 1982 provided for an "incidental take" of enlisted species, if "a landowner provides a long-term commitment to species conservation through development of a Habitat Conservation Plan (HCP)." These amendments have opened the door to a series of market-based transactions, described as conservation banking, which permits land containing a natural resource (such as wetlands, forests, rivers, or watersheds) that is conserved and maintained for specified enlisted species to be used to offset impacts occurring elsewhere to the same natural resource.[1] A private landowner may request an "incidental take" permit and mitigate it by purchasing "species credits" from preestablished conservation banks. Credits are administered according to individuals, breeding pairs, acres, nesting sites, and family units. Conservation banking has maximized the value of underutilized commercial real estate and given private landowners incentive to conserve habitat.

California was the first state to authorize the use of conservation banking and has established 50 conservation banks since 1995. Other states, including Alabama, Colorado, and Indiana, have followed suit. In April 2002, the Indiana Department of Transportation, the Federal Highway Association Indiana Division, and four local government agencies finalized an HCP for the endangered Indiana bat as part of the improvement of transportation facilities around Indianapolis International Airport. These highway improvements will occur in an area of known Indiana bat habitat that is predicted to experience nearly $1.5 billion in economic development during the next ten years. Under the HCP, approximately 3,600 acres will be protected, including 373 acres of existing bat habitat.

Notes

1. A. Davis, "Conservation Banking," presentation to the Katoomba Group-Lucarno Workshop, Lucarno, Switzerland, November 2003.

direct payments for easements, land leases, and management contracts.

> The fastest-growing compo-
> nent of future market demand
> for biodiversity services is
> likely to be in eco-labeling of
> crop, livestock, timber, and fish
> products for export and for ur-
> ban consumers.

A rough back-of-the-envelope estimate suggests that the current value of international, national, and local direct payments and trading markets for ecosystem services from tropical forests alone could be worth several hundred million dollars per year, while the value of certified forest and tropical tree crop products may reach as much as a billion dollars. While this is a large and significant amount, it represents a small fraction of the value of conventional tropical timber and other forest product markets. For example, by comparison, the total value of tropical timber exports is $8 billion (including only logs, sawnwood, veneer, and plywood), which is a small fraction of the total exports and domestic timber, pulpwood, and fuelwood markets in tropical countries. NTFP markets are far larger still.[14] The total value of international trade for NTFPs is $7.5 billion–$9 billion per year, with another $108 billion in processed medicines and medicinal plants.[15] Domestic markets for NTFPs are many times larger (for example, domestic consumption accounted for 94 percent of the global output of fresh tropical fruits 1995–2000.[16] Nonetheless, these rough figures are quite interesting when compared with the scale of public and donor forest conservation finance summarized in Table 1.

Scaling Up Payments for Biodiversity: Next Steps

Markets for ecosystem services are steadily growing and can be expected to grow even more rapidly in the next decade. Yet they predominate as pilot projects. What will it take to transform these markets to impact ecosystem conservation on the global scale? The four most strategic and catalytic areas for policy and action are to

- structure emerging markets to support community-driven conservation;
- mobilize and organize buyers for ecosystem services;
- connect global and national action on climate change to biodiversity conservation; and
- invest in the policy frameworks and institutions required for functioning ecosystem service payment systems.

Supporting Community-Driven Conservation

The benefits of investments in ecosystem services will be maximized over the long term if markets reward local participation and utilize local knowledge. In community forests and agroforestry landscapes, communities have already established sophisticated conservation strategies. Studies of indigenous timber enterprises document conservation investments on the order of $2 per hectare per year apart from other management activities and investments of community time and labor; this is equal to the average available budget per hectare for protected areas worldwide. Conservation policies must recognize the role that local people are playing in the conservation of forest ecosystems worldwide and support them (either with cash or in-kind support) to continue to be good environmental stewards.

To enable conservation-oriented management to remain or become economically viable, it is important that ecosystem service payments and markets are designed so that they strategically channel financial payments to rural communities. Such payments can be used to develop and invest in new production systems that increase productivity and rural incomes, and enhance biodiversity at a landscape scale—an approach referred to as "ecoagriculture."[17] Ecosystem service payments to poor rural communities that are providing stewardship services of national or international value can help to meet multiple Millennium Development Goals. For any semblance of a sustainable future to be realized, it is crucial that our long-term vision includes biodiversity and natural ecosystems as part of the "natural

infrastructure" of a healthy economy and society.

Mobilizing and Organizing Buyers for Ecosystem Services

Turning beneficiaries into buyers is the driving force of ecosystem service markets. Because beneficiaries are often hesitant to pay for goods previously considered free, "willingness to pay" for ecosystem services must be organized on a greater scale. The private sector must be called upon to engage in responsible corporate behavior in conserving biodiversity. For example, Insight Investment, a major financial firm, has developed a biodiversity policy that uses conservation as a screen for investment. Voluntary payments by consumers, retail firms, and other actors can be encouraged through social advertising. This approach is growing rapidly now for eco-labeling programs (labeling of some personal care products and foods) and voluntary carbon emission offset programs involving investment in reforestation. Stockholder pressure is beginning to influence some firms to avoid investments and activities that harm biodiversity, and this is evolving to positive action. Civil society campaigns can also mobilize willingness to pay for biodiversity offsets and payments to local partners for conservation.

Connecting Climate Action with Biodiversity Conservation

Far more aggressive action must and will be taken to mitigate and adapt to climate change. Land use and land-use change currently contribute more than 20 percent of carbon emissions and other greenhouse gases. Action to reduce these emissions must be a central part of our response, and it is critical that action to sequester carbon through improved land uses accompanies strategies to reduce industrial emissions. There is thus an unprecedented opportunity at this time to structure our responses to climate change so that actions related to land use are also designed to protect and restore biodiversity. Moreover, such actions can be designed in ways that enhance and protect livelihoods, especially for those most vulnerable to the impacts of climate change. Indeed, it is imperative that they do so.

As a result of the deliberations at the Conference of the Parties of the United Nations Framework Convention on Climate Change last year, payments for forest carbon through the Clean Development Mechanism (CDM) of the Kyoto Protocol can be used to finance forest restoration and regeneration projects that conserve biodiversity while providing an alternative income source for local people.[18] But the scale of forest carbon under CDM is very small—too small to have a major impact on climate, biodiversity, or livelihoods. It is critical that we aim for a much larger program in the second commitment period, and it is crucial that nations affiliated with the Organisation for Economic Co-operation and Development (OECD) create initiatives to utilize carbon markets for biodiversity conservation in their own internal trading programs. It is imperative to develop a new principle of international agreements on climate response and carbon trading, one that builds a system that encourages overlap of the major international environmental agreements and the Millennium Development Goals. This could mobilize demand by creating an international framework for investing in good ecosystem service markets. It is also important that emerging private voluntary markets for carbon (that is, with actors who do not have a regulatory obligation) are encouraged to pursue such biodiversity goals as well. The Climate, Community and Biodiversity Alliance, for example, is seeking to develop guidelines and indicators for private investments in carbon projects that will achieve these multiple goals. The Forest Climate Alliance of The Katoomba Group is seeking to mobilize the international rural development community to advocate for such approaches.[19]

Investing in Policy Frameworks and Institutions for Biodiversity Markets

Ecosystem service markets are genuinely new—and biodiversity markets are the newest and most challenging. Every market requires basic rules and institutions in order to function, and this is equally true of biodiversity markets. The biodiversity

conservation community needs to act quickly and strategically to ensure that as these markets develop, they are effective, equitable, and operational and are used sensibly to complement other conservation approaches.

Policymakers and public agencies play a vital role in creating the legal and legislative frameworks necessary for market tools to operate effectively. This includes establishing regulatory rules, systems of rights over ecosystem services, and mechanisms to enforce contracts and settle ownership disputes. Ecosystem service markets pose profound equity implications, as new rules may fundamentally change the distribution of rights and responsibilities for essential ecosystem services. Forest producers and civil society will need to take a proactive role to ensure that rules support the public interest and create development opportunities.

New institutions will also be needed to provide the business services required in ecosystem service markets. For example, in order for beneficiaries of biodiversity services to become willing to pay for them, better methods of measuring and assessing biodiversity in working landscapes must be developed, as well as the institutional capacity to do so. New institutions must be created to encourage transactions and reduce transaction costs. Such institutions could include "bundling" biodiversity services provided by large numbers of local producers, as well as investment vehicles that have a diverse portfolio of projects to manage risks. Registers must be established and maintained, to record payments and trades. For example, The Katoomba Group is developing a Web-based "Marketplace" to slash the information and transaction costs for buyers, sellers, and intermediaries in ecosystem service markets.[20]

Conclusion

Conservation of biodiversity and of the services biodiversity provides to humans and to the ecological health of the planet requires financing on a scale many times larger than is feasible from public and philanthropic sources. It is essential to find

PROTECTING BRAZIL'S ATLANTIC FOREST: THE GUARAQUEÇABA CLIMATE ACTION PROJECT

Due to excessive deforestation, the Atlantic Forest of Brazil has been reduced to less than 10 percent of its original size. The Guaraqueçaba Climate Action Project has sought to regenerate and restore natural forest and pastureland.[1] Companies such as American Electric Power Corporation, General Motors, and Chevron-Texaco have invested US$18.4 million to buy carbon emission offset credits from the approximately 8.4 million metric tons of carbon dioxide that the project is expected to sequester during its lifespan. The project has initiated sustainable development activities both within and outside the project boundary, including ecotourism, organic agriculture, medicinal plant production, and a community craft network. The project has made significant contributions toward enhancing biodiversity in the area, creating economic opportunities for local people (such as jobs), restoring the local watershed, and substantially mitigating climate change.

Notes

1. The Nature Conservancy (TNC), *Climate Action: The Atlantic Forest in Brazil* (Arlington, VA: TNC, 1999).

new mechanisms by which resource owners and managers can realize the economic values created by good stewardship of biodiversity. Moreover, private consumers, producers, and investors can financially reward that stewardship. New markets and payment systems, strategically shaped to deliver critical public benefits, are showing tremendous potential to move biodiversity conservation objectives to greater scale and significance.

NOTES

1. M. Cernea and K. Schmidt-Soltau, 2003. "Biodiversity Conservation versus Population Resettlement, Risks to Nature

and Risks to People," paper presented at CIFOR (Center for International Forestry Research) Rural Livelihoods, Forests and Biodiversity Conference, Bonn, Germany, 19–23 May 2003.

2. See S. Wood, K. Sebastian, and S. J. Scherr, *Pilot Analysis of Global Ecosystems: Agroecosystems* (Washington, DC: International Food Policy Research Institute and the World Resources Institute, 2000), 64; and E. W. Sanderson et al., "The Human Footprint and the Last of the Wild," *Bioscience* 52, no. 10 (2002): 891–904.

3. K. Chomitz, *Forest Cover and Population Density in Latin America*, research notes to the World Bank (Washington, DC: World Bank, 2003).

4. R. P. Cincotta and R. Engelman, *Nature's Place: Human Population and the Future of Biological Diversity* (Washington, DC: Population Action International, 2000).

5. Wood, Sebastian, and Scherr, note 2 above.

6. J. McNeely and S. J. Scherr, *Ecoagriculture: Strategies to Feed the World and Conserve Wild Biodiversity* (Washington, DC: Island Press, 2003).

7. A. Khare et al., *Joint Forest Management: Policy Practice and Prospects* (London: International Institute for Environment and Development, 2000).

8. S. J. Scherr, A. White, and D. Kaimowitz, *A New Agenda for Forest Conservation and Poverty Reduction: Making Markets Work for Low-Income Communities* (Washington, DC: Forest Trends, CIFOR, and IUCN-The World Conservation Union, 2004).

9. E. Corcuera, C. Sepulveda, and G. Geisse, "Conserving Land Privately: Spontaneous Markets for Land Conservation in Chile," in S. Pagiola et al., eds., *Selling Forest Environmental Services: Market-Based Mechanisms for Conservation and Development* (London: Earthscan Publications, 2002).

10. J. W. Clay, *Community-Based Natural Resource Management within the New Global Economy: Challenges and Opportunities*, a report prepared by the Ford Foundation (Washington, DC: World Wildlife Fund, 2002).

11. M. Arnold and M. Jenkins, "The Business Development Facility: A Strategy to Move Sustainable Forest Management and Conservation to Scale," proposal to the International Finance Corporation (IFC) Environmental Opportunity Facility from Forest Trends, Washington, DC, 2003.

12. N. Landell-Mills, and I. Porras. 2002. *Markets for Forest Environmental Services: Silver Bullet or Fool's Gold? Markets for Forest Environmental Services and the Poor, Emerging Issues* (London: International Institute for Environment and Development, 2002).

13. International Federation of Organic Agriculture Movements (IFOAM), "Cultivating Communities," 14th IFOAM Organic World Congress, Victoria, BC, 21–28 August 2002.

14. S. J. Scherr, A. White, and A. Khare, *Current Status and Future Potential of Markets for Ecosystem Services of Tropical Forests: A Report for the International Tropical Timber Organization* (Washington, DC: Forest Trends, 2003).

15. M. Simula, *Trade and Environment Issues in Forest Protection*, Environment Division working paper (Washington, DC: Inter-American Development Bank, 1999).

16. Food and Agricultural Organization of the United Nations (FAO), *FAOSTAT* database for 2000, accessible via `http://www.fao.org`.

17. For more information on ecoagriculture, see the Ecoagriculture Partners' Web site at `http://www.ecoagriculturepartners.org`.

18. S. J. Scherr and M. Inbar, *Clean Development Mechanism Forestry for Poverty Reduction and Biodiversity Conservation: Making the CDM Work for Rural Communities* (Washington, DC: Forest Trends, 2003).

19. For more information on this project, see `http://www.katoombagroup.org/Katoomba/forestcarbon`.

20. The Katoomba Group is a unique network of experts in forestry and finance companies, environmental policy and research organizations, governmental agencies and influential private, community, and nonprofit groups. It is dedicated to advancing markets for some of the ecosystem services provided by forests, such as watershed protection, biodiversity habitat, and carbon storage. For more information on the Katoomba Group, see `http://www.katoombagroup.org`. Forest Trends serves as the secretariat for the group. More information on Forest Trends can be found at `http://www.forest-trends.org`.

Michael Jenkins is the founding president of Forest Trends, a nonprofit organization based in Washington, D.C., and created in 1999. Its mission is to maintain and restore forest ecosystems by promoting incentives that diversify trade in the forest sector, moving beyond exclusive focus on lumber and fiber to a broader range of products and services. Previously he worked as a senior forestry advisor to the World Bank (1998), as associate director for the Global Security and Sustainability Program of the MacArthur Foundation (1988–1998), as an agroforester in Haiti with the U.S. Agency for International Development (1983–1986), and as technical advisor with Appropriate Technology International (1981–1982). He has also worked in forestry projects in Brazil and the Dominican Republic and was a Peace Corps volunteer in Paraguay. He speaks Spanish, French, Portuguese, Creole, and Guaraní, and can be contacted by telephone at (202) 298-3000 or via e-mail at mjenkins@forest-trends.org. Sara J. Scherr is an agricultural and natural resource economist who specializes in the economics and policy of land and forest management in tropical developing countries. She is presently director of the Ecosystem Services program at Forest Trends, and also director of Ecoagriculture Partners, the secretariat of which is based at Forest Trends. She previously worked as principal researcher at the International Center for Research in Agroforestry, in Nairobi, Kenya, as (senior) research fellow at the International Food Policy Research Institute in Washington, D.C., and as adjunct professor at the Agricultural and Resource Economics Department of the University of Maryland, College Park. Her current work focuses on policies to reduce poverty and restore ecosystems through markets for sustainably grown products and environmental services and on policies to promote ecoagriculture—the joint production of food and environmental services in agricultural landscapes. She also serves as a member of the Board of the World Agroforestry Centre, and as a member of the United Nations Millennium Project Task Force on Hunger. Scherr can be reached by telephone at (202) 298-3000 or via e-mail at sscherr@forest-trends.org. Mira Inbar is program associate with Forest Trends. She works in the Ecosystem Services program, supporting efforts to establish frameworks and instruments for emerging transactions in environmental services worldwide. Before joining Forest Trends, she worked with communities in the Urubamba River Valley of Peru to initiate a forest conservation plan. She has worked with the National Fishery Department in Western Samoa, the Marie Selby Botanical Gardens, and Environmental Defense. Inbar may be reached by telephone at (202) 298-3000 or via e-mail at minbar@forest-trends.org. This article is © The Aspen Institute and is published with permission.

From *Environment,* July/August 2004, pp. 32-42. Reprinted by permission of the Aspen Institute. Copyright © 2004.

UNIT 5

Resources: Land and Water

Unit Selections

Key Points to Consider

- How has ecosystem management in the drylands region south of the Sahara produced some hopeful mitigation of the trend toward desertification in that area? What lessons can the management techniques developed for this area be expanded into other parts of the world?

- What are the conflicting uses of water that prompt such environmental problems as those recently experienced in the Klamath River basin in Oregon? Are there ways to resolve conflicts over the use of water that have not yet been implemented?

- Is there a relationship between water pollution and the cost of water? How can conservation methods alter the price of water for commercial, domestic, industrial, and agricultural needs?

- How serious are claims that increasing demand for water places limits on agricultural production? What might be some of the contributing factors to a diminishing global water supply in either quantitative or qualitative terms?

Student Website

www.mhcls.com/online

Internet References

Further information regarding these websites may be found in this book's preface or online.

Global Climate Change
http://www.puc.state.oh.us/consumer/gcc/index.html

National Oceanic and Atmospheric Administration (NOAA)
http://www.noaa.gov

National Operational Hydrologic Remote Sensing Center (NOHRSC)
http://www.nohrsc.nws.gov

Terrestrial Sciences
http://www.cgd.ucar.edu/tss/

The worldwide situations regarding reduction of biodiversity, scarcity of energy resources, and pollution of the environment have received the greatest amount of attention among members of the environmentalist community. But there are a number of other resource issues that demonstrate the interrelated nature of all human activities and the environments in which they occur. One such issue is the declining quality of agricultural land. In the developing world, excessive rural populations have forced the overuse of lands and sparked a shift into marginal areas, and the total availability of new farmland is decreasing at an alarming rate of 2 percent per year. In the developed world, intensive mechanized agriculture has resulted in such a loss of topsoil that some areas are experiencing a decline in food production. Other natural resources, such as minerals and timber, are declining in quantity and quality as well; in some areas they are no longer usable at present levels of technology. The overuse of groundwater reserves has resulted in potential shortages beside which the energy crisis pales in significance. And the very productivity of Earth's environmental systems—their ability to support human and other life—is being threatened by processes that derive at least in part from energy overuse and inefficiency and from pollution. Many environmentalists believe that both the public and private sectors, including individuals, are continuing to act in a totally irresponsible manner with regard to the natural resources upon which we all depend.

Uppermost in the minds of many who think of the environment in terms of an integrated whole, as evidenced by many of the selections in this unit, is the concept of the threshold or critical limit of human interference with natural systems of land and water. This concept suggests that the environmental systems we occupy have been pushed to the brink of tolerance in terms of stability and that destabilization of environmental systems has consequences that can only be hinted at, rather than predicted. Although the broader issue of system change and instability, along with the lesser issues such as the quantity of agricultural land, the quality of iron ore deposits, the sustained yield of forests, or the availability of fresh water seems to be quite diverse, all are closely tied to a pair of concepts—that of resource marginality and of the globalization of the economy that has made marginality a global rather than a regional problem.

Not all the utilizations of marginal resources turns out quite so badly. In the first article in this unit, independent researcher Michael Mortimore describes recent successes in developing the highly marginal environments of dryland West Africa. In "Dryland Development: Success Stories from West Africa," Mortimore describes the fragility (or marginality) of the African drylands that encompass 40% of the continent's surface. Away from river valleys, dryland soils are easily eroded by wind and water, lose their fertility quickly under cultivation, and are subject to desertification—the conversion of soils and accompanying vegetation from merely a dryland condition to one of absolute desert. While the general trend of human use of drylands for farming has produced degradation, experimental farming techniques begun in Kenya in the 1930s have recently been applied to West African dryland farming regions in Nigeria, Niger, and

Senegal. While farmers in these areas continue to live in poverty, their lot has been improved by the new farming techniques adapted specifically to drylands and involving conservation of water and soil. The results have been striking: sustainable ecosystem management, increasing farm investment, stable outputs per areal unit, and increasing farm incomes. The use of water also figures importantly—but in entirely different ways—in the third selection of this unit. Journalist Bruce Barcott, writing in *Mother Jones*, discusses the conflicts between various water users, from farmers to fishermen, that have erupted in the drought-stricken American West. In "What's a River For?" Barcott focuses his attention on the Klamath River basin of northern California and southern Oregon where a struggle for an increasingly scarce resource has erupted between farmers who need water for irrigation, factories, and cities that need water for manufacturing and domestic purposes, ranchers whose livestock depend upon water for their very existence, and fishermen—both recreational and subsistence—who depend upon the river's salmon habitat for nourishment for both body and soul. The Klamath simply could not and cannot support all of these conflicting uses and, in 2002, a decision was made to release irrigation water from the river, resulting in one of the largest fish die-backs in American history. Use of water for irrigation impacts not only on fish populations, however. We now know that the loss of topsoil to accelerated erosion, the loss of living and farming space to reservoirs, and increasing the salinity in irrigation waters and soil are problems greater than the dryness that prompts irrigation to begin with.

In the next articles in the unit, the struggle between different stakeholders in the land and water use wars also takes center stage. Journalist Jim Morrison in "How Much Is Clean Water Worth?" asks the essential question in water management: can we put a dollar value not just on the price of water per quart, gallon, or acre-foot, but also on the ecosystems that water sustains? Morrison answers his own question in an affirmative but non-quantitative way. The watershed that supplies the city of New York with its basic water supply is easy to calculate: $1.3 billion per year. But calculating the dollar value of the Catskills in terms of recreation, scenic value, wildlife habitat is more difficult, although even a cursory look at the region would suggest that habitat and wildlife are powerful economic engines for both the region and for a wider area. If the ecosystem service values could be more accurately calculated, then wiser choices could be made about how to use the Catskill's water supply—is it best to reserve the water for use of residents of New York City or to leave it as a part of an integrated ecosystem that has its own inherent value? A similar set of questions on a wider scale is asked in "A Human Thirst," by UN consultant Don Hinrichsen. Hinrichsen reports that human uses now consume more than half the world's freshwater, often leaving other animal species to go thirsty. As water is withdrawn from rivers and other sources to feed the demands of agriculture, industry, and over six billion humans, the aquatic ecosystems, the plants and animals they support, and the very surface of the land itself suffers. Competition for water on a global scale goes beyond that of regional competition for the watershed of the Catskill Mountains of New York State. On the wider stage of the entire world, freshwater resources are coming increasingly into a competitive system involving humans and other species.

There are two possible solutions to all these problems posed by the use of increasingly marginal and scarce resources and by the continuing pollution of the global atmosphere. One is to halt the basic cause of the problems—increasing population and consumption. The other is to provide incentives and techniques for the conservation and management of existing resources and for the discovery of alternative resources to eliminate the demand for more marginal resources and the use of heavily polluting ones.

Dryland Development

Success Stories from West Africa

Michael Mortimore

The African continent spans a huge range of climates, from equatorial to tropical, desert, and subtropical on both sides of the equator. Much of Africa—around 40 percent—is dryland, receiving less than 1,000 millimeters (mm) of mean annual rainfall in short rainy seasons, the remaining months being relatively or absolutely rainless.[1] High temperatures during the tropical rainy season cause much of the rainfall to be lost in evaporation, and the high intensity of storms may cause much of it to run off in floods.

For securing human livelihoods, the principal constraints of dryland climates are aridity and the variability of the rainfall. In areas receiving less than 250 mm of rainfall during the growing season, farming is more or less impossible without irrigation, and most livelihoods are based on livestock. Yet in some parts of semiarid Africa, with rainfall of only 400–800 mm, very large and dense rural populations (up to 400 people per square kilometer (km²)) are found. The variability of the rainfall compounds the effects of aridity and tends to increase the drier the location. Frequent and unpredictable droughts (and occasional floods) introduce high levels of risk into farming and livestock production. Rainfall variability also occurs over the long term. In the Sahel, Africa's biggest dryland, there was a decline of up to 33 percent in the average rainfall between the periods 1931–1960 and 1961–1990.[2]

Dryland soils—away from river valleys—have low natural fertility, measured in terms of the key plant nutrients.[3] Many soils are derived from desert sands and are incompletely formed. Scarce organic matter reflects the poor natural vegetation that grows with such a low rainfall.[4] Thus, nutrients limit plant growth where rainfall is higher, and aridity limits growth where rainfall is lower.[5] The economic output of cropping systems therefore depends on inputs and management.[6]

The African drylands are home to 268 million people—40 percent of the continent's population. That many of these people are very poor is not in doubt. However, because drylands form subregions within nations, statistics on dryland incomes are rare. Access to health and education services is poor in rural areas; infant, child, and maternal mortality are high; and average life expectation at birth is low. Poor nutrition reduces available energy, and the high temperatures in early summer make farm work arduous. Given the large numbers of people, failure to reduce poverty in the drylands will prejudice the achievement of the Millenium Development Goals (MDGs).[7]

For the past five decades, the threat of desertification has dominated dryland development policy and debate.[8] The term "desertification" describes a set of land degradation processes, which include soil fertility decline, erosion by wind or water, dune formation, hydrological decline, biodiversity loss, deforestation, and declining bioproductivity. Some changes are irreversible and extend beyond the normal climatic oscillations of the desert edge and the effects of random droughts to forms of land degradation that are commonly attributed to the actions of humans.[9]

The mainstream view of dryland management continues to be that widespread and "inappropriate" land use practices require transformation. This view is reflected in the Convention to Combat Desertification, which was formulated after the Rio Earth Summit in 1992. In neo-Malthusian interpretations, unsustainable practices are blamed on population growth, which has driven human and animal populations beyond the rather low carrying capacity of the drylands.[10]

Challenging Malthus: The Machakos Story

In an article published in *Environment* in 1994,[11] based on a study of long-term change in Machakos District, Kenya, it was shown that degradation is not inevitable in African drylands. As long ago as the 1930s, the Machakos Reserve acquired some infamy among conservationists, who thought they saw "every phase of misuse of the land," leading to soil erosion and deforestation on a large scale, with its inhabitants consequently "rapidly drifting to a state of hopeless and miserable poverty and their land to a parching desert of rocks, stones and sand."[12]

By the 1990s, the district's population had multiplied sixfold, while expanding into previously uninhabited areas (most of them dry and risky). The study found, however, that erosion had been largely brought under control on private farmlands. This was achieved through innumerable small investments in terracing and drainage, advised by the extension services but carried out by voluntary work groups, hired laborers, or the farmers themselves. On some grazing land, significant improvements in management were also taking place. The

Table 1. Defining "success" in drylands, based on the Machakos experience

Domain	Outcome	Indicators
Ecosystem management	Stabilization or reversal of degradation	Soil erosion controlled Soil water holding-capacity improved Nutrient losses minimized or compensated Trees managed sustainably[a] Useful biodiversity maintained
Land investments	Viability and sustainability in economic and/or social terms[b]	Private farm investments Cross-sectoral financial flows Acceptable economic rate of return on public investments
Productivity	Maintenance or increase	Stable or increasing crop yields or livestock production per hectare (ha) Increasing value of output per ha Increasing market participation
Incomes and welfare	Maintenance or increase in real terms	Increasing value of output per capita Strengthened access to off-farm incomes Rising achievement in education Asset accumulation on- and off-farm

[a] The major factor driving forest clearance in dryland Africa is agricultural expansion. In Europe and South and East Asia, it is accepted that the historical benefits of agricultural expansion exceed the value of lost woodland. In African drylands, where deforestation is often declaimed, woodland clearance is often later followed by conservation of trees on farms. M. Mortimore and B. Turner, "Does the Sahelian Smallholder's Management of Woodland, Farm Trees and Rangeland Support the Hypothesis of Human-Induced Desertification?" *Journal of Arid Environments,* forthcoming, 2005. It is not practicable to propose a reversal of agricultural expansion.

[b] Investments are made for social as well as financial benefits (such as houses for retirement, domestic water catchments, and gardens).

value of agricultural production per km² increased between 1930 and 1987 by a factor of six and doubled on a per-capita basis.[13] At the same time, a rapid change in agricultural technology occurred, with a switch from an emphasis on livestock production to increasingly intensive farming, close integration of crops with livestock production, and increased marketing of higher value commodities (such as fruit, vegetables, and coffee). A social transformation also occurred with the enthusiastic pursuit of education, giving increased access to employment opportunities outside the district and intensifying rural-urban linkages.[14]

The Machakos story upset the Malthusian scenario for drylands, suggesting in its place the hypothesis that positive linkages between population growth and environmental management may occur under the right conditions.[15] As a widely cited "success story," is it a model for other African drylands?

What is a "Success Story"?

Dryland communities live and work on an intersection between managed ecosystems and human systems, which are coevolving as time goes on.[16] Seen this way, such interactions may take negative forms, leading to irreversible damage to the ecosystem, and/or impoverishment in the human system. Such a model means that change in the natural ecosystems should be in harmony with the course of human development. If not, negative impacts such as those characterized as desertification must occur. To guide policy, therefore, it is necessary to define what might be called a "success story." Farmers' and pastoralists' knowledge and achievements are central to success: Unlike outsiders, local peoples possess accumulated experience of specific dryland environments.[17]

Based on the Machakos experience, it is proposed that "success" may be evident in one or more of the following four interconnected domains: ecosystem management (soil and biological resources), land investments, productivity, and personal incomes or welfare (see Table 1). For the conservationist, the first of these may be considered preeminent. However, without a healthy level of investment, ecosystem stability, functions, and services (economic products, biodiversity, hydrological cycles, microclimates, and so on) cannot be sustained under increasing exploitation. Likewise, if productivity fails, new investments cease, and human incomes or welfare stagnate. Conservation cannot be achieved by edict, because it is socially constructed in a given situation.

Integrated approaches to understanding and managing ecosystems are seen to be essential for the future relations between societies and nature.[18]

In the Machakos story, long-term evidence suggests success in all four domains but not in all of the indicators (Table 1). A transition was achieved from acute risk of further degradation to a pathway characterized instead by conserving the ecosystem, with a strong narrative of soil and water conservation and terracing—on arable land in particular—that now extends to every corner of the old district.[19] These works present impressive evidence of investments at the farm level, and there is a range of other investments such as water catchments and storage and livestock structures. In the individual household, financial flows occurred between sectors, whereby farm profits financed education and off-farm employment, while off-farm incomes were themselves used to finance capital developments on-farm.[20] Farm output indicators show increased value of output per km², including crop and livestock production. This fed into income effects, with an improving trend in farm incomes per capita until 1987. Rising educational achievement supported off-farm employment and the diversification of livelihoods.

However, all success is relative and incomplete. Thus it is unlikely that any system, including that of Machakos, can score highly in all four domains at one time or perform equally well at all times. Incomes in particular are at the mercy of macroeconomic changes. For example, in Machakos, increasing value of output per capita may have hesitated after 1987, with adverse trends in global markets.[21] The indicators are not all relevant everywhere. This descriptive model, however, suggests key variables in very complex systemic interactions and measurable indicators that are amenable to policies.

Are There Success Stories in West Africa?

Success stories have been documented from a variety of locations in African drylands.[22] In West Africa, three of these stories are told in studies that were designed specifically to test the Machakos model under different conditions. Findings suggest that success is by no means unqualified.[23] However, evidence has been found of significant achievements in the first three domains—ecosystem management, land investments, and productivity—identified in Table 1. Such evidence, which emerges from analyses of long-term data and from well-supported inferences, runs counter to some current perceptions.

The Kano Close-Settled Zone, Nigeria

The significance of the Kano story lies in the scale and longevity of the area's intensive farming system. As originally delimited,[24] the zone now has a population of more than 5 million, excluding the urban population of the city of Kano, which exceeds 1.5 million.[25] More than 85 percent of the land surface is occupied by farmland. Intensive farming of small holdings under annual cultivation, with less than 0.5 hectares (ha) per person, a dense scattering of trees, and livestock, is centuries old in the zone and is strongly oriented toward the conservation and care of land resources.[26]

It is possible to analyze ecosystem management as follows:

- Soil organic matter is protected by applying manure from domestic animals, dry compost, and waste at rates of 4–6 tons per ha per year. In the dry season, animals roam the fields during the day but are penned at night, where they stay during the growing season, feeding on cut-and-carried fodder. The manure and bedding are mixed and returned to the fields by cart, donkey's back, or even head loading.

- Plant nutrients are recycled through feeding crop residues, tree browse, and weeds to the animals; growing nitrogen-fixing crops such as cowpea or groundnuts that fix nitrogen with the grain (millet or sorghum); protecting leguminous trees (*Faidherbia albida*); and taking advantage of dust deposition in the dry season and low rates of leaching in the wet. When farm budgets permit, inorganic fertilizers are applied, but in small quantities—only about a third of recommended doses even when they were subsidized in the 1980s. Quantities of nitrogen, phosphorus, potassium, calcium, and magnesium in the soil vary from field to field and from year to year and tend to be low; nevertheless, no evidence was found of a general decline over a period of 13 years.[27]

- Soil moisture control is important where rainfall is scarce but intensive. It is achieved through field ridging (formerly by hand-hoeing but increasingly with ox-ploughing). This conserves runoff in furrows and maximizes infiltration between rainfall events. Competition between crops and weeds for scarce moisture is reduced by three or even four weeding operations during the short growing season (which is only 12–15 weeks long).

- Biodiversity is conserved in a functional sense through protecting or planting a variety of tree and shrub species on farms or along field boundaries and through on-farm selection and breeding of seed for at least 76 cultivars.[28] Biodiversity occurring naturally in the ecosystem is protected for its contribution to food security.[29]

- Erosion by wind or water is controlled on the gentle slopes of the farmlands (usually at grades less than five degrees) by planted field boundaries (including the henna bush, *Lawsonia inermis*, and the grasses *Andropogon gayanus* and *Vetiveria nigritana*), by planting spreading intercrops (cowpea, groundnut) among the grain, and by maintaining densities of mature farm trees at 7–15 per ha.

The result is an intricate, anthropogenic landscape of permanent fields, farmed parkland, and hamlets. A misleadingly sterile appearance in the dry season is confounded by a startling luxuriance of biomass during the wet.

The investments that made this possible in the past were mainly in human labor (manuring, weeding, and animal tending), but during the twentieth century, groundnut exports and increasing opportunities for employment or trading in the cities provided cash. After groundnut production failed in 1975 (on account of rosette disease and drought), farmers switched to selling some of their grain. Monetary transactions are extremely common, and during the 1990s farm investments that were visibly evident, especially near Kano, included ox-ploughs and teams, livestock kept for fattening, and "modern" building materials such as metal roofing sheets.

For long-term yield trends it is necessary to depend on inferences, as compatible and accurate data are not available. At least one-third of households in inner, high-density villages were estimated to be food-insufficient in the 1960s, although whether this was from poverty or from specializing in groundnuts for export is unclear. Poverty was noticeable on highly subdivided holdings in a very densely populated village close to the walls of Kano City in 1970.[30] Yet in the 1990s, except after a drought,[31] farm holdings with only one-third of a hectare per person could still be self-sufficient in basic grains.[32] The population of Kano Emirate, which roughly coincides with the zone, increased at a rate of 1.9 percent per year from 1931 to 1991.[33] Certainly there are food-insecure households today. Nevertheless, the information available does not suggest a general decline in yields but rather a sustained effort to increase the output of food in line with increasing consumption needs.

The livestock population, the integration of which into the farming system is essential for maintaining organic inputs to cultivated soils, increases in density in line with the human population. People like to invest their savings in animals, and fattening the animals for market has become a thriving industry. Sheep and goats, being easier to manage on the crowded farmlands, are increasing at the expense of cattle, the owners of which have to rely on transhumance (grazing away from home during the cropping season).

With their sales of animals, higher-value crops (cowpea; sesame; and new, disease-resistant groundnuts), and labor and skills in the rapidly growing urban markets of Nigeria, farming families are participating in more monetary transactions than ever before. Kano City's markets are supplying a population six times larger than in the 1960s, and although the zone contributes only a small part of its needs, what it does sell is very important to local livelihoods.[34] During the same period, increased retail activity and increasing domestic capital, such as bicycles, motorcycles, radios, clothing, equipment, and furnishings, suggested a slow, incremental growth for better-off families until the 1990s.[35] However, there is much anecdotal evidence of rural poverty.

Long-term change in the zone suggests significant successes in ecosystem management, many small-scale investments, and no evidence of a decline in yields per hectare. But trends in rural incomes depend on impressionistic evidence that is ambivalent. Recently, positive trends have been held ransom by economic recession, inflation, and adjustment policies.[36]

Diourbel Region, Senegal

According to land-cover maps of the administrative departments of Bambey and Diourbel, the Diourbel Region, like the Kano Close-Settled Zone, has more than 85 percent of its surface under cultivation. But rural population densities are lower (46–150 per km²); therefore, there is more land per capita (0.6–0.7 ha). More than a century of production for export, promoted by strong policy incentives, reduced the priority given to grain production.

Farm expansion, the demarcation of boundaries with perennial grasses or shrubs, the protection of economic farm trees, and the scattering of small settlements have produced a landscape resembling that seen around Kano. Consequently, in the face of repeated cultivation and the export of plant nutrients, maintaining soil fertility is a focal theme in ecosystem management. But the soil fertility equation in Diourbel differs from that in Kano. Less rainfall (500 mm compared with 650 mm), labor, and animal manure mean that fertility, measured in terms of chemical soil properties, can only be sustained on about a fifth of cultivated land, the *champs de case* (the fields nearest to the house).[37] On these, familiar strategies are employed: recycling organic matter and available plant nutrients, erosion control, and biodiversity protection. The remaining fields, the *champs de*

brousse, used to be fallowed regularly, but economic pressures to bring them under annual cultivation led to the substitution of inorganic fertilizers for fallowing. This strategy was promoted by fertilizer subsidies under the national agricultural program from 1960 to 1980. When the subsidy was removed, fertilizer use fell, and soil fertility faced a crisis of economic sustainability.

The Senegalese economy depended on groundnut exports from the inception of French rule in the 1880s until the 1980s.[38] In the late 1960s, a million tons per year were produced, making Senegal (with Nigeria) a leading world exporter. Under the French colonial policy and the independent government's *programme agricole*, the state took over responsibility not only for fertilizer promotion and supply but also for credit, groundnut seed supply, technical advice, marketing, processing, and export. From the sales of groundnuts, it deducted credit repayments, thus reducing the producers to a high level of dependency. It even subsidized rice imports so that Senegal could exploit its comparative advantage in growing groundnuts. Food commodity markets developed only slowly and late, as the Senegalese developed a strong preference for eating rice (mostly imported). Because the state was supplying much of the need for farm investments, private funds tended to be channelled into urban investments instead of agriculture.

The system proved to be financially unsustainable, as costs increased and global prices fell during the 1970s. At the same time, the Sahel Drought of 1968–1974 was followed by a persistent decline in annual rainfall until the 1990s. Structural adjustment policies, introduced in 1984, included an ending to credit and subsidies, the withdrawal of some agricultural services, late payments and stagnant, low producer prices. There was a dramatic fall in groundnut production, a diversion of output to local consumption, and agricultural stagnation. The amount of fallow land increased, and some farm tree populations may have declined.[39] In some areas, there was a slowing in the growth of the population and even absolute decline.[40]

However, not all the evidence on productivity is negative.[41] Long-term improvements in millet yields per ha and per mm of rainfall (though not per capita) are visible in official data.[42] Expanding production of cowpeas, which command a good price, and of crops for urban niche

markets appear in the statistics or have been widely observed. Livestock populations have increased in response to buoyant meat prices, and fattening animals is a popular sideline for farming families even though they often have to purchase supplementary feed. Rural markets, for many years frustrated by the state's dominant role in commodity movements, are being reinvigorated.

The search for alternative incomes from outside farming is proportionately most important for households that, owing to small holdings or underinvestment, only produce a fraction of their annual cereal needs. On the other hand, households that are food-secure throughout the year can generate five times as much agricultural income from crops and livestock.[43] Income diversification thus plays a critical role in the adjustment to changing conditions.

The evidence of success in Diourbel, therefore, is very different from that in Machakos and Kano. With regard to managing the ecosystem, there has been partial success in stabilizing or reversing degradation, but the collapse of state investment has left farmers unable to compensate, and only a proportion of the soil benefits from organic fertilization. Crop production and farm tree populations in some areas are said to be languishing.[44] But private investment is bouncing back in response to market signals, and the livestock sector is growing. Trends in rural incomes cannot be established from the data available, but much energy is devoted to widening the household's portfolio—whether from sheer necessity or in response to opportunity. The case for success rests essentially in the resilience that has characterized the responses of farming families, both to the failure of the government's state-led agricultural policy and the concurrent decline in rainfall.

Maradi Department, Niger

The evidence of success so far assembled is found in densely populated regions where, according to Danish economist Ester Boserup's theory of agricultural change, land scarcity drives farmers to use labor-intensive practices and adopt new technologies, thereby increasing their output per hectare.[45] It is, however, in low-density regions that rapid population growth and agricultural expansion often draw attention to largescale erosion and deforestation. One such region is Maradi Department in Niger, where until the nineteenth century, large

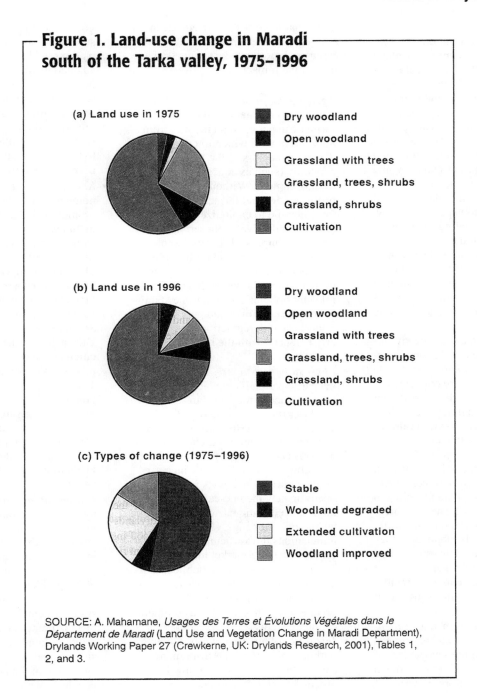

Figure 1. Land-use change in Maradi south of the Tarka valley, 1975–1996

(a) Land use in 1975

Dry woodland
Open woodland
Grassland with trees
Grassland, trees, shrubs
Grassland, shrubs
Cultivation

(b) Land use in 1996

Dry woodland
Open woodland
Grassland with trees
Grassland, trees, shrubs
Grassland, shrubs
Cultivation

(c) Types of change (1975–1996)

Stable
Woodland degraded
Extended cultivation
Woodland improved

SOURCE: A. Mahamane, *Usages des Terres et Évolutions Végétales dans le Département de Maradi* (Land Use and Vegetation Change in Maradi Department), Drylands Working Paper 27 (Crewkerne, UK: Drylands Research, 2001), Tables 1, 2, and 3.

areas lay uninhabited and where farming villages were mostly concentrated in the extreme south.

French colonization encouraged rapid migration from the south and from places further afield, taking advantage of relatively good rainfall during the 1950s and extensive cultivable land that stood unclaimed in the department's central *arrondissements*. The pace of change was rapid. The cultivated fraction in four representative village territories averaged 31 percent in 1957, according to aerial photograph interpretation.[46] In 1999, all four were virtually entirely cultivated, and between 1974 and 1996, satellite imagery of the whole department south of the effective limit of rainfed cultivation indicated an increase from 59 percent to 73 percent.[47] This confirms that little uncultivated land (and even less unclaimed land) remained (see Figure 1).

The Sahel Drought exposed the risk associated with farming in such marginal areas, even as a growing population sought to appropriate more land for millet and groundnut production. Deforestation, soil fertility decline, erosion, falling yields, overgrazing, food insecurity, poverty, and increasing dependence on resources outside the area (food aid and urban employment) seemed to threaten not only the environment but also economic sustainability of rural communities.[48] A major rural development project was put in place. The demand for cultivable land was driven by demographic, market, and institutional forces, as well as a threat of declining productivity.[49]

This was not an auspicious setting for a success story. However, there are some surprises to be found in long-term data and recent field investigations.[50] Two changes in the management of farmland have been detected. The first of these is a growing use of animal manure—especially in the southern arrondissements where land is most scarce—as well as inorganic fertilizer when it can be afforded. The second is the practice, called *défrichement amélioré*, of protecting trees that are regenerating naturally and that are considered to have economic value. Promoted by the development program, this indigenous practice is now firmly established, creating an increasingly wooded appearance in the once bare farmland around the villages. Individuals rather than patrilineal families are tending to take charge of the natural resources, and conservationist attitudes toward biodiversity resources are being reasserted.[51]

Farm investments are thus already significant, and permanent fields and vegetated boundaries are appearing as private rights to the use of land are confirmed.[52] Livestock numbers in the department trended upward throughout the 1990s. Especially in southern areas, ox-drawn ploughs, seeders, and carts are becoming commonplace.

Crop yield data are available for each arrondissement and when compared, these suggest a positive trend for millet in the south (in response to the need to intensify farming practice) while in the north, where land is relatively abundant, they are stagnant.[53] In the department as a whole, cereal production kept ahead of the nutritional requirements of a rapidly growing population from 1964 until 1998, except in drought years.[54]

There are no long-term data on incomes and welfare, and recent surveys confirmed the continuing existence of poverty in Maradi.[55] Not too much should be claimed, therefore, in terms of success. But Niger as a whole is now less dependent on food aid than it was in the 1980s, despite its larger population. Maradi Department lies across the border from Katsina and Kano in Nigeria and is increasingly incorporated into buoyant transborder marketing chains in food commodities.[56]

Policies to Promote Success

Other regions in West Africa have also yielded counterintuitive evidence of positive trends in environmental management and productivity.[57] From the studies that have been carried out in all these regions, we can learn that the following policy-dependent agents have had positive impacts:

- *Agricultural product markets.* Without effective demand, no dryland farmer can generate an income from agriculture with which to cross-subsidize other sectors of the household economy (such as education fees and trading capital). Without a prospect of profit, no dryland farmer will use off-farm income to capitalize agriculture (such as soil or water conservation and purchased inputs). It is remarkable that there is no dryland region of any size that lacks linkages with one or more markets, however distant. The relevant policies concern price stabilization, the control of competition from imports, and deregulation that is properly balanced with the needs of local producers.

- *Physical infrastructure.* Increased market participation is unavoidable for dryland producers; they need to sell surpluses, buy their way out of deficit, achieve food security, and access labor and commodity markets. The provision of roads and transport systems reduces marketing costs and thereby increases producer prices, increases interaction, opens up new income-earning opportunities and educational access. With technological change, electronic communications infrastructure is also critical for accessing market information. The relevant policies are public investments in roads and telephones and regulatory frameworks that facilitate private sector investments in communications systems.

- *Institutional infrastructures.* Given a withdrawal of the state from many areas (including produce buying and processing and agricultural extension support), a failure of private sector enterprises to adequately fill the gap, and failures of local governance, attention is shifting to building the capacity of decentralized and autonomous institutions. These can restore local ownership of natural resources, negotiate on behalf of local communities with higher levels of government, control access to endangered resources, defend the common interests of local producers in wider markets, and supply needed information of all kinds to communities. The relevant policies concern genuine decentralization of government powers, recognition and support of producers' or traders' associations, and the facilitation of new or adapted institutions for ecosystem management (such as common resources) and of information networks. An area offering major potential for poverty reduction is institutional frameworks to manage risk, which go to the heart of the uncertainty of living in dryland environments.[58]

- *Knowledge management.* Useful technologies, rather than being spread by promotional blueprint, tend to be taken up in response to users' particular circumstances. Discriminating users in rainfed farming systems need a variety of options from which to choose. There have been no "miracle" technologies so far that resemble the high-yielding crop varieties used in humid or irrigated South and East Asia. Farmers' own knowledge, exchanges, and experimentation have been as important to the process as experimental agriculture. The relevant policies concern education provision and partnering local producers in extension systems. As knowledge is a public good, the state still has a role to play in putting new knowledge into the public domain and facilitating access to it by poor rural people.

- *Investment incentives.* The assumption that all dryland families are too poor to invest is shown to be false in the cumulative growth of small on- and off-farm investments in the longer-settled and successful systems. Meanwhile, many of the problems of recently occupied, sparsely settled, and highly dynamic systems stem from a lack of investment. Attention was drawn to this priority in Maradi as long ago as 1980,[59] and the evidence suggests that the capitalization of the landscape using local peoples' own resources does indeed take decades rather than years. Clearly, secure title to use of and benefits from ecosystem resources is a necessary condition for investing scanty, private financial resources. The remarkable landscape of Machakos is a witness to what can be achieved in a half-century of incremental growth based on inputs of personal finance and family labor. The relevant policies, therefore, concern an enlightened adjustment by central government of variable incentives (such as taxation and exchange rates) and a prioritization of enabling incentives (such as secure resource tenure, dispute settlement, and conflict prevention) and efficient markets.[60]

- *Income diversification incentives.*
Rather than posing a threat to urban areas, seasonal migrants from rural areas provide a range of services at minimal cost, in a dynamic and adaptive informal sector (which tends to defy government regulation). Urban informal sectors offer low average incomes, insecurity, squalor and sometimes illhealth. Paradoxically, perhaps, having such options available provides some of the flexibility necessary in rural drylands to cope with risk and recurrent food insecurity. They also provide investment funds for raising productivity and incomes in drylands, thereby helping to stabilize rural populations. Dryland peoples not only help themselves, they also help to integrate the national economy through consumption and investment streams. The relevant policies are relaxing controls on the movement of people, commodities, and capital around the country and ensuring the personal security of migrants. In an economic union such as the Economic Community of West African States (ECOWAS), these considerations apply to international movements, as drylands are interdependent with more humid or urbanized regions.

Policies, of course, evolve in response to articulated interests. To plead a special case for drylands, even if evidence-based, may have little impact on such a process. However, given the recent advances in democratic government in many countries of Africa, there are new opportunities to empower dryland people to plead their own cause with local and central government. This is especially true where evidence is available from targeted research. This type of empowerment relies on alliances between local people, researchers, and the private and voluntary sectors and on resources to support interactive debate and negotiation. Villagers can express their demands effectively when barriers of language and isolation are removed, as shown by experience in Kenya, Senegal, Niger, and northern Nigeria.[61]

Conclusion

Is "success" a justified term in dryland environments where risk, low productive potentials, and high levels of poverty are endemic? Some areas can show

evidence of achievements that run counter to the expectations generated by some orthodox models of desertification. There are four components of success: a movement toward more sustainable ecosystem management, evidence of increasing investment both on and off the farm, stable or improving output or output value per hectare, and evidence of improving incomes and welfare. Not all of these components are evident in every story, and the last in particular is difficult to pin down in the absence of data. However, a positive trajectory sustained over a timescale of decades is evidence too important to ignore, even though it may conflict with "expert" opinions based on shallow timeframes. It is remarkable that many African smallholders have sustained such achievements against policy failure and urban bias.[62]

The dryland cases show that rural development should be seen in a broader context than that of agriculture alone. Currently, efforts are being made to re-emphasize the role of science and technology for improving productivity and food security in Africa. But it is acknowledged that if the potential of technology is to be fully realized, a market-led strategy is required to raise productivity.[63] The dryland success stories support the view that productivity trends respond to economic incentives. The capacities of resource-poor farmers to invest in on-farm improvements should not be underestimated, notwithstanding many constraints.

Basing dryland development policy on evidence of poor peoples' achievements offers a radical alternative to the doomsday scenarios that have a profound though not always admitted influence on policymakers. In starting with a perceived failure in ecosystem management—as implied in the idea of desertification—governments or donors drive themselves into a blind alley. Because the diagnosed "mismanagement" is defined and assessed in quantitative, biophysical terms, it calls for technical solutions that have to be imposed through an assertion of outsider knowledge over insider experience, devaluing the indigenous resource. Participatory rhetoric cannot on its own correct this distortion. On the other hand, knowledge partnerships, constructed on local peoples' achievements, imply equality between the agents of intervention and the intended beneficiaries, an equality that is too often denied in authoritarian governmental structures. This di-

lemma is widely recognized. It is especially acute in drylands, where the "knowledge gap" between insiders and outsiders is all the greater: The outsider is often unfamiliar with the harsh realities of managing relatively unproductive natural resources at high levels of risk.

By providing evidence of real achievements and internal potentials, success stories can therefore point the way toward laying a new foundation for evidence-led policies for dryland development: Rather than aiming to transform "inappropriate" local practices, such policies instead aim to build on local experience, suggesting a more organic model for development.

NOTES

1. The United Nations (UN) Food and Agricultural Organization (FAO) defines drylands by the average length of the growing period, while the UN Environment Programme (UNEP) defines drylands by an index of aridity based on precipitation over potential evapotranspiration. See FAO, *Land, Food and People*, FAO Economic and Social Development Series 30 (Rome: FAO, 1984); and UNEP, *World Atlas of Desertification* (Nairobi: Arnold for UNEP, 1992). According to the first, drylands have fewer than 180 growing days per year, and according to the second, an aridity index between 0.05 and 0.65. M. Mortimore, *Roots in the African Dust: Sustaining the Sub-Saharan Drylands* (Cambridge, UK: Cambridge University Press, 1998), 12.

2. M. Hulme, "Rainfall Changes in Africa: 1931–1960 to 1961–1990," *International Journal of Climatology* 12 (1992): 685–99.

3. For example, in Maradi Department, Niger, levels of carbon are less than 0.2 percent, nitrogen less than 0.02 percent, and available phosphorus less than 4 parts per million (ppm) on uncultivated soils. M. Issaka, *Évolution à Long Terme de la Fertilité de la Sol dans la Région de Maradi* (Long-Term Change in Soil Fertility in the Maradi Region), Drylands Research Working Paper 30 (Crewkerne, UK: Drylands Research, 2001). These values for phosphorus are exceptionally low; in Kano, Nigeria, and

Machakos, Kenya, 11–16 ppm are found. F. Harris and M. A. Yusuf, "Manure Management by Smallholder Farmers in the Kano Close-Settled Zone, Nigeria," *Experimental Agriculture* 37 (2001): 319–32; and J. P. Mbuvi, *Makueni District Profile: Soil Fertility Management*, Drylands Research Working Paper 6 (Crewkerne, UK: Drylands Research, 2000).

4. Organic soils comprise less than 0.4 percent of Maradi soils (see Issaka, ibid.).

5. F. W. T. Penning de Vries and M. A. Djiteye, eds., *La Productivité des Pâturages Sahèliens. Une Etude des Sols, des Végétations et l'Exploitation de Cette Ressource Naturelle* (The Productivity of Sahelian Pastures. A Study of Soils, Vegetation, and the Exploitation of this Natural Resource) (Wageningen, Netherlands: Pudoc, 1991).

6. Issaka, note 3 above; and F. M. A. Harris, "Farm-level Assessment of the Nutrient Balance in Northern Nigeria," *Agriculture, Ecosystems and Environment* 71 (1998): 201–14.

7. The Millenium Development Goals are: halving the proportion of people in extreme poverty and suffering hunger between 1990 and 2015; achieving universal primary education by 2015; eliminating gender disparity in education by 2015; reducing the under-five child mortality ratio by two thirds by 2015; reducing the maternal mortality ratio by three quarters by 2015; halting and beginning to reverse the spread of HIV/AIDS and the incidence of malaria and other diseases by 2015; ensuring environmental stability; and creating a global partnership for development.

8. The most widely used characterization is the revised definition used for the 1992 UN Rio Earth Summit: "Desertification means land degradation in arid, semi-arid and sub-humid areas resulting from various factors, including climatic variations and human activities." UN, *UN Earth Summit. Convention on Desertification*, UN Conference on Environment and Development, Rio de Janeiro, Brazil, 3–14 June 1992, DPI/SD/1576 (New York: United Nations, 1994). Definition is not, however, straightforward: More than 100 attempts have been recorded. M. Glantz and N. Orlovsky, "Desertification: A Review of the Concept," *Desertification Control Bulletin* no. 9 (1983): 15–20.

9. J. F. Reynolds and D. M. Stafford Smith, eds., *Global Desertification. Do Humans Cause Deserts?* (Berlin: Dahlem University Press, 2002).

10. Thomas Malthus argued, with respect to eighteenth-century England, that human populations outgrow their capacity to feed themselves. In his own time, his theory was answered by the industrial revolution, which financed the English in importing food from a growing world market and led to technical innovations that increased agricultural productivity. Since World War II, the inability of many poor countries to follow the same path led to a revival of "neo-Malthusian" thinking. Evidence of land degradation under pressure from rapidly growing farming populations, considered as a system closed to the larger economy, appears to offer only impoverishment and starvation. A summary is provided in M. Mortimore, "Technological Change and Population Growth," in P. Demeny and G. McNicoll, eds., *Encyclopedia of Population* (New York: Macmillan Reference, 2003), 932–35; and for a new discussion of some issues, see Q. Gausset, M. Whyte, and T. Birch Thomsen, eds., *Beyond Territory and Scarcity: Social, Cultural and Political Aspects of Conflicts on Natural Resource Management* (Uppsala, Sweden: Nordic African Institute, 2004).

11. M. Mortimore and M. Tiffen, "Population Growth and a Sustainable Environment: The Machakos Story," *Environment*, October 1994, 10–20, 28–32.

12. C. Maher, *Soil Erosion and Land Utilisation in the Ukamba Reserve (Machakos)*, Report to the Department of Agriculture, Mss, Afr.S.755, Rhodes House Library, Oxford (Nairobi: Department of Agriculture, 1937).

13. M. Tiffen, M. Mortimore, and F. Gichuki, *More People Less Erosion: Environmental Recovery in Kenya* (Chichester, UK: John Wiley & Sons, 1994), 93–96.

14. The Machakos story is told in great detail elsewhere: ibid.; J. English, M. Tiffen, and M. Mortimore, *Land Resource Management in Machakos District, Kenya: 1930–1990*, World Bank Environment Paper No. 5, (Washington, DC: World Bank, 1994); and M. Tiffen and M. Mortimore, "Malthus Controverted: The Role of Capital and Technology in Growth and Environment Recovery in Kenya," *World Development* 22, no. 7 (1994): 997–1010.

15. E. Boserup, *The Conditions of Agricultural Growth: The Economics of Agricultural Change under Population Pressure* (London: Allen and Unwin, 1965); and E. Boserup, *Economic and Demographic Relationships in Development* (Baltimore, MD: Johns Hopkins University Press, 1990).

16. P. F. Robbins et al., "Desertification at the Community Scale. Sustaining Dynamic Human-Environment Systems," in Reynolds and Stafford Smith, note 9 above, pages 326–56.

17. M. Mortimore, *Adapting to Drought, Farmers, Famines and Desertification in West Africa* (Cambridge, UK: Cambridge University Press, 1989); and Mortimore, note 1 above.

18. R. D. Smith and E. Maltby, *Using the Ecosystem Approach to Implement the Convention on Biological Diversity. Key Issues and Case Studies*, Ecosystem Management Series No. 2 (Gland, Switzerland: IUCN-The World Conservation Union, 2003).

19. The former Machakos District is now divided between Machakos and Makueni Districts. F. N. Gichuki, S. G. Mbogoh, M. Tiffen, and M. Mortimore, *District Profile: Synthesis*, Drylands Research

Working Paper 11 (Crewkerne, UK, Drylands Research, 2000).

20. J. Nelson, *Makueni District Profile: Income Diversification and Farm Investment, 1989–1999*, Drylands Research Working Paper 10 (Crewkerne, UK: Drylands Research, 2000).

21. Tiffin, Mortimore, and Gichuki, note 13 above, 93–95.

22. Global Mechanism of the Convention to Combat Desertification (GM-CCD), *Why Invest in Drylands? A Study Carried out for the Global Mechanism of the Convention to Combat Desertification* (Rome: GM-CCD, 2004); and C. Reij and D. Steeds, *Success Stories in Africa's Drylands: Supporting Advocates and Answering Skeptics* (Amsterdam: CIS/Centre for International Cooperation, Vrije Universiteit Amsterdam, 2003).

23. The studies are reported in Drylands Research Working Papers 1–41, available at www.drylandsresearch.org.uk. See also M. Tiffen and M. Mortimore, "Questioning Desertification in Dryland Sub-Saharan Africa," *Natural Resources Forum* 26, no. 3 (2002): 218–33.

24. The Kano Close-Settled Zone was defined by Mortimore as having local population densities of 350 per square mile (147 per square kilometer (km²) or more, according to the census of 1962. M. Mortimore, "Land and Population Pressure in the Kano Close-Settled Zone, Northern Nigeria," *The Advancement of Science* 23 (1967): 677–88. Over an area of about 10,000 km², densities decrease regularly outward to distances of 40–100 km from Kano City. This threshold density is now exceeded over a wider area. In the innermost periurban districts, densities of more than 800 per km² are found. M. Tiffen, *Profile of Demographic Change in the KanoMaradi Region, 1960–2000*, Drylands Research Working Paper 24 (Crewkerne, UK: Drylands Research, 2001).

25. Tiffin, ibid.

26. Harris, note 6 above; and M. Mortimore, "The Intensification of Peri-Urban Agriculture: The Kano Close-Settled Zone, 1964–86," in B. L. Turner II, R. W. Kates, and H. L. Hyden, eds., *Population Growth and Agricultural Change in Africa* (Gainesville, FL: University Press of Florida, 1993), 358–400.

27. The exception was potassium, which declined significantly in the 59 samples analyzed. M. Mortimore, "Northern Nigeria: Land Transformation under Agricultural Intensification," in C. L. Jolly and B. B. Torrey, eds., *Population and Land Use in Developing Countries. Report of a Workshop* (Washington, DC: National Academy Press, 1993), 42–69.

28. M. Mortimore and W. Adams, *Working the Sahel: Environment and Society in Northern Nigeria* (London: Routledge, 1999), 78.

29. F. M. A. Harris and S. Mohammed, "Relying on Nature: Wild Foods in Northern Nigeria," *Ambio*, 32, no. 1 (2003): 24–29.

30. P. Hill, *Population, Prosperity and Poverty. Rural Kano, 1900 and 1970* (Cambridge, UK: Cambridge University Press, 1977).

31. Droughts have occurred unpredictably throughout the history of Kano and recently in the great Sahel Drought that hit Kano in 1972–1974, 10 years later in 1982–1984, and in several individual years during the 1990s (M. Mortimore, *Adapting to Drought: Farmers, Famines and Desertification in West Africa* (Cambridge, UK: Cambridge University Press, 2001); and M. Mortimore, *Profile of Rainfall Change in the Kano-Maradi Region, 1960–2000*, Drylands Research Working Paper 25 (Crewkerne, UK: Drylands Research, 2000)).

32. Grain yields ranged between 0.5 and 3 tons per hectare (depending on rainfall, fertilization, and weeding) on a small but accurately measured sample located 40 km from Kano. Some output might be sold, lent, or given away, rendering the household food-insecure. F. Harris, *Intensification of Agriculture in Semi-arid Areas: Lessons from the Kano Close-Settled Zone, Nigeria*, Gate-keeper Series No. 59 (London: International Institute for Environment and Development, 1996).

33. Tiffen, note 24 above.

34. J. A. Ariyo, J. P. Voh, and B. Ahmed, *Long-Term Change in Food Provisioning and Marketing in the Kano Region*, Drylands Research Working Paper 34 (Crewkerne, UK: Drylands Research, 2001).

35. An absence of long-term compatible data on real income is frustrating the analysis of income and welfare trends. Meanwhile, general price inflation (linked to a depreciation of the national currency by 200 percent since the 1970s) makes people strongly aware of hardship, especially since structural adjustment policies were introduced in 1986. A. R. Mustapha and K. Meagher, *Agrarian Production, Public Policy and the State in Kano Region, 1900–2000*, Drylands Research Working Paper 35 (Crewkerne, United Kingdom, Drylands Research, 2000).

36. Ibid.

37. A. N. Badiane, M. Khouma, and M. Sène, *Région de Diourbel: Gestion des Sols* (Diourbel Region: Soil Management), Drylands Research Working Paper 15 (Crewkerne, UK: Drylands Research, 2000); and P. Garin, B. Guigou, and A. Lericollais, "Les Pratiques Paysannes dans le Sine" (Farming Practices in Sine), in A. Lericollais, ed., *Paysans Sereer: Dynamiques Agraires et Mobilité au Sénégal* (Sereer Farmers: Agrarian Dynamics and Mobility in Senegal) (Paris: Editions Institut de Recherches en Développement, 1999), 211–98.

38. M. Gaye, *Région de Diourbel: Politiques Nationales Affectant l'Investissement chez les Petits Exploitants* (Diourbel Region: National Policies Affecting Small Farmers' Investments), Drylands Research Working Paper 12 (Crewkerne, UK: Drylands Research, 2000).

39. Farm tree populations are said to be under threat in the well-studied village of Sob, just outside the boundary of Diourbel Region. (See Garin,

Guigou, and Lericollais, note 37 above.) Data gathered in 1999 in three villages indicated that the status of farmed parkland is more ambiguous than in Kano, with large variations between villages, abundant regeneration in places, but decline in some species and scarcities of middle-aged trees. S. Sadio, M. Dione, and S. Ngom, *Région de Diourbel: Gestion des Ressources Forestières et de l'Arbre* (Diourbel Region: Management of Forest Resources and Trees), Drylands Research Working Paper 17 (Crewkerne, UK: Drylands Research, 2000).

40. A. Barry, S. Ndiaye, F. Ndiaye, and M. Tiffen, *Région de Diourbel: Les Aspects Démographiques* (Diourbel Region: Demographic Aspects), Drylands Research Working Paper 13 (Crewkerne, UK: Drylands Research, 2000).

41. A. Faye, A. Fall, M. Mortimore, M. Tiffen, and J. Nelson, *Région de Diourbel: Synthesis*, Drylands Research Working Paper 23e (Crewkerne, UK: Drylands Research, 2001).

42. Official data at region and department level were used in the long-term analysis (1960–2000). They are based on a mixture of censuses and estimates. Small trends were only considered potentially significant if sustained over many years.

43. Such a difference is too large to be fully explained by the size of household, which averaged 10 members, 13 and 12 in types 1, 2, and 3 shown in Figure 2.

44. Garin, Guigou, and Lericollais, note 37 above; and Sadio, Dione, and Ngom, note 39 above.

45. See Boserup 1965 and 1990, note 15 above.

46. C. Raynaut, *Recherches Multidisciplinaires sur la Région de Maradi: Rapport de Synthèse* (Multidisciplinary Research in the Maradi Region: Synthesis) (Bordeaux: Maradi Region Research Programme, Université de Bordeaux II, 1980), 10.

47. A. Mahamane, *Usages des Terres et Évolutions Végétales dans le Département de Maradi* (Land Use

and Vegetation Change in Maradi Department), Drylands Working Paper 27 (Crewkerne, UK: Drylands Research, 2001).

48. Raynaut, note 46 above; C. Raynaut, "Le Cas De La Région de Maradi (Niger)" (The Maradi Region Case, Niger), in J. Copans, ed., *Sécheresses et Famines du Sahel. Tome II. Paysans et Nomads* (Droughts and Famines in the Sahel. Farmers and Nomads) (Paris: Librairie Francoix Maspero, 1975), 5–43; and C. Raynaut, J. Koechlin, C. Cheung, and M. Stigliano, *Le Dévelopement Rural de la Région au Village—Analyser et Comprendre la Diversité* (Rural Development from Region to Village—Analysing and Understanding Diversity) (Bordeaux, France: Maradi Region Research Programme, Université de Bordeaux II, 1988).

49. The rural population of Maradi Department grew at more than 3.0 percent per year between 1960 and 1988 (Tiffin, note 24 above). Maradi Department was a breadbasket of Niger: It produced the staple food crop, bulrush millet and was the main center of groundnut production until 1975. It was also a cotton-producing area. During the 1990s, the development of Niger's Code rural (which was based on the principle of "land to the tiller") motivated farmers to claim as much free land as they could plant.

50. M. Mortimore, M. Tiffen, Y. Boubacar, and J. Nelson, *Department of Maradi: Synthesis*, Drylands Research Working Paper 39e (Crewkerne, UK: Drylands Research, 2001).

51. A. Luxereau, "Usages, Représentations, Évolutions de la Biodiversité Végétales chex les Haoussa du Niger" (Use, Perceptions, and Change in Plant Biodiversity among the Hausa of Niger), *Journal d'Agriculture et de Botanique Appliqué NS* (Journal of Agriculture and Applied Botany NS) 36, no. 2 (1994): 67–85; and A. Luxereau and B. Roussel, *Changements Écologiques et Sociaux au Niger*

(Ecological and Social Change in Niger) (Paris: L'Harmattan, 1997).

52. Y. Boubacar, *Évolution des Régimes de Propriété et d'utilisation des Ressources Naturelles dans la Région de Maradi* (Changes in the Tenure and Use of Natural Resources in the Maradi Region), Drylands Research Working Paper 29 (Crewkerne, UK: Drylands Research, 2000).

53. Tiffin and Mortimore, note 23 above.

54. The nutritional requirements are estimated at 200 kilograms per person per year. This finding does not exclude the possibility that cereals were exported from the department, leaving some families food-insecure.

55. Cooperative for Assistance and Relief Everywhere (CARE), *Evaluation de la Sécurité des Conditions de Vie dans le Département de Maradi* (Assessment of Living Conditions in Maradi Department) (Niamey, Niger: CARE Niger, 1997).

56. Ariyo, Voh, and Ahmed, note 34 above; and K. Meagher, *Current Trends in Cross-Border Grain Trade between Nigeria and Niger* (Paris: IRAM, 1997).

57. V. Mazzucato and D. Niemeijer, *Rethinking Soil and Water Conservation in a Changing Society*, Tropical Resource Management Papers, No. 32 (Wageningen, Netherlands: Wageningen University and Research Centre, 2000); D. Niemeijer and V. Mazzucato, "Soil Degradation in the West African Sahel: How Serious Is It?" *Environment*, March 2002, 20–31; and C. Reij and T. Thiombiano, *Développement Rural et Environnement au Burkina Faso: La Réhabilitation de la Capacité Productive des Terroirs sur la Partie Nord du Plateau Central entre 1980 et 2000* (Rural Development and the Environment in Burkina, Faso: The Restoration of the Productive Capacity of Village Lands in the North of the Central Plateau between 1980 and 2000) (Amsterdam: CIS, Vrije Universiteit Amsterdam, 2003).

58. Grain banks and other attempts to counter the insecurity of living under drought risk have enjoyed only chequered success. It is arguable that this problem will be found as challenging as achieving sustainable natural resource management. Drought management will be the focus of a new initiative by UNDP's Dryland Development Centre, Nairobi.

59. Raynaut, note 46 above, page 66.

60. D. Knowler, G. Acharya, and T. van Rensburg, *Incentive Systems for Natural Resources Management*, (Rome: FAO Investment Centre, 1998).

61. M. Mortimore and M. Tiffen, "Introducing Research into Policy: Lessons from District Studies of Dryland Development in Sub-Saharan Africa," *Development Policy Review* 22, no. 3 (2004): 259–86.

62. Drylands Research, *Livelihood Transformations in Semi-Arid Africa 1960–2000: Proceedings of a Workshop Arranged by the ODI with Drylands Research and the ESRC*, in the series "Transformations in African Agriculture," Drylands Research Working Paper 40 (Crewkerne, UK: Drylands Research, 2001).

63. InterAcademy Council, *Realizing the Promise and Potential of African Agriculture. Science and Technology Strategies for Improving Agricultural Productivity and Food Security in Africa* (Amsterdam: InterAcademy Council, 2004).

Michael Mortimore is an independent researcher with Drylands Research, Crewkerne, United Kingdom. His research focuses on policy studies of environmental management by small-scale farmers in Africa's drylands. Previously, he carried out research studies as a senior research associate in the Department of Geography, Cambridge University, the Overseas Development Institute, and as an honorary fellow of the Centre of West African Studies, University of Birmingham. Mortimore is the author or coauthor of several books, including *Working the Sahel: Environment and Society in Northern Nigeria* (Routledge, 1999), *Roots in the African Dust: Sustaining the Sub-Saharan Drylands* (Cambridge University Press, 1998), *More People, Less Erosion: Environmental Recovery in Kenya* (John Wiley, 1994), and *Adapting to Drought: Farmers, Famines and Desertification in West Africa* (Cambridge University Press, 1989). He may be reached at (44) (0) 1963-250-198 or via e-mail at mikemortimore @compuserve.com. The author wishes to thank Bill Adams, Abdou Fall, Adama Faye, Francis Gichuki, Mary Tiffen, and Boubacar Yamba, who, together with other colleagues in Africa and the United Kingdom, made valuable contributions to the ideas expressed in this article. The article is based on work mainly funded by the U.K. Department for International Development.

WHAT'S A RIVER FOR?

Thousands of dead salmon, acres of dying crops, pesticide-poisoned birds: How the Klamath River became the first casualty in the West's new water wars

By Bruce Barcott

ON THE MORNING of September 19, 2002, the Yurok fishermen who set their gill nets near the mouth of the Klamath River arrived to find the largest salmon run in years fully under way. The fish had returned from the ocean to the Klamath, on the Northern California coast, to begin their long trip upstream to spawn; there were thousands of them, as far as the eye could see. And they were dying. Fully-grown 30-pounders lay beached on shoreline rocks. Smaller fish floated in midriver eddies. Day after day they kept washing up; by the third day, biologists were estimating that 33,000 fish had been killed in one of the largest salmon die-offs in U.S. history.

The Yurok knew immediately what had happened. For months they, along with state experts and commercial fishermen, had been pleading with the federal government to stop diverting most of the river's water into the potato and alfalfa fields of Oregon's upper Klamath Basin. But the Bureau of Reclamation, the agency in charge of federal irrigation projects, refused to intervene. No one had proved, it argued, that the fish really needed the water.

When the die-off was discovered, federal authorities raced to send a flush of water downriver. But they didn't change their long-term policy—using the Klamath's water to support farmers over fish, fishermen, and some of the most environmentally critical wildlands in the nation.

From Montana to New Mexico, the Bush administration is favoring the water rights of farmers, ranchers, and developers over endangered species, Indian tribes—and the federal government itself.

FOR THE BUSH ADMINISTRATION, and for the advocacy groups that have joined the battle, the fight over the Klamath is more than a regional dispute. It's a bellwether signaling a key shift in the federal government's stance in a new generation of Western water wars. From Montana to New Mexico, conflicts over rivers, wetlands, and irrigation projects are pitting federal water rights against local and state governments and private interests. And in each case, the Bush administration is favoring farmers, ranchers, and developers over the rights of endangered species, Indian tribes, and the federal government itself. In the past year alone, administration officials have backed away from water policies designed to protect fish and birds along the Rio Grande and in California's Central Valley, given up their claim to protect thousands of acres of wetlands from being filled in for subdivisions and shopping malls, and moved toward ceding federal rights to water in several national parks and wildlife refuges.

"The administration sees water rights as property rights that come before other rights, including the right of ecosystems to exist," says Steve Malloch, executive director of the Western Water Alliance, a regional environmental advocacy group. "When faced with the choice between [environmental protection] and what they perceive to be inviolable property rights, they've made it clear what side they're going to come down on."

Nowhere have these issues played out more dramatically than on the 250-mile course the Klamath takes from Oregon's arid high country to the redwood forests of California's northern coast. When the federal government was forced by the courts in 2001 to withhold irrigation water and keep the Klamath flowing, national property-rights groups rallied to support struggling farmers. A year later, when the administration took water from the fish and gave it to farmers, the salmon kill made the river a cause célèbre for environmental groups. At least four separate lawsuits, with plaintiffs ranging from fishermen to property-rights advocates to environmentalists, are challenging government policies on the river. And a National Marine Fisheries Service biologist has filed for whistleblower protection, arguing that, under pressure from "a very high level," his agency changed a key scientific report to justify withholding water from the fish.

Decisions in some of the cases are expected soon, just as another season of trouble gets under way on the Klamath. After a winter of meager snowfalls, another drought is likely. Someone will have to go without water again. And that, in this part of the world, is the one thing no one can afford.

THE KLAMATH IS BORN A CRIPPLE. The river begins in Klamath Falls, Oregon, at the southern outlet of Upper Klamath Lake, a body of water so shallow that a tall man could nearly cross it without wetting his hat. Before the river becomes a river, the U.S. Bureau of Reclamation's "A" Canal diverts half its water into a labyrinthine system of aqueducts that irrigate more than 1,400 farms and ranches. The falls? There are no falls, never have been. The town's founders invented the name to attract homesteaders.

This was what you found when you arrived in Klamath Falls a century ago: swamps everywhere, and birds thick as flies. The Klamath Basin, a high arid plateau the size of Connecticut, didn't get much rain—about as much, on average, as the Texas Panhandle—but what it got it kept, draining the runoff from surrounding mountains into three shallow lakes and a vast system of wetlands. Ten million birds paused to feed here during their migration from Canada to Mexico. Egrets, terns, mallards, pelicans, eagles, tundra swans, and herons browsed amid thickets of 10-foot-tall bulrushes known as tule (too-lee). The Klamath Indians pulled tens of thousands of sucker fish from the lakes every year. This was the Everglades of the American West.

Where some saw a thriving wetland, the U.S. Bureau of Reclamation saw farmland drowning under water. Starting in 1906, the Bureau drained and replumbed nearly the entire basin, building a complex set of canals that shrank Lower Klamath Lake and Tule Lake to one-quarter their original size and sending the water to irrigate thousands of acres of crops and pasture. After the farmers had their share, some of the water would drain into the Tule Lake and Lower Klamath National Wildlife refuges, and some would be pumped back into the Klamath River. Over time, a map of the project came to resemble a guide to the London Underground.

A few weeks after low river levels caused one of the largest salmon kills in U.S. history, a federal biologist said his superiors had changed a key scientific assessment under pressure from "a very high level."

Thanks to irrigation, things hummed along for nearly a century. Homesteaders built farm towns along the Oregon-California border: Merrill, Malin, Tulelake. Teenage girls were crowned queen of the potato festival. Teenage boys hung out at the Tulelake bowling alley. A local historian proclaimed the Klamath Basin Project "one of the most successful government projects in American history."

But over those 100 years, any life that didn't grow in neat harrowed rows began to drain out of the basin.

The Klamath Indians noticed first. The shortnose sucker, a staple of the tribal diet, began growing scarce in Upper Klamath Lake. Farther downstream, the coho salmon that provided sustenance to the Yurok tribe—and to thousands of commercial fishermen along the coast—all but vanished. Environmentalists successfully fought to have both fish protected under the Endangered Species Act. The tribes, whose treaties guaranteed them the right to harvest sucker and salmon in perpetuity, filed lawsuits demanding protection for the fish. By 1995, the Interior Department's regional solicitor—the department's in-house lawyer—issued an opinion reflecting the new legal reality: Tribes, and the endangered species they sought to protect, would have first right to disputed federal waters. Irrigators would have to come second.

The Bureau of Reclamation, an agency created not to protect nature but to overcome it, was slow to implement the new priorities. But by the spring of 2001, it had no choice. A drought had struck the Klamath and a new lawsuit, from commercial fishermen along the coast, required that river levels be kept high enough for the fish. That meant withholding some of the farmers' irrigation water and sending it downriver—at least until the drought broke.

The drought did not break.

"WE WERE LOOKING GOOD AND GOING into 2001," says Dick Carleton. "We had our three tractors and potato equipment nearly paid off. Then we got word that the water wouldn't be coming." The 59-year-old farmer steers his red F-250 truck down a muddy track separating fields of alfalfa stubble. Carleton's farm sits three days' flow from the Klamath headwaters. He and his son Jim, 34, grow alfalfa and potatoes on 1,500 acres in Merrill, Oregon, 15 miles south of Klamath Falls. If you've eaten Campbell's cream of potato soup or snacked on Frito-Lay potato chips you may have tasted their crop. But you might not again. The Carletons are bankrupt.

"It's Chapter 12," Dick explains. "You don't lose everything, but you still owe your debts. Gives us a chance to regroup."

Without irrigation water, the Carletons lost their 2001 potato crop. Their bills piled up. A typical potato farmer can carry a $300,000 loan just on his equipment. Multiply Dick Carleton by 1,400 (the number of local growers who depend on Bureau irrigation) and you end up with a lot of angry farmers.

By the summer of 2001, those farmers were making national news. They paraded their tractors through Klamath Falls and carried signs that said "Klamath Basic Betrayed." National property-rights advocates flocked to the basin, accusing the government of engaging in "rural cleansing." Western members of Congress held hearings in Klamath Falls and vowed to rewrite the Endangered Species Act. The farmers started a vigil at the "A" Canal headgate. When somebody surreptitiously opened the canal—and local police refused to make arrests—federal marshals were called in.

"I wasn't a political person before 2001," says Dick Carleton. "But I spent a lot of time up at the headgate." For the farmers, things grew desperate. Some started selling off equipment for 10 cents on the dollar.

Many farmers lashed out at the Bush administration for locking up their water. But behind the scenes, top officials at the Interior Department, which oversees the Bureau of Reclamation, were scrambling to keep the farmers in business. "The department was in no position to say, 'We're not going to comply with the law,'" recalls Sue Ellen Wooldridge, deputy chief of staff for Interior Secretary Gale Norton. "We tried to see if there was any flex"—any wiggle room in the law that would allow delivery of irrigation water—"and there wasn't."

Officials knew that the only way to change the irrigation plan was a new scientific assessment of how much water the fish needed. In October 2001, Norton asked a committee of the National Research Council to review the available research on the Klamath. Four months later, the committee delivered a draft report: There wasn't enough evidence, it concluded, to know exactly how much water was enough. Environmentalists argue that the document left out key studies of the river; the council says it is still working on a final report.

But if the committee's findings were ambivalent, the Bureau of Reclamation's response was anything but. As long as science hadn't proved that low flows would harm the salmon, the agency announced, farmers would receive their full allocation of water. Wildlife experts—especially those at the National Marine Fisheries Service, which is charged with protecting the salmon—objected, but they were overruled. In a ceremony on March 29, 2002, with national TV cameras in attendance and farmers chanting, "Let the waters flow," Norton herself helped open the Klamath headgate.

The bureau's allocation to the farmers was so generous that by summer, irrigation ditches in Klamath County were overflowing with water, prompting Oregon state officials to complain about flooding. By September, dead salmon were washing up on the Yurok reservation. And the following month, Michael Kelly—the National Marine Fisheries Service's lead biologist in the Klamath case—filed a claim for protection as a whistleblower, charging that his superiors had changed a key scientific assessment, known as a biological opinion, in response to "political pressure." Environmentalists are now challenging the biological opinion—the foundation for the Klamath irrigation plan—in court. Kelly isn't speaking to the press while the claim is pending.

Throughout the controversy, the Interior Department has argued that its policy on the Klamath is driven by science; if the science changes, says Wooldridge, so will the irrigation plan. But the department's stance also reflects Interior Secretary Norton's own long-standing philosophy. The secretary cut her teeth working for the Rocky Mountain Legal Foundation, a property-rights group in Colorado where her boss was Reagan's controversial Interior Secretary James Watt, who once proposed selling off public lands, including some national parks. As Colorado attorney general in the 1990s, Norton frequently challenged the federal government's land and water rights in her state. Among her first actions at Interior was to appoint Colorado attorney Bennett Raley, who has represented cities and farmers in water disputes, as the assistant secretary overseeing the Bureau of Reclamation.

"There's a new wind blowing out of Washington," says Janet Neuman, an environmental law professor at Lewis & Clark College who, in the '90s, served on a federal water commission charged with evaluating Bureau of Reclamation policies. "At one point, the bureau was trying to move away from being perceived as being captive to [private interests]. Now they are pulling back from that."

In Colorado, for example, the administration has announced that it will not insist on its right to keep water flowing in the Gunnison River—a move that would endanger the unique ecosystems of Black Canyon of the Gunnison National Park, but could allow the water to be diverted toward Denver's booming suburbs. And on the Rio Grande, Norton is backing the city of Albuquerque in its attempts to withhold water from the river, even if that means destroying the habitat of an endangered fish called the silvery minnow.

Similar conflicts are likely to erupt throughout the West, experts believe, as drought and booming development exacerbate the pressure on already overtaxed water systems. "The Klamath is looked at as symptomatic," says Neuman. "There but for the grace of God go many, many other basins where there has been decades of over-commitment of the water resources. Endangered species listings and irrigation demands and unmet tribal demands—those circumstances exist in lots of places around the West, and all it takes is a particularly low water year to bring them to a head. It's just a matter of how low, and how soon."

"THERE—IN THAT TREE," says Bob Hunter. "Northern harrier." The marsh hawk perches in a bare cottonwood tree, scowling at a flock of bufflehead ducks bobbing on the marsh. A flock of white swans flaps overhead, their long necks slanted like 737s at takeoff.

Hunter is a lawyer for the Oregon environmental group Water Watch, and today he is guiding me through some of the most bizarre wetlands in the nation's wildlife refuge system. "What you're looking at is the only wildlife refuge in America that grows potatoes and hay," he says as we drive down a gravel levee road. To our left is the water of Tule Lake. To our right is an overwintering alfalfa field. Both are within the refuge's borders.

The peculiar setup goes back to the days of Teddy Roosevelt, who created both the National Wildlife Refuge System and the Bureau of Reclamation. The Klamath Basin became their battlefield. Reclamation engineers wanted it for farmland. Conservationists claimed it as a bird sanctuary. Over the years a compromise emerged that allowed local farmers to lease about one-third of the land in the refuges at Lower Klamath Lake and Tule Lake. To grow crops, they need irrigation water—the "excess" water that would otherwise keep the wetlands wet.

For years, environmentalists have argued for an end to farming in the refuge: "If you take those 32,000 acres of lease land out of the irrigation loop," says Hunter, "that's more than

15 percent of the Klamath Project put back into the river—at no cost to the government." In the '90s, the refuge's manager began doing just that, withholding water from the lease lands in drought years. But the Bush administration scrapped that policy last year.

If you imagine the Pacific flyway as an hourglass stretching from Alaska to South America, the Klamath marshes sit at its waist; 80 percent of all the migratory birds in the West stop here at some point on their journeys. "We don't really know if there's any other place in the United States that has the same significance for wildlife," says Wendell Wood, southern Oregon field representative for the Oregon Natural Resources Council. But since irrigation began a century ago, the number of birds in the basin has dropped by 90 percent, and those that remain—including nearly 1,000 bald eagles—find their breeding grounds drained and polluted. During the summer, many of the wetlands near Tule Lake turn into fields of cracked mud, even as the alfalfa fields next to them sprout a lush crop. At least 46 different insect- and weed-killing chemicals are regularly applied to the fields in the refuge, sometimes mixed in with irrigation water in a process called "chemigation"; the Bureau of Reclamation itself uses several toxic herbicides to keep irrigation canals weed-free. The refuge's longtime manager, Phil Norton, once said that when he first came to Tule Lake, he "couldn't believe they called it a wildlife refuge."

NOT FAR FROM THE REFUGE'S DUCK PONDS, on about 600 acres of what used to be Tule Lake, John Anderson grows alfalfa and mint. His father, Robert, won the land in a 1947 government lottery that offered World War II veterans the chance to be America's last homesteaders. "This was a thriving community back then," recalls Anderson, now 50. "Lot of neighbors, lot of kids around. Prices were better. Lot of folks got out of the business since then. At my church I'm one of the only farmers younger than 75."

Sure enough, a stroll down Tulelake's dusty streets reveals a town whose decline began long before the 2001 water crisis. The bowling alley is shuttered, and the hardware store closes all winter. The only signs of life come from a small grocery store, a county ag extension office, and Tulelake High School.

Years ago, subsidizing water-intensive crops like potatoes in a region that gets only 18 inches of rain each year seemed like an efficient use of federal resources. But, John Anderson says, farmers can tell that the winds have shifted. "Americans have changed their priorities," he says. "Now they want rivers, wetlands, clean water, wildlife. I can understand. But the American people should be willing to pay for it."

Anderson is among a growing number of farmers who support a seemingly simple solution to the water dilemma: What if the government paid some of the farmers to quit irrigating? A buyout of, say, 30 percent of the Klamath Projects' 200,000 acres of irrigated land, at roughly $2,500 an acre, would cost $150 million. The scheme would remove a lot of claims on the river and could leave enough water for both the fish and the remaining farmers. Variations of the plan have been floated by numerous conservation groups and some members of Oregon's congressional delegation.

When it comes to water in the West, it's appropriate to borrow from Faulkner: The past isn't dead here. It's not even past.

But when it comes to water in the West, it's appropriate to borrow from Faulkner: The past isn't dead here. It's not even past. Decades-old contracts and double crosses are recalled as if they happened last week. A century ago, farmers in California's Owens Valley took a water buyout and saw the lifeblood of their valley diverted to Los Angeles. Roman Polanski immortalized the scam in *Chinatown*. "Some of the Klamath pioneers," says Dan Keppen, head of the organization that represents the valley's irrigators, "moved here from the Owens Valley."

Keppen's group, the Klamath Water Users Association, is dead set against a buyout, which it says would simply usher in the end of farming in the basin. Last year the group successfully lobbied to block a $175 million congressional aid package for the Klamath because some of the money could have been used to buy out farms. National property-rights groups, who oppose turning private land into public property, have resisted the idea, and so has the Bush administration. As one official privately notes, Secretary Norton feels that the federal land portfolio "is quite large enough, thank you very much."

AS IT LEAVES the Klamath plateau, the river drops into Northern California's woolly Siskiyou country, home to Bigfoot sightings, marijuana patches, off-the-grid rednecks, and long-toothed hippies. State Highway 96 follows its course past a string of abandoned mines, through former mill towns that now get by on fly-fishing and rafting tours. As the Klamath Mountains segue into the Coast Range, moist Pacific air creeps up the river valley in cottony mists. Moss overcoats wrap around trunks of toyon, the California holly that inspired the name Hollywood. Strengthened by dozens of winter streams, the Klamath throws up rapids and surfable waves that draw kayakers from hundreds of miles away. At the village of Weitchpec (Witchpeck), the green Trinity River joins the muddy Klamath, and the combined channel veers away from the highway. A one-lane road on the Hoopa Indian Reservation continues to shadow the river until finally it, too, gives out and the Klamath rolls on through the woods, for the first time truly wild.

But the water wars don't stop when you leave the river. A few miles from the Klamath-Trinity confluence, I knocked at a house that had a sign posted in its yard. "DYING 4 WATER," it said, over a drawing of a salmon. Duane Sherman Sr., the 33-year-old former Hoopa tribal chairman, invited me in.

"I'm dividing up this deer we killed a week ago," Sherman said, offering me a bite of venison jerky. "You see, Indians are nothing but extended family. This deer feeds my grandmother, my sister, my aunt," he said, nodding to three women chatting at the kitchen table, "and my nephews too." Two boys watched cartoons in the next room. "The salmon's the same way. Those fish aren't just our livelihood. If we don't fish, we don't eat."

Most of the 33,000 fish lost in last September's salmon kill were headed upriver toward the Trinity, where the Hoopa have fishing rights. (Their Yurok neighbors have rights along the lower 40 miles of the Klamath, from the mouth to the Trinity confluence.) Which means there's a lot more deer than salmon on the Sherman family table these days.

For more than a decade the Hoopa, along with the Yurok, state and municipal governments, and landowners up and down the river, have been working to restore the Klamath fishery, once among the most productive on the West Coast. Last year was supposed to be the first season it all paid off. Instead, the salmon kill left local communities with what one study estimates will be about $20 million in losses, roughly as much as the farmers lost during the water crisis of 2001.

"I understand that the irrigators have a contract with the government," Susan Masten, the Yurok tribal chair, tells me over a plate of eggs at Sis' diner, near the mouth of the Klamath. "The government also has a contract with us. And it was the first contract."

This is Masten's second water war. The Yuroks' federal agreements, which date back to the mid-19th century, guarantee them access fish in perpetuity. But during the 1970s, tribal members battled the federal government and commercial fishermen for the right to set their nets. Federal agents with M-16s and riot gear occupied the reservation to keep tribal fishermen out of the river. A popular bumper sticker off the reservation read, "Can an Indian, Save a Salmon."

But over the years, the commercial fishermen have become the Yuroks' strongest allies. "After years of fighting with the tribes, we looked at each other and realized we had fewer and fewer fish to fight over," says Glen Spain, the Northwest regional director of the Pacific Coast Federation of Fishermen's Associations. "It's the same thing with the farmers. We're all workaday people trying to make a living. We understand the market forces driving the farmers down."

Like the farm towns farther upstream, the fishing communities on this stretch of coast are dotted with "For Sale" signs and shuttered family businesses. "From Fort Bragg, California, to Coos Bay, Oregon—the range of Klamath River salmon—we've lost 3,700 jobs and an $80 million-a-year economy," says Spain.

But the fishermen haven't been cast as victims in this crisis—not by the media, and not by the government. Their decline has come slowly, invisibly; they lack the ag industry's national lobbying muscle, or the property-rights movement's clout with the Bush administration. They are no more a political force than the ducks that land in the Tule Lake refuge.

SOME RIVERS ARE SO BIG you can't see them meet the sea. They simply fan out and merge with the tide. The amazing thing about the Klamath is that you can actually watch it pour into the Pacific. In Yurok country the river is a quarter-mile wide and quite; at its mouth it narrows into a frightening rush that cuts between a sandbar and a rocky cliff and then it's gone, lost in the foam. Standing there in the teeth of a January storm, watching a flock of gulls wheel in the cold wind, I thought of the classic closing line in *Chinatown*: "Forget it, Jake, it's Chinatown." Some things, it suggests, are simply too murky to understand.

But the Klamath water war comes down to a single, very simple equation: too many takers, not enough water.

"When I was a boy," John Anderson told me as we watched dusk settle over fields north of Tule Lake, "the ducks and geese came here by the millions. You could hear the flocks roar at night. We've given up a lot of the life that used to be here. Sometimes I question whether it's been a good trade-off."

It wasn't a wholly bad deal. A nation got fed, the West settled, families raised. But the math never did add up, and it's now becoming clear what got shortchanged in the bargain.

Additional reporting by Stephen Baxter.

Based in Seattle, **Bruce Barcott** is a contributing writer for *Outside*. In addition to this story about the battle over water rights on the Klamath River, Barcott covered the Pacific Northwest's salmon-fishing industry in "Aquaculture's Troubled Harvest" (November/December 2001).

HOW MUCH IS CLEAN WATER WORTH?

A lot, say researchers who are putting dollar values on
wildlife and ecosystems—and providing that conservation pays

Jim Morrison

The water that quenches thirsts in Queens and bubbles into bathtubs in Brooklyn begins about 125 miles north in a forest in the Catskill Mountains. It flows down distant hills through pastures and farmlands and eventually into giant aqueducts serving 9 million people with 1.3 billion gallons daily. Because it flows directly from the ground through reservoirs to the tap, this water—long regarded as the champagne of city drinking supplies—comes from what's often called the largest "unfiltered" system in the nation.

But that's not strictly true. Water percolating through the Catskills is filtered naturally—for free. Beneath the forest, fine roots and microorganisms break down contaminants. In streams, plants absorb nutrients from fertilizer and manure. And in meadows, wetlands filter nutrients while breaking down heavy metals.

New York City discovered how valuable these services were 15 years ago when a combination of unbridled development and failing septic systems in the Catskills began degrading the quality of the water that served Queens, Brooklyn and the other boroughs. By 1992, the U.S. Environmental Protection Agency (EPA) warned that unless water quality improved, it would require the city to build a filtration plant, estimated to cost between $6 and $8

billion and between $350 and $400 million a year to operate.

Instead, the city rolled the dice with nature in a historic experiment. Rather than building a filtration plant, officials decided to restore the health of the Catskills watershed, so it would do the job naturally.

What's this ecosystem worth to the city of New York? So far, $1.3 billion. That's what the city has committed to build sewage treatment plants upstate and to protect the watershed through a variety of incentive programs and land purchases. It's a lot of money. But it's a fraction of the cost of the filtration plant—a plant, city officials note, that wouldn't work as tirelessly or efficiently as nature.

"It was a stunning thing for the New York City council to think maybe we should invest in natural capital," says Stanford University researcher Gretchen Daily.

Daily is one of a growing number of academics—some from economics, some from ecology—who are putting dollar figures on the services that ecosystems provide. She and other "ecological economists" look not only at nature's products—food, shelter, raw materials—but at benefits such as clean water, clean air, flood control and storm mitigation, irreplaceable services that have been taken for granted throughout history. "Much of Mother Nature's labor has enormous and obvious

value, which has failed to win respect in the marketplace until recently," Daily writes in the book *The New Economy of Nature: The Quest to Make Conservation Profitable.*

Ecological economist Geoffrey Heal, a professor of public policy and business responsibility at Columbia University, became interested in the field as an economist who was concerned about the environment. "The idea of ecosystem services is an interesting framework for thinking why the environment matters," says Heal, author of *Nature and the Marketplace: Capturing the Value of Ecosystem Services.* "The traditional argument for environmental conservation had been essentially aesthetic or ethical. It was beautiful or a moral responsibility. But there are powerful economic reasons for keeping things intact as well."

Daily notes that beyond providing clean water, the Catskills ecosystem has value for its beauty, as wildlife habitat and for recreation, particularly trout fishing. Such values are not inconsequential. While no one has assessed the total worth of the watershed, even a partial look reveals that habitat and wildlife are powerful economic engines.

Restored habitat for trout and other game fish, for example, attracts fishermen, and angling is big business in this state. According to a report by the U.S. Fish and Wildlife Service (FWS), more than 1.5 million people fished in New York during 2001, yielding an economic benefit to the state of more than $2 billion and generating the equivalent of 17,468 full-time jobs and more than $164 million in state, federal, sales and motor fuel taxes. Though not as easily measured, individual Catskills species also have value. Beavers, for instance, create wetlands that are vital to filtering water and to biodiversity.

Ecological economists maintain that ecosystems are capital assets that, if managed well, provide a stream of benefits just as any investment does. The FWS report, for example, notes that 66 million Americans spent more than $38 billion in 2001 observing, feeding or photographing wildlife. Those expenditures resulted in more than a million jobs with total wages and salaries of $27.8 billion. The analysis found that birders alone spent an estimated $32 billion on wildlife watching that year, generating $85 billion of economic benefits. In Yellowstone National Park, the reintroduction of gray wolves that began in 1995 has already increased revenues in surrounding communities by $10 million a year, with total benefits projected to reach $23 million annually as more visitors come to catch a glimpse of these charismatic predators.

When it comes to water quality, EPA projects that the United States will have to spend $140 billion over the next 20 years to maintain minimum required standards for drinking water quality. No wonder, then, that 140 U.S. cities have studied using an approach similar to New York's. Under that agreement, finalized in 1997, the city promised to pay farmers, landowners and businesses that abided by restrictions designed to protect the watershed. (The city owns less than 8 percent of the land in the 2,000-square-mile watershed; the vast majority is in private hands.) "In the case of the Catskills, it was a matter of coming up with a way to reward the stewards of the natural asset for something they had been providing for free," Daily says. "As soon as they got paid even a little bit,

---

NATURE'S SERVICES

HOW THE EXPERTS CATEGORIZE THEM

• Ecosystem goods, the traditional measure of nature's products such as seafood, timber and agriculture
• Basic life support functions such as water purification, flood control, soil renewal and pollination
• Life-fulfilling functions, the beauty and inspiration we get from nature, including activities such as hiking and wildlife watching
• Basic insurance, the idea that nature's diversity contains something—like a new drug—the value of which isn't known today, but may be large in the future

---

they were much happier and inclined to go about their stewardship." There's no guarantee this experiment will work, of course; it may be another decade before the city finds out.

Elsewhere, other governing bodies are also recognizing the value of ecosystem services. The U.S. Army Corps of Engineers, for example, bought 8,500 acres of wetlands along Massachusetts' Charles River for flood control. The land cost $10 million, a tenth of the $100 million the Corps estimated it would take to build the dam and levee originally proposed. To fight floods in Napa, California, county officials spent $250 million to reconnect the Napa River to its historical floodplains, allowing the river to meander as it once did. The cost was a fraction of the estimated $1.6 billion that would have been needed to repair flood damage over the next century without the project. Within a year, notes Daily, flood insurance rates in the county dropped 20 percent and real estate prices rose 20 percent, thanks to the flood protection now promised by nature.

Even insects supply vital ecosystem services. More than 218,000 of the world's 250,000 flowering plants, including 70 percent of all species of food plants, rely on pollinators for

reproduction—and more than 100,000 of these pollinators are invertebrates, including bees, moths, butterflies, beetles and flies. Another 1,000 or more vertebrate species, including birds, mammals and reptiles, also pollinate plants. According to University of Arizona entomologist Stephen Buchmann, author of *The Forgotten Pollinators,* one of every three bites of food we eat comes courtesy of a pollinator.

A Cornell University study estimated the value of pollination by honeybees in the United States alone at $14.6 billion in 2000. Yet honeybee populations are dropping everywhere, as much as 25 percent since 1990, according to one study. Now many farms and orchards are paying to have the bees shipped in.

Today's interest in assigning dollar values to pollination and other ecosystem services was spawned by publication of a controversial 1997 report in *Nature* that estimated the total global contribution of ecosystems to be $33 trillion or more each year—roughly double the combined gross national product of all countries in the world. The study became a lightning rod. Detractors scoffed at the idea that one could put a dollar value on something people weren't willing to purchase. One report by researchers at the University of Maryland, Bowden College and Duke University called the estimate "absurd," noting that if taken literally, the figure suggests that a family earning $30,000 annually would pay $40,000 annually for ecosystem protection.

Other researchers, including Daily and Heal, charged that the $33 trillion figure greatly underestimates nature's value. "If you believe, as I do, that ecosystem services are necessary for human survival, they're invaluable really," Heal says. "We would pay anything we could pay."

Daily doesn't believe the absolute value of an ecosystem can ever be measured. Heal agrees, yet both scientists say that pricing ecosystem services is an important tool for making decisions about nature—and for making the case for conservation.

NATURAL CAPITAL

WHAT'S THE ANNUAL DOLLAR VALUE OF ...?

- Recreational saltwater fishing in the United States: $20 billion
- Wild bee pollinators to a single coffee farm in Costa Rica: $60,000
- Tourism to view bats in the city of Austin, Texas: $8 million
- Wildlife watching in the United States: $85 billion
- U.S. employment income generated by wildlife watching: $27.8 billion
- State and federal tax revenues from wildlife watching: $6.1 billion
- Natural pest control services by birds and other wildlife to U.S. farmers: $54 billion

"Valuation is just one step in the broader politics of decision making," she says. "We need to be creative and innovative in changing social institutions so we are aligning economic forces with conservation."

Indeed, as dollar values for nature's services become available, environmentalists increasingly use them to bolster arguments for conservation. One high-profile example is the contentious dispute over whether to tear down four dams on the lower Snake River in southeastern Washington to restore salmon habitat, and thus the region's lucrative salmon fishery. Ed Whitelaw, a professor of economics at the University of Oregon, notes that estimates of the economic impact of breaching the dams range from $300 million in net costs to $1.3 billion in net benefits, largely due to the wide range of projections about recreational spending.

A 2002 report by the respected, nonprofit think-tank RAND Corporation concluded the dams could be breached without hurting economic growth and employment. Energy lost as a result of the breaches could be replaced with more efficient sources, including natural gas, resulting in 15,000 new jobs. Further, the report noted that recreation, retail, restaurants and real estate would experience a marked growth. Recreational activities alone would increase by an estimated $230 million over 20 years.

There's no question that returning the salmon runs would have a major impact on the region. When favorable ocean conditions increased the runs in 2001, Idaho's Department of Fish and Game estimated the salmon season that year alone generated more than $90 million of revenue in the state, most of it in rural communities that badly needed the funds.

"Some people think it sounds crass to put a price tag on something that's invaluable, careening down the slippery slope of the market economy," says Daily. "In fact, the idea is to do something elegant but tricky: to finesse the economic system, the system that drives so much of our individual and collective behavior, so that without even thinking it makes natural sense to invest in and protect our natural assets, our ecosystem capital."

What Daily and other ecological economists want is to insinuate consideration of ecosystem services into daily decision making, whether it takes the form of financial incentive or penalty. "At a practical level, decisions are made at the margin, not at the 'should we sterilize the Earth' level," she says. "It's in all the little decisions—whether to farm here or leave a few trees, whether to build the shopping mall there or leave the wetland, whether to buy an SUV or a Prius—that ecosystem service values need to be incorporated."

Heal agrees. "Although ecosystem services have been with us for millennia," he says, "the scale of human activity is now sufficiently great that we can no longer take their continuation for granted."

*Virginia journalist **JIM MORRISON** wrote about polar bears and global warming in the February/March 2004 issue.*

A **Human** THIRST

Humans now appropriate more than half of all the freshwater in the world.
Rising demands from agriculture, industry, and a growing population have
left important habitats around the world high and dry.

by Don Hinrichsen

On March 20, 2000, a group of monkeys, driven mad with thirst, clashed with desperate villagers over drinking water in a small outpost in northern Kenya near the border with Sudan. The Pan African News Agency reported that eight monkeys were killed and 10 villagers injured in what was described as a "fierce two-hour melee." The fight erupted when relief workers arrived and began dispensing water from a tanker truck. Locals claimed that a prolonged drought had forced animals to roam out of their natural habitats to seek life-giving water in human settlements. The monkeys were later identified as generally harmless vervets.

The world's deepening freshwater crisis—currently affecting 2.3 billion people—has already pitted farmers against city dwellers, industry against agriculture, water-rich state against water-poor state, county against county, neighbor against neighbor. Inter-species rivalry over water, such as the incident in northern Kenya, stands to become more commonplace in the near future.

"The water needs of wildlife are often the first to be sacrificed and last to be considered," says Karin Krchnak, population and environment program manager at the National Wildlife Federation (NWF) in Washington, D.C. "We ignore the fact that working to ensure healthy freshwater ecosystems for wildlife would mean healthy waters for all." As more and more water is withdrawn from rivers, streams, lakes and aquifers to feed thirsty fields and the voracious needs of industry and escalating urban demands, there is often little left over for aquatic ecosystems and the wealth of plants and animals they support.

The mounting competition for freshwater resources is undermining development prospects in many areas of the world, while at the same time taking an increasing toll on natural systems, according to Krchnak, who co-authored an NWF report on population, wildlife, and water. In effect, humanity is waging an undeclared water war with nature.

"There will be no winners in this war, only losers," warns Krchnak. By undermining the water needs of wildlife we are not just undermining other species, we are threatening the human prospect as well.

Pulling Apart the Pipes

Currently, humans expropriate 54 percent of all available freshwater from rivers, lakes, streams, and shallow aquifers. During the 20th century water use increased at double the rate of population growth: while the global population tripled, water use per capita increased by six times. Projected levels of population growth in the next 25 years alone are expected to increase the human take of available freshwater to 70 percent, according to water expert Sandra Postel, Director of the Global Water Policy Project in Amherst, Massachusetts. And if per capita water consumption continues to rise at its current rate, by 2025 that share could significantly exceed 70 percent.

As a global average, most freshwater withdrawals—69 percent—are used for agriculture, while industry accounts for 23 percent and municipal use (drinking water, bathing and cleaning, and watering plants and grass) just 8 percent.

The past century of human development—the spread of large-scale agriculture, the rapid growth of industrial development, the construction of tens of thousands of large dams, and the growing sprawl of cities—has profoundly altered the Earth's hydrological cycle. Countless rivers, streams, floodplains, and wetlands have been dammed, diverted, polluted, and filled. These components of the hydrological cycle, which function as the Earth's plumbing system, are being disconnected and plundered, piece by piece. This fragmentation has been so extensive that freshwater ecosystems are perhaps the most severely endangered today.

Left High and Dry

Habitat destruction, water diversions, and pollution are contributing to sharp declines in freshwater biodiversity. One-fifth of all freshwater fish are threatened or extinct. On continents where studies have been done, more than half of amphibians are in decline. And more than 1,000 bird species—many of them aquatic—are threatened.

More than 40,000 large dams bisect waterways around the world, and more than 500,000 kilometers of river have been dredged and channelized for shipping. Deforestation, mining, grazing, industry, agriculture, and urbanization increase pollution and choke freshwater ecosystems with silt and other runoff.

Water diversion for irrigation, industry, and urban use has increased 35-fold in the past 300 years. In some cases, this increased demand has deprived entire ecosystems of water. Sprawl is an increasing concern, as the spread of urban areas is destroying important wetlands, and paved-over area is reducing the amount of water that is able to recharge aquifers.

Consider the plight of wetlands—swamps, marshes, fens, bogs, estuaries, and tidal flats. Globally, the world has lost half of its wetlands, with most of the destruction having taken place over the past half century. The loss of these productive ecosystems is doubly harmful to the environment: wetlands not only store water and transport nutrients, but also act as natural filters, soaking up and diluting pollutants such as nitrogen and phosphorus from agricultural runoff, heavy metals from mining and industrial spills, and raw sewage from human settlements.

In some areas of Europe, such as Germany and France, 80 percent of all wetlands have been destroyed. The United States has lost 50 percent of its wetlands since colonial times. More than 100 million hectares of U.S. wetlands (247 million acres) have been filled, dredged, or channeled—an area greater than the size of California, Nevada, and Oregon combined. In California alone, more than 90 percent of wetlands have been tilled under, paved over, or otherwise destroyed.

Destruction of habitat is the largest cause of biodiversity loss in almost every ecosystem, from wetlands and estuaries to prairies and forests. But biologists have found that the brunt of current plant and animal extinctions have fallen disproportionately on those species dependent on freshwater and related habitats. One fifth of the world's freshwater fish—2,000 of the 10,000 species identified so far—are endangered, vulnerable, or extinct. In North America, the continent most studied, 67 percent of all mussels, 51 percent of crayfish, 40 percent of amphibians, 37 percent of fish, and 75 percent of all freshwater mollusks are rare, imperiled, or already gone.

The global decline in amphibian populations may be the aquatic equivalent of the canary in the coal mine. Data are scarce for many species, but more than half of the amphibians studied in Western Europe, North America, and South America are in a rapid decline.

Around the world, more than 1,000 bird species are close to extinction, and many of these are particularly dependent on wetlands and other aquatic habitats. In Mexico's Sonora Desert, for instance, agriculture has siphoned off 97 percent of the region's water resources, reducing the migratory bird population by more than half, from 233,000 in 1970 to fewer than 100,000 today.

Pollution is also exacting a significant toll on freshwater and marine organisms. For instance, scientists studying beluga whales swimming in the contaminated St. Lawrence Seaway, which connects the Atlantic Ocean to North America's Great Lakes, found that the cetaceans have dangerously high levels of PCBs in their blubber. In fact the contamination is so severe that under Canadian law the whales actually qualify as toxic waste.

Waterways everywhere are used as sewers and waste receptacles. Exactly how much waste ends up in freshwater systems and coastal waters is not known. However, the UN Food and Agriculture Organization (FAO) estimates that every year roughly 450 cubic kilometers (99 million gallons) of wastewater (untreated or only partially treated) is discharged into rivers, lakes, and coastal areas. To dilute and transport this amount of waste requires at least 6,000 cubic kilometers (1.32 billion gallons) of clean water. The FAO estimates that if current trends continue, within 40 years the world's entire stable river flow would be needed just to dilute and transport humanity's wastes.

The Point of No Return?

The competition between people and wildlife for water is intensifying in many of the most biodiverse regions of the world. Of the 25 biodiversity hotspots designated by Conservation International, 10 are located in water-short regions. These regions—including Mexico, Central America, the Caribbean, the western United States, the Mediterranean Basic, southern Africa, and southwestern China—are home to an extremely high number of endemic and threatened species. Population pressures and overuse of resources, combined with critical water shortages, threaten to push these diverse and vital ecosystems over the brink. In a number of cases, the point of no return has already been reached.

China

China, home to 22 percent of the world's population, is already experiencing serious water shortages that threaten both people and wildlife. According to China's former environment minister, Qu Geping, China's freshwater supplies are capable of sustainably supporting no more than 650 million people—half its current population. To compensate for the tremendous shortfall, China is draining its rivers dry and mining ancient aquifers that take thousand of years to recharge.

As a result, the country has completely overwhelmed its freshwater ecosystems. Even in the water-rich Yangtze River Basin, water demands from farms, industry, and a giant population have polluted and degraded freshwater and riparian ecosystems. The Yangtze is one of the longest rivers in Asia, winding

6,300 kilometers on its way to the Yellow Sea. This massive watershed is home to around 400 million people, one-third of the total population of China. But the population density is high, averaging 200 people per square kilometer. As the river, sluggish with sediment and laced with agricultural, industrial, and municipal wastes, nears its wide delta, population densities soar to over 350 people per square kilometer.

The effects of the country's intense water demands, mostly for agriculture, can be seen in the dry lake beds on the Gianghan Plain. In 1950 this ecologically rich area supported over 1,000 lakes. Within three decades, new dams and irrigation canals had siphoned off so much water that only 300 lakes were left.

China's water demands have taken a huge toll on the country's wildlife. Studies carried out in the Yangtze's middle and lower reaches show that in natural lakes and wetlands still connected to the river, the number of fish species averages 100. In lakes and wetlands cut off and marooned from the river because of diversions and drainage, no more than 30 survive. Populations of three of the Yangtze's largest and more productive fisheries—the silver, bighead, and grass carp—have dropped by half since the 1950s.

Mammals and reptiles are in similar straits. The Yangtze's shrinking and polluted waters are home to the most endangered dolphin in the world—the Yangtze River dolphin, or Baiji. There are only around 100 of these very rare freshwater dolphins left in the wild, but biologists predict they will be gone in a decade. And if any survive, their fate will be sealed when the massive Three Gorges Dam is completed in 2013. The dam is expected to decrease water flows downstream, exacerbate the effects of pollution, and reduce the number of prey species that the dolphins eat. Likewise, the Yangtze's Chinese alligators, which live mostly in a small stretch near the river's swollen, silt-laden mouth, are not expected to survive the next 10 years. In recent years, the alligator population has dropped to between 800 and 1,000.

The Aral Sea

The most striking example of human water demands destroying an ecosystem is the nearly complete annihilation of the 64,500 square kilometer Aral Sea, located in Central Asia between Kazakhstan and Uzbekistan. Once the fourth largest inland sea in the world, it has contracted by half its size and lost three-quarters of its volume since the 1960s, when its two feeder rivers—the Amu Darya and the Syr Darya—were diverted to irrigate cotton fields and rice paddies.

The water diversions have also deprived the region's lakes and wetlands of their life source. At the Aral Sea's northern end in Kazakhstan, the lakes of the Syr Darya delta shrank from about 500 square kilometers to 40 square kilometers between 1960 and 1980. By 1995, more than 50 lakes in the Amu Darya delta had dried up and the surrounding wetlands had withered from 550,000 hectares to less than 20,000 hectares.

The unique *tugay* forests—dense thickets of small shrubs, grasses, sedges and reeds—that once covered 13,000 square kilometers around the fringes of the sea have been decimated. By

Alien Invaders

"Rapidly growing populations place heavy demand on freshwater resources and intensify pressures on wildlands," concludes a combined World Resources Institute and Worldwatch report called "Watersheds of the World." But increasingly, the introduction of exotic or alien species is playing a large role in wreaking havoc on freshwater habitats.

The spread of invasive species is a global phenomenon, and is increasingly fostered by the growth of aquaculture, shipping, and commerce. Whether introduced by accident or on purpose, these alien invaders are capable of altering habitats and extirpating native species en masse.

The invasion and insidious spread of the zebra mussel in the U.S. Great Lakes highlights the tremendous costs to ecosystems and species. A native of Eastern Europe, the zebra mussel arrived in the Great Lakes in 1988, released most likely through the discharge of ballast waters from a cargo ship. Once established, it spread rapidly throughout the region.

The mussels have crowded out native species that cannot compete with them for space and food. A study of the mussels in western Lake Erie found that all of the native clams at each of 17 sampling stations had been wiped out. Moreover, the last known population of the winged maple leaf clam, found in the St. Croix River in the upper Mississippi River basin, is now threatened by advancing ranks of the zebra mussel.

1999 less than 1,000 square kilometers of fragmented and isolated forest remained.

The habitat destruction has dramatically reduced the number of mammals that used to flourish around the Aral Sea: of 173 species found in 1960, only 38 remained in 1990. Though the ruined deltas still attract waterfowl and other wetland species, the number of migrant and nesting birds has declined from 500 species to fewer than 285 today.

Plant life has been hard hit by the increase in soil salinity, aridity, and heat. Forty years ago, botanists had identified 1,200 species of flowering plants, including 29 endemic species. Today, the endemics have vanished. The number of plant species that can survive the increasingly harsh climate is a fraction of the original number.

Most experts agree that the sea itself may very well disappear entirely within two decades. But the region's freshwater habitats and related communities of plants and animals have already been consigned to oblivion.

Lake Chad

Lake Chad, too, has shrunk—to one-tenth of its former size. In 1960, with a surface area of 25,000 square kilometers, it was the second-largest lake in Africa. When last surveyed, it was down to only 2,000 square kilometers. And here, too, massive water withdrawals from the watershed to feed irrigated agriculture have reduced the amount of water flowing into the lake to a trickle, especially during the dry season.

Lake Chad is wedged between four nations: populous Nigeria to the southwest, Niger on the northwest shore, Chad to the northeast, and Cameroon on a small section of the south shore. Nigeria has the largest population in Africa, with 130 million inhabitants. Population-growth rates in these countries average 3 percent a year, enough to double human numbers in one generation. And population growth rates in the regions around the lake are even higher than the national averages. People gravitate to this area because the lake and its rivers are the only sources of surface water for agricultural production in an otherwise dry and increasingly desertified region.

Although water has been flowing into the lake from its rivers over the past decade, the lake is still in serious ecological trouble. The lake's fisheries have more or less collapsed from over-exploitation and loss of aquatic habitats as its waters have been drained away. Though some 40 commercially valuable species remain, their populations are too small to be harvested in commercial quantities. Only one species—the mudfish—remains in viable populations.

As the lake has withered, it has been unable to provide suitable habitat for a host of other species. All large carnivores, such as lions and leopards, have been exterminated by hunting and habitat loss. Other large animals, such as rhinos and hippopotamuses, are found in greatly reduced numbers in isolated, small populations. Bird life still thrives around the lake, but the variety and numbers of breeding pairs have dropped significantly over the past 40 years.

A Blue Revolution

As these examples illustrate, the challenge for the world community is to launch a "blue revolution" that will help governments and communities manage water resources on a more sustainable basis for all users. "We not only have to regulate supplies of freshwater better, we need to reduce the demand side of the equation," says Swedish hydrologist Malin Falkenmark, a senior scientist with Sweden's Natural Science Research Council. "We need to ask how much water is available and how best can we use it, not how much do we need and where do we get it." Increasingly, where we get it from is at the expense of aquatic ecosystems.

If blindly meeting demand precipitated, in large measure, the world's current water crisis, reducing demand and matching supplies with end uses will help get us back on track to a more equitable water future for everyone. While serious water initiatives were launched in the wake of the World Summit on Sustainable Development held in Johannesburg, South Africa, not one of them addressed the water needs of ecosystems.

There is an important lesson here: just as animals cannot thrive when disconnected from their habitats, neither can humanity live disconnected from the water cycle and the natural systems that have evolved to maintain it. It is not a matter of "either or" says NWF's Krchnak. "We have no real choices here. Either we as a species live within the limits of the water cycle and utilize it rationally, or we could end up in constant competition with each other and with nature over remaining supplies. Ultimately, if nature loses, we lose."

By allowing natural systems to die, we may be threatening our own future. After all, there is a growing consensus that natural ecosystems have immense, almost incalculable value. Robert Costanza, a resource economist at the University of Maryland, has estimated the global value of freshwater wetlands, including related riverine and lake systems, at close to $5 trillion a year. This figure is based on their value as flood regulators, waste treatment plants, and wildlife habitats, as well as for fisheries production and recreation.

The nightmarish scenarios envisioned for a water-starved not too distant future should be enough to compel action at all levels. The water needs of people and wildlife are inextricably bound together. Unfortunately, it will probably take more incidents like the one in northern Kenya before we learn to share water resources, balancing the needs of nature with the needs of humanity.

Don Hinrichsen *is a UN consultant. He is former editor-in-chief of* Ambio *and was a news correspondent in Europe for 15 years.*

From *World Watch*, January/February 2003. © 2003 by Worldwatch Institute, www.worldwatch.org.

UNIT 6

The Hazards of Growth: Pollution and Climate Change

Unit Selections

Key Points to Consider

- Why are agricultural pesticides that have been banned in the United States and other developed countries still used in the lesser-developed regions of the world? Do the benefits of increased food production outweigh the costs of using agricultural pesticides?

- How can monitoring of chemical concentrations in human tissues give clues as to environmental quality? Should such monitoring become a regular part of the health care system?

- What are some of the potential relationships between the beginnings of agriculture and the first human alterations of global climates? Is there a connection between the possible climate-altering practices of farmers thousands of years ago and the potential global climate change of the present and future?

- Since carbon dioxide is the chief atmospheric component in the process of enhancing the global greenhouse effect, what new ways might be developed to remove a portion of this gas from the atmosphere? What are the benefits and potential risks of some of the suggested new technologies of carbon sequestration?

Student Website

www.mhcls.com/online

Internet References

Further information regarding these websites may be found in this book's preface or online.

Persistent Organic Pollutants (POP)
http://www.chem.unep.ch/pops/

School of Labor and Industrial Relations (SLIR): Hot Links
http://www.lir.msu.edu/hotlinks/

Space Research Institute
http://arc.iki.rssi.ru/eng/index.htm

Worldwatch Institute
http://www.worldwatch.org

Of all the massive technological changes that have combined to create our modern industrial society, perhaps none has been as significant for the environment as the chemical revolution. The largest single threat to environmental stability is the proliferation of chemical compounds for a nearly infinite variety of purposes, including the universal use of organic chemicals (fossil fuels) as the prime source of the world's energy systems. The problem is not just that thousands of new chemical compounds are being discovered or created each year, but that their long-term environmental effects are often not known until an environmental disaster involving humans or other living organisms occurs. The problem is exacerbated by the time lag that exists between the recognition of potentially harmful chemical contamination and the cleanup activities that are ultimately required.

A critical part of the process of dealing with chemical pollutants is the identification of toxic and hazardous materials, a problem that is intensified by the myriad ways in which a vast number of such materials, natural and man-made, can enter environmental systems. Governmental legislation and controls are important in correcting the damages produced by toxic and hazardous materials such as DDT, PCBs, or CFCs; in limiting fossil fuel burning; or in preventing the spread of living organic hazards such as pests and disease-causing agents. Unfortunately, as evidenced by most of the articles in this unit, we are losing the battle against harmful substances regardless of legislation, and chemical pollution of the environment is probably getting worse rather than better.

The first article in this unit deals with one of the most serious of all international pollution problems: that of soil and water pollution from the widespread application of agricultural pesticides. In "Agricultural Pesticides in Developing Countries" Sylvia Karlsson of Yale University's Center for Environmental Law and Policy notes that it was the use of agricultural pesticides in North America and Europe, along with their undesired and unanticipated side effects, that produced the first alerts (largely through Rachel Carson's *Silent Spring*) that modern society can have major impacts on Earth's ecological systems. While many of the most dangerous chemicals have been banned in countries like the United States, their benefits in terms of increased food production outweighs their perceived dangers in countries in the lesser-developed world. As a consequence, increases rather than decreases in biological pollution from pesticides have been noted in Africa and Asia, posing massive challenges for governments that now, whether they like it or not, operate in a large, interconnected system.

In "A Little Rocket Fuel with Your Salad?" science writer Gene Ayres begins with a brief description of the California reaction to the appearance of an insect pest that might damage crops in the state's important agribusiness areas—full coverage spraying with a toxic chemical, malathion. This introduction provides a segue into his real story, the dumping into groundwater and wastewater systems of perchlorates, a primary component of rocket fuel, by both industrial and government entities. Perchlorates are persistent chemicals—that is they stay within the envi-

ronmental system for a long time—that eventually end up, in concentrated form, in crops irrigated with water that has been contaminated. Consumption of food items containing perchlorates is damaging to organic systems, including humans, and often causes significant organic damage.

The section's next two articles also deal with the unwanted injection of harmful substances into the environment—but at the other end of the scale. Where the two previous articles dealt with local contamination, these selections deal with global contamination and what it might mean at all scales from local to worldwide. In "How Did Humans First Alter Global Climate," climate historian William Ruddiman ponders whether the recent acknowledgement that human activities over the last century or two have had a warming effect on the global climate goes far enough. Ruddiman suggests that farming practices begun as early as 8000 years ago began increasing global carbon dioxide and contributing to a global warming trend that began slowly and, as industrialization increased the use of fossil fuels and as human populations increased, gathered momentum to the point where it became noticeable over the last few decades. Ruddiman suggests anomalies in atmospheric carbon dioxide that are large enough to have prevented the world from entering a new glacial state several thousand years ago. However, the warming trend of the past—which has proven generally beneficial for humankind—now runs the risk of accelerating into a course of increasing global temperatures that will lead us into completely unknown territory. The fourth article in this section also deals with the issue of global warming produced by pollution. In "Can We Bury Global Warming?" Robert H. Socolow of Princeton University acknowledges the fact of global warming as a function of an overabundance of atmospheric carbon dioxide. For Socolow, an engineer, problems beg solutions and he advances a novel solution to the world's most important pollution problem: pump carbon dioxide deep into the ground in order to prevent it from entering the atmosphere. While this novel solution is feasible, its feasibility depends upon a series of key challenges: most important are the mechanisms of carbon dioxide capture and carbon dioxide storage. While it may be economically impossible to capture the released carbon dioxide from an automobile, capturing it from an existing industrial plant or power-generating facility may prove more workable. Once it is captured, then the challenge becomes one of storage. The best destination seems to be deep underground formations of sedimentary rock loaded with pores now filled with salty water. These formations must be deep enough to eliminate contamination of drinking or irrigation water, preventing the re-release of CO_2 back into the atmosphere. Socolow contends that these challenges can be overcome; indeed, he concludes, they must be overcome if we are to reduce the risk of global warming.

In the final article of the section, we move to a combined pollution scale: global pollution that concentrates naturally in localized areas. Similar to the mechanisms that moved chlorofluorocarbons into the Earth's ozone layer but concentrated it in polar areas, atmospheric and oceanic dynamics are currently moving industrial pollutants from the world's factories (most of them concentrated in the northern hemisphere) into the food chain of the world's high latitude oceans—especially the Arctic. Native inhabitants of the Arctic, with a diet dependent upon the resources of the sea, have long been among the world's healthiest people. But that is changing as pollutants—PCBs, chlorinated pesticides such as DDT, mercury, and other toxic substances—are moved by ocean currents into the northern seas where they enter the systems of marine animals and, eventually, of those animals at the top of the food chain: humans. If there were ever the perfect object lesson as to how events in one part of the world impact peoples half a world away, the increasingly toxic diet of Arctic populations is that lesson. Pollution problems such as the increasingly toxic Arctic Ocean might appear nearly impossible to solve. But solutions do exist: massive cleanup campaigns to remove existing harmful chemicals from the environment and to severely restrict their future use; strict regulation of the production, distribution, use, and disposal of potentially hazardous chemicals; the development of sound biological techniques to replace existing uses of chemicals for such purposes as pest control; the adoption of energy and material resource conservation policies; and the use of more conservative and protective agricultural and construction practices. We now possess the knowledge and the tools to ensure that environmental cleanup is carried through. It will not be an easy task, and it will be terribly expensive. It will also demand a new way of thinking about humankind's role in the environmental systems upon which all life forms depend. If we do not complete the task, however, the support capacity of the environment may be damaged or diminished beyond our capacities to repair it. The consequences would be fatal for all who inhabit this planet.

AGRICULTURAL PESTICIDES IN DEVELOPING COUNTRIES

A Multilevel Governance Challenge

By Sylvia I. Karlsson

The use of agricultural pesticides in North America and Europe and their non-desired side effects were among practices and subsequent effects that set alarm bells ringing—bells that, in the 1960s and 1970s, awakened the public and, eventually, government agencies to the impact our modern human life can have on the Earth's life-sustaining ecological systems. For the past 30-40 years in developing countries, agricultural pesticide use has set off a continuously ringing alarm, alerting us to a heavy toll on human health and the environment. It is an issue that clearly demonstrates links—between the local and the global and between developed and developing countries—often associated with environmental and societal problems. This creates a complex picture with many challenges to address in terms of governance.[1]

Following the trajectory of modern agriculture in developed countries, developing countries have, over the past half century, increasingly adopted a pest management approach that centers on the use of chemical pesticides. The pesticide world market's total value surpasses US$26 billion. Developing countries' share of this is approximately one-third.[2] It is reasonable to expect that developing countries will continue to experience similar—if not identical—patterns of negative human-health and environmental side effects that motivated industrialized countries to develop the pesticide regulatory apparatus in the 1970s. Looking at potential effects on human health, a variety of factors is likely to make local populations in developing countries—in particular farmers and agricultural workers but in general all food consumers—more vulnerable to the toxicological effects of pesticides.[3] Examples of such factors include

- low literacy and education levels;
- weak or absent legislative frameworks;
- climatic factors (which make the use of protective clothing while spraying pesticides uncomfortable);
- inappropriate or faulty spraying technology; and
- lower nutritional status (less physiological defense to deal with toxic substances).

The first four of these factors increase the likelihood of higher exposure; the last factor increases the toxic effects from that exposure on the human body. In addition, it has been found that when most organochlorine pesticides are banned or restricted, developing countries have often turned to substances that exert higher toxicity.[4]

According to a World Health Organization/United Nations Environment Programme (WHO/UNEP) working group report from 1990, unintentional acute poisonings with severe symptoms exceed 1 million cases each year, out of which 20,000 are fatal.[5] Additionally, there are estimated to be 2 million intentional poisonings (mainly suicide attempts) resulting in 200,000 deaths per year.[6] The vast majority of both unintentional and intentional intoxications occur in developing countries. However, good intoxication data are sparse in developing countries and these global estimates are often contested: Some stakeholders argue that they are too high, others insist that they are too low.[7] The data for estimating the toll of less acute effects, including long-term effects such as cancer, are even more sparse.[8]

Looking at potential environmental impacts of pesticide use in developing countries, many of the same factors that increase the vulnerability to health impacts are likely to exacerbate the release of the chemicals into the environment. Furthermore, in these regions there are unique ecosystems and species of ecological and economic importance and a range of managed systems (agricultural, silvicultural, and aquacultural) that are only marginally present in developed countries.[9] Because most species in developing countries are not subject to any tests before an industrialized country approves a particular pesticide, they may exhibit previously unknown sensitivities to pesticide exposure. Studies of pesticide impact on local environments in developing countries are few and far between: Data collection and research in these regions is extremely limited.[10] However, observed effects include contaminated ground, river, and coastal waters; fish kills; and impacts on cattle.[11] Environmental effects from organochlorine pesticides, a few of which are still in use, may also extend well beyond the region of use.[12] In the last decade, a hypothesis regarding the transport of such substances has received stronger support: It is believed that, due to their chemical characteristics, organochlorine pesticides are partially transported via the atmosphere from warmer tropical regions toward colder regions (the Arctic and Antarctica). According to this theory, the polar regions are thought to function as sinks for these substances as they condense and accumulate in ecosystems and food chains—ultimately affecting human health as well.[13]

Governance Challenges, from Local to Global

Pesticide use in developing countries is seen to produce benefits to society. At the same time, however, it has the potential to produce negative consequences on different scales, ranging from immediate health effects for those who spray them to environmental effects in remote regions. The reasons, or driving forces, behind these negative effects can be found in policies and actions covering levels of governance from the local to the global. The responses by stakeholders to reduce the negative effects are also found at each governance level. It is thus an issue that well illustrates an interconnected, globalized world and a multilevel governance challenge.[14]

To search for appropriate responses to such a globalized problem, it is necessary to examine the human activity—the use of pesticides in agriculture—that acts as the direct driving force for various undesired effects. In addition, it is helpful to analyze governance at the local, national, and global levels. It is equally important to determine how the diversity of stakeholders at these levels understand and structure the "pesticide problem" and address the problem through strategies of risk reduction.

Compared with other nations, Costa Rica produces the highest amount of coffee per hectare. Along with a number of other factors (many having to do with the banana trade), this has contributed to relatively high levels of pesticide use.

Such research—examining many aspects of pesticide use and governance in developing countries—was carried out in association with Linköping University, Sweden, in 1997-1999.

A close look at the stakeholders included in the study reveals three major categories: organizations (government and intergovernmental organizations (IGOs)), civil society (nongovernmental organizations (NGOs), private companies, and academia), and individuals (farmers and workers).[15]

At the global level of governance, some key stakeholders involved in pesticide use in developing countries (and which were examined in the 1997-1999 study) include IGOs such as the Food and Agriculture Organization of the United Nations (FAO) in Rome and the United Nations Environment Programme (UNEP) in Nairobi and Geneva (a more complete list appears below in the discussion of the activities of such organizations).

On the national level, two countries, Kenya and Costa Rica, served as important case studies of the issue. Both countries have enacted relatively ambitious pesticide legislation and other governance measures compared with other countries on their respective continents. Taken together, the countries' heavy dependence on—and the involvement of large businesses in—export agriculture, the type of export crops grown, and a number of other factors have made Kenya and Costa Rica substantial users of agricultural pesticides. The trend in this regard has been increasing for both nations in the past two decades.[16] In terms of product category, fungicides dominate import figures by volume in Costa Rica, followed by insecticides and fumigants.[17] In Kenya, fungicides account for half the market, insecticides 20 percent, and herbicides 18 percent.[18]

The high and increasing use of pesticides in Costa Rica has been attributed to such factors as the switch to horticultural crops; the expansion of land under banana production, a particularly difficult problem in banana plantations with the leaf spot disease *Mycosphaerella fijiensis*; the highest production of coffee per hectare (ha) in the world; and the fact that some transactions with pesticides have been exempted from taxes.[19] The volume of pesticide use is usually closely linked to the type of crop. In Costa Rica the average amount of active ingredient applied per ha was, according to one study, 0.25 kilograms (kg) for pasture, 3.5 kg for sugarcane, 6.5 kg for coffee, 10 kg for rice, 20 kg for vegetables/fruits, and 45 kg for bananas.[20]

Kenyan agricultural exports have been traditionally dominated by coffee and tea, but nontraditional crops such as pineapples, vegetables, and ornamentals have expanded quickly in recent years.[21] This expansion has contributed to increasing pesticide use. Historically, pesticide use per ha on the large agricultural farms has been much higher than on small farms. However, a substantial percentage of smallholder farmers in Kenya use pesticides, mainly on their export crops.[22]

When comparing approaches to pesticide use at the local level in different countries, it is helpful to look at the same crop. In the case of Kenya and Costa Rica, coffee is a good choice, and two prominent coffee growing regions are the Meru District in Kenya and the Naranjo *cantón* (district) in Costa Rica. Kenya's Meru District, which is located about 300 kilometers northeast of Nairobi (in the Eastern Province, covering the slopes of Mount Kenya), is one of the most fertile areas of the country with a range of subsistence and cash crops. Naranjo *cantón*, in Costa Rica's Alajuela province, is located in the hills northwest of San José. Coffee is the dominant crop, but there is also sugarcane and livestock production.[23] The land distribution is similar in both districts, with many small-scale landholders and a few medium- or large-scale landowners. However, the differences are substantial: Naranjo has significantly larger-sized farms, a higher level of modernization, and much higher literacy rates and general education levels than does Meru. Although the same variety of coffee, *Coffee arabica*, is grown in Meru and Naranjo, the answers farmers in the two districts gave to inquiries about pest problems share only some similarities. (Farmers in Meru were interviewed in early 1999; Naranjo

farmers were interviewed later that year.) In both areas, farmers said that the dominating pest problems had been fungi. However, Meru was also beset by a number of insect pests. In Naranjo, farmers said that insect pests were largely absent, but they mentioned that a number of nematodes (commonly known as roundworms) and different fungi have affected coffee production. In Meru, at least a quarter of the farmers responding to interviews were not spraying anything on their coffee. This was a new situation that had occurred for a few years prior to the interviews and were due primarily to falling coffee prices. Farmers in Meru have been spraying their coffee trees since the crop was introduced in the area in the 1950s. The interviews in Naranjo found that virtually all of the farmers there sprayed pesticides. Coffee was introduced in Costa Rica in the early 1900s—long before pesticides were available—but since the arrival of these chemicals in the 1950s, they have been applied ubiquitously.[24]

Which Problems and Whose Problems?

The stakeholders mentioned above raised five problem categories from pesticide use in developing countries: economic, production, human health, environment, and trade.[25] At the global level, pesticides are primarily seen as a health issue, especially for the poor, uneducated farmers and workers in developing countries who apply them. The pesticide-related trade problems for developing countries emerge in some agencies involved with regulations for pesticide residues in food crops. Persistent categories of pesticides—organochlorines—have emerged as a transboundary environmental issue in multilateral discussions and agreements, but there has been little mention of potential local and regional environmental effects in developing countries themselves. While the lack of data and research makes it impossible to give solid numbers and fig-

ures on health and environmental impacts of pesticides on a large scale in developing countries, the discussion above showed that there is enough isolated data coupled with higher risk factors in developing countries for IGOs to be taking the problem seriously.

At the national level in Kenya and Costa Rica, stakeholders involved in the study tended to either associate the entire assortment of pesticides with health and environmental effects or claimed there was no evidence for substantive negative effects. The government in Kenya tilted toward the latter position; various environmental NGOs in the country referred to substantial problems.[26] Costa Rican stakeholders made more references to significant health effects, and although environmental effects were usually discussed in general terms, the situation in banana plantations was often lifted out as potentially more serious. The trade problem has definitely remained high on the agenda for both countries, but stakeholders largely referred to it as a problem of the past, before appropriate policies had been put in place.[27]

There is little data on pesticide-related health effects in Kenya to substantiate opinions expressed regarding their magnitude. There was (at the time of the study) no working system to report intoxications to the authorities. In fact, one of the main areas of concern regarding chemical use generally was the absence of documentation on the risks in the country.[28] Figures surfaced in a national debate in the early 1990s claiming that 7 percent of the people in the agricultural sector—about 350,000 people—suffered pesticide poisoning each year.[29] Other sources reported that, in 1985, three major hospitals treated an average of two cases of pesticide poisoning every week.[30] Kenya's Ministry of Health estimated in 1996 that 700 deaths per year were caused by pesticide-related poisonings.[31] However, there were virtually no data at all on subacute poisonings from

Table 1. Toxicity classification of imported pesticides, Kenya and Costa Rica, 1993

World Health Organization (WHO) toxicity class	Percent total Kenya	Percent total Costa Rica
1a-Extremely hazardous	-	10
1b-Highly hazardous	3	8
Highly hazardous volatile fumigants	11	-
II-Moderately hazardous	22	24
Ill-Slightly hazardous	24	8
Unlikely to present acute hazard	25	40
Not classified by WHO	3	-
Unidentified	10	10

NOTE: It is very difficult to compare and interpret these kinds of import figures from different countries: Some of the products are imported as technical-grade material (concentrated) and others as ready-to-use formulations. For example, after formulation in Kenya the proportion of highly hazardous pesticides will be significantly higher, because a substantial portion of those substances are imported as technical-grade material. The figures show that nearly one-quarter of the pesticides used are moderately hazardous. More than 10 percent are highly hazardous.

SOURCE: H. Partow, *Pesticide Use and Management in Kenya* (Geneva: Institut Universitaire D'Etudes du Développement (Graduate Institute of Dévelopement Studies), 1995); and F. Chaverri, and J. Blanco, *Importación, Formulación y Use de Plaguicidas en Costa Rica.* Periodo 1992-1993 (Importation, Formulation, and Uso of Pesticides in Costa Rica between 1992-1993) (San José, Costa Rica: Programa de Plaguicidas: Desarollo, Salud y Ambiente, Escuela de Ciencias Ambientales Universidad Nacional (Pesticide Program: Development, Health, and Environment, School of Environmental Sciences, National University, 1995).

pesticides.[32] A few studies investigated residue levels of organochlorine pesticides in food and human tissue, but none of these looked at possible symptoms from such chronic, low-level exposure.[33]

Farmers and workers interviewed in Kenya's Meru District said the biggest problem with pesticides was their cost: None mentioned the possibility of long-term health effects.

Costa Rica had better data on pesticide-related health effects. In 1991, a new law there made it obligatory to report intoxications from pesticides to the Ministerio de Salud (Ministry of Health).[34] In the period 1980-1986, 3,347 cases of intoxications were treated in hospitals.[35] The figures for the years 1990-1996 indicate a significant increase of intoxications until 1995, followed by a slight decrease in 1996—when the reported number was 792.[36] Out of all the reported intoxications in 1996, farmers and workers associated with banana cultivation suffered the highest number (64 percent), followed by those working with ornamental plants (3.7 percent).[37] In terms of chemical classes, insecticides and nematicides were attributed to more than one-half of the intoxication cases, followed by herbicides.[38] One study identified young workers (under 30) and women as groups particularly affected by occupational poisonings.[39] The same study concluded that on a yearly basis, 1.5 percent of the agricultural workforce is medically treated for occupational poisonings.[40] Intoxications from cholinesterase-inhibiting pesticides (organophosphates and carbamates) and the herbicide paraquat together accounted for a majority of poisonings identified in hospitalization and fatality records in the 1980s in Costa Rica.[41] Unique studies—in a developing country context—have been done in Costa Rica on subacute health effects such as cancer. The organochlorines that were used in agriculture until the 1980s have shown a rather strong association with breast cancer in areas where rice—a crop that has been intensely sprayed with such products—is grown.[42] Other associations between pesticides and cancer include paraquat and lead arsenate, which have been linked to skin-related cancers; formaldehyde and leukemia; and dibromochloropropane (DBCP), which has been associated with lung cancer and melanoma.[43] Banana workers in particular showed increased incidence of some cancer types, specifically melanoma in men and cervical cancer in women.[44] Table 1 above shows pesticide import figures for Kenya and Costa Rica in the early 1990s, classified in categories according to acute toxicity. (Such classification gives an indication for assessing potential health risks.) Nearly one-quarter of the pesticides used are moderately hazardous, and more than 10 percent are highly hazardous.

Unfortunately, the lack of data on environmental effects is the rule and not the exception. The few studies that were found in Kenya only look at pesticide levels in the environment and do not indicate significant problems.[45] In Costa Rica, the Ministry of Environment and Energy had commissioned a report on the envi-

Table 2. Pesticides most frequently mentioned by local-area coffee farmers

Product Name	Pesticide category	Active ingredient	WHO hazard classification[1]
Meru			
Copper (various)	Fungicide	Copper	III
Lebaycid®	Insecticide	Fenthion	II
Sumithion®	Insecticide	Fenitrothion	II
Gramoxone®	Herbicide	Paraquat	II
Ambush®	Insecticide	Permethrin	II
Karate®	Insecticide	Lamba-cyhalothrin	II
Naranjo			
Atemi®	Fungicide	Cyproconazole	III
Silvacur®	Fungicide	Triadimenol	III
Gramoxone®	Herbicide	Paraquat	II
Roundup®	Herbicide	Glyphosphate	[2]
Counter®	Nemtaicide	Terbufos	1a
Furudan®	Nematicide	Carbofuran	1b

[1]The World Health Organization (WHO) classification places products' active ingredients into the following categories: 1a-extremely hazardous; 1b-highly hazardous; highly hazardous volatile fumigants; II-moderately hazardous; III-slightly hazardous; and unlikely to present acute hazard.

[2]Unlikely to present hazard under normal use.

NOTE: Coffee farmers in Meru, Kenya, and Naranjo, Costa Rica, responded to questions about which pesticides they apply with a few product names: The table above lists the most commonly mentioned. In total, farmers in Meru identified more than 20 different products; in Naranjo more than 30 were mentioned. Fungicides, the products sprayed most frequently, belong to the less acutely toxic categories (II and III). The nematicides—used only in Costa Rica—stand out as having highly or even extremely toxic ingredients. The three pesticides most commonly involved with intoxications in Costa Rica during the years 1994-1996 were paraquat (bipyridylium), carbofuran (carbamate), and terbufos (organophosphate).

SOURCE: S. Karlsson, *Multilayered Governance. Pesticide Use in the South: Environmental Concerns in a Globalised World* (Linköping, Sweden: Linköping University, 2000); and R. Castro Córdoba, N. Morera González, and C. Jarquín Núñez, *Sistema de Vigilancia Epidemiologica de Intoxicaciones con Plaguicidas, la Experiencia de Costa Rica, 1994-1996* (Epidemiological Monitoring System of Pesticide Intoxications, Costa Rica's Experience, 1994-1996) (San José, Costa Rica: Departamento de Registro y Control de Sustancias y Medicina del Trabajo del Ministerio de Salud (Ministry of Health Department of Registration and Control of Occupational Substances and Medicine, 1998).

ronmental effects of pesticide use on banana plantations in three regions in the Atlantic zone. The report, however, could also mainly refer to pesticide residues detected in the environment that have unknown effects: The only concrete direct negative effects reported were some incidences of fish killed in rivers and declining fish populations.[46]

At the local level, the chief complaint farmers in Meru expressed was an economic problem: They were unable to purchase pesticides due to low coffee prices. Interview revealed that farmers in neither Meru nor Naranjo saw production problems from pesticides. Respondents said they had not seen pests develop resistance, and with few exceptions, all pests could be adequately controlled. Health effects that workers associated with using pesticides varied substantially. Some said they suffered problems such as dizziness, nausea, headache, skin problems, eye problems, or fever; others said they never had any problems.[47] There was hardly any mention in Meru of the possibility of more long-term health effects. However, it was not uncommon for respondents in Naranjo to mention such effects, including cancer. The general environment as a possible victim of negative effects was practically absent in the interviews in Meru, while respondents in Naranjo were slightly more aware of its vulnerability. Table 2 above shows some of the most commonly used pesticides in the districts. Many are moderately hazardous; some have notorious records as intoxicants in developing countries.

The upshot of all this is that there were, at the time of the study, substantial divergences in the understanding of the pesticide-linked problems among different stakeholder groups and at varying governance levels. This lack of common understanding is not surprising considering the diversity of priorities among stakeholders and prevalent lack of knowledge on so many aspects of the potential impacts.

Table 3. Strategies to reduce health and environmental risks, by governance level

Level	Country/district	Use	Type	Mode
Global		Integrated pest management Organic farming[a]	Phase-out Information exchange Risk assessment Toxicity classification	Codes of conduct Guidelines
National	Kenya	Integrated pest management Organic farming[a]	Registration Banning	Regulation and training
	Costa Rica	Integrated pest management Organic farming	Registration Banning	Regulation and training
Local	Meru, Kenya	Resistant coffee variety Getting others to spray	——	Safe-use training
	Naranjo, Costa Rica	Integrated pest management	——	Safe-use training

(a) Organic farming exists as a strategy on this level but is not strongly stressed.

NOTE: Strategies to reduce health and environmental risks are categorized under "use," "type," or "mode;" that is, they reduce pesticide use, address certain types of pesticides use related to their individual characteristics, or they address modes of the chemicals' application.

SOURCE: S. Karlsson, *Multilayered Governance. Pesticides in the South: Environmental Concerns in a Globalised World* (Linköping, Sweden: Linköping University, 2000).

Which Risk-Reduction Efforts, Where?

Three basic risk-reduction strategies emerge for addressing the risks with pesticides: reducing their use, using less-toxic types of substances, and using them in a more precautionary fashion or mode?[48]

These approaches can be linked to three factors of pesticides that contribute to their health and environmental risks—the volume of use, the type of pesticide used, and the mode in which they are used. Table 3 on page 6 shows which kind of risk-reduction efforts existed at each governance level and to which strategy they belong. At the global level, the predominant risk-reduction efforts focused on the type of pesticides, applying a chemical-by-chemical approach in many activities. Several IGOs have assisted developing countries to regulate and control the use of pesticides. This guidance has come in the form of facilitating the development of national legislation, helping to build capacity in chemical management, and providing scientific data on individual pesticides. Several international agreements

have been developed to address problems associated with pesticides and other chemicals by targeting individual substances—by phasing them out, for example.[49] Some IGOs have also made significant efforts to support a better mode of use. For example, FAO developed the International Code of Conduct on the Distribution and Use of Pesticides, and there have been several efforts promoting its universal observance.[50] In addition, a number of technical guidelines have been made on how pesticides should be used.[51] Finally, there have been some efforts to reduce the overall use of pesticides. For instance, some IGO programs have encouraged the implementation of integrated pest management (IPM) techniques rather than supporting increased use of pesticides in agricultural projects.[52] However, there has been very low support for organic farming.[53] Table 4 on pages 7 and 8 gives more details on some of the relevant organizations and their activities as well as agreements on the international level.[54]

Most risk-reduction efforts by governments and industry at the national level in Kenya and Costa Rica

belong to the type and mode categories, although IPM efforts have been increasingly encouraged. To address risks arising from the types of pesticides used, both countries have developed extensive laws and regulations on pesticides. The core of these have been registration processes.[55] Within these processes, pesticide companies submit each product for approval before it can be used in the country.[56] Both countries have banned a number of primarily organochlorine pesticides over the last two decades.[57] In both countries, training programs for the safe use of pesticides has involved government agencies as well as the pesticide industry.[58] The pesticide laws in both countries cover, for example, conditions for storage of pesticides by retailers, the training of pesticide retailer staff, and the application of pesticides by farmers (prescribing that they must be used safely).[59] In Kenya, there have been small efforts to establish IPM as a national policy, but there has been very little implementation in this regard. However, it appears that IPM has received more official support inCosta Rica.[60] On the margin, some NGOs in both

Table 4. Global-level actors, activities, and agreements on developing country pesticide risks

Actor/organization	Activities and international agreements
Intergovernmental Forum on Chemical Safety (IFCS) A noninstitutional arrangement where governments meet with intergovernmental and nongovernmental organizations every three years. Smaller meetings are held every year. Geneva, Switzerland (small secretariat) Established in 1994 www.who.int/ifcs	• Provides advice to governments, international organizations, intergovernmental bodies, and nongovernmental organizations on chemical risk assessment and environmentally sound management of chemicals. • Sets priorities and promotes coordination mechanisms at the national and international level.
Food and Agricultural Organization of the United Nations (FAO) Specialized agency Rome, Italy Established in 1945 www.fao.org	• Supports integrated pest management (IPM) projects with Farmer Field Schools in Asia and Africa (for example) at national and regional levels for various crop systems. • Assists countries with development of national pesticide legislation. • Produces technical guidelines on various aspects of pesticide risks. • Runs projects to clean up obsolete stocks of pesticides (wastes) in developing countries. • Facilitated the negotiation of the International Code of Conduct on the Distribution and Use of Pesticides (the FAO Code of Conduct) and monitors its observance. www.fao.org/ag/agp/agpp/pesticid/
Global IPM Facility Cosponsored by FAO, UNEP, United Nations Development Programme (UNDP), and the World Bank. FAO Headquarters, Rome, Italy Established in 1995 www.fao.org/globalipmfacility/home.htm	• Assists governments and nongovernmental organizations to initiate, develop, and expand IPM. • Strengthens IPM programs through, for example, initiation of pilot projects around the world. Programs include policy development and capacity building. • Applies a farmer-led, participatory approach to IPM.
United Nations Environment Programme (UNEP) Nairobi, Kenya Established in 1972	• Hosts the Interim Secretariat for the Stockholm Convention on Persistent Organic Pollutants (POPs), which initially targets 12 chemicals (out of which 9 are pesticides) for reduction and eventual elimination. It also sets up a system for identifying further chemicals for action. Signed by 151 countries, it will enter into force 17 May 2004. www.pops.int
UNEP Chemicals Geneva, Switzerland www.unep.org www.chem.unep.ch	• Hosts (with FAO) the Interim Secretariat for the Rotterdam Convention on Prior Informed Consent for Certain Hazardous Chemicals and Pesticides in International Trade, which prevents export of harmful pesticides (and industrial chemicals) unless the importing country agrees to accept them. Signed by 73 countries, it was entered into force 24 February 2004. www.pic.int • Produces the Legal File, which contains information on regulatory actions on hazardous chemicals in 13 countries and 5 international organizations. www.chem.unep.ch/irptc/legint.html

table continues on next page

Table 4, continued	
Actor/organization	**Activities and international agreements**
International Programme on Chemical Safety (IPCS) A joint program of the World Health Organization (WHO), FAO, and the International Labour Organization (ILO). WHO Headquarters, Geneva, Switzerland Established in 1980 www.who.int/pcs/	• Produces and disseminates evaluations of the risk to human health and the environment from exposure to chemicals (including pesticides) and produces guideline values for exposure. • Carries out projects with governments to support their capacity in chemical safety. • Since the early 1990s, has carried out activities to develop a project for collecting data on pesticide poisoning. Several countries are testing the harmonized approach of data collection.
Pesticide Action Network (PAN) A network of more than 600 participating nongovernmental organizations, institutions, and individuals in more than 60 countries. Five regional centers (San Francisco, California; Santiago, Chile; London, United Kingdom; Dakar, Senegal; and Penang, Malaysia) www.pan-international.org/	• Works to replace the use of hazardous pesticides with ecologically sound alternatives. • Working from five autonomous regional centers, the network's programs include research, policy development, and media and advocacy campaigns.
CropLife International (formerly (pre-2000) Global Crop Protection Federation (GCPF)) Represents the crop-protection product manufacturers and their regional associations. Brussels, Belgium www.gcpf.org	• Supports the FAO Code of Conduct. • Initiated three pilot projects in 1991 to promote the safe use of pesticides (in Guatemala, Kenya, and Thailand). Now continuing to support safe-use programs through its National Associations. • Endorses and supports IPM. • Has carried out several projects addressing obsolete pesticide stocks. Supports a nongovernmental organization-initiated program to eliminate stocks of POPs in Africa.

SOURCE: Sylvia I. Karlsson, 2004.

countries—and even the Costa Rican government—have encouraged organic farming.[61]

On the local level in the two coffee-growing districts, safe-use training has been the only explicit approach to reduce the risks of pesticides. More than 20,000 farmers in Meru were trained in a project sponsored by the pesticide industry in the first half of the 1990s, but this number is still a small part of Meru's agricultural population: The project neglected groups such as women and casual and permanent farm workers, and it is likely that such farm workers are the group most heavily exposed to pesticides. In Naranjo, training on protective measures has been present for a number of years, although not in the form of an explicit safe-use project: It was incorporated in the general training from the national agricultural extension system. Despite these training efforts, the adoption of the safe-use message among interview respondents was low in both districts. In Meru, a few said they used improvised or partial protective clothing, but most respondents neither owned nor had access to these. Farmers and workers interviewed in Meru said they were too expensive. In Naranjo, access was generally not a problem, but farmers said they were uncomfortable: Few used the protective wear, or if they did they wore it only in the early morning before it got too hot. When asked, most farmers said they believed there were no alternatives to pesticides. The few references to nonchemical alternatives to combat pests were primarily made by farmers in Naranjo and included pruning, hand weeding, and general soil conservation measures. In Meru, the extension system had provided a coffee variety that was resistant to the two most prominent fungal pests, but only some farmers at the time of the interviews had planted them. Interviewees had not applied any of these measures to address either health-risk or environmental problems. Accumulated experience and research in Costa Rica had, at the end of the 1990s, begun to show the negative consequences of herbicide use both on the productivity of the coffee plants and erosion levels. Consequently, the extension service had begun to discourage farmers from using herbicides. Many coffee farms were de facto organic: They could afford neither pesticides nor chemical fertilizers. But because there was no market infrastructure for organic coffee, they could not receive a higher price for their crop. In Naranjo, some farmers were aware that it was possible to receive a significant premium price for growing certified organic coffee.

However, even the promoters of organic coffee farming did not encourage existing coffee farms to switch to organic farming: Organic certification requires a farm to be without agrochemical inputs for three years, and the harvest slumps in the meantime. (Generally, long-abandoned coffee farms are preferred instead; they can most successfully be developed into an organic farming system.) There were no efforts at the local level to reduce risks by avoiding certain types of pesticides. Farmers bought and applied the products that were recommended by companies, cooperatives, or the extension system.

Which Institutions, at What Level?

Despite large uncertainties in the precise nature and levels of risks, the description above of the scope of the health and environmental problems, globally and on a smaller scale in Kenya and Costa Rica, shows that there is by no means an effective governance system for pesticides in developing countries. As evidenced by the risk-reduction measures described above, considerable efforts of governance at different levels are not only insufficient but are also characterized by fragmentation and incoherence. One way to identify the reasons for this situation as well as potential options for change is to focus on institutions—here defined as formal and informal rules of human interactions—and their role in governance at various levels.[62] Institutions are at the core of governance: They influence who has access to what information, shape the incentives for various courses of action, and affect who has the capacity to act. In a multilevel governance context, the following question arises: Is it possible to identify some criteria to determine the types of institutions that should preferably be established, enforced, or changed at particular levels? Some suggestions for elements of such criteria can be found in theories of the management of collectively owned natural re-

sources. To identify current gaps in the governance system, it is useful to explore two sets of criteria. The first relate to the potential effectiveness of institutions if they were to match up the level of effects, driving forces and capacity; the second set relates to how the possibility to change institutions may vary across levels. Tables 5a and b illustrate these criteria.

Matching Institutions

It is particularly instructive to determine criteria that can address how well institutions and governance "match" the level where most negative effects occur, the driving forces behind the problems originate, and where there is capacity to take action.

• *Matching effects.* If institutions are not in place at the level to correspond with, or "fit" the geographical scope of the negative effects, their effectiveness can be limited.[63] The potential for persistent organic pollutants (POPs)—which include a number of organochlorine pesticides—to act as transboundary pollutants made countries like Canada and Sweden push for an international agreement to ban them. The development of the Stockholm Convention can thus be seen as an effort to match the global scope of the problem with global governance. The efforts to establish a harmonized registration system for pesticides in Central America can be seen in a similar light. Banning a pesticide in Costa Rica while it is allowed in a neighboring country invites smuggling, black market sales, and potential for cross-border pollution via rivers.[64] However, pesticides also exert very local and context-dependent effects on health and environment—which by aggregation can be seen as global problems—and these are not at all well matched with institutions. Consider, for example, the common use of paraquat in both Meru and Naranjo despite its well-known (on the international level) intoxication record, and the use of category la and lb (extremely and

highly hazardous, respectively) pesticides by small farmers in Naranjo (see Table 2).

The banana trade is a huge market in Costa Rica for pesticides. Unfortunately, data show that banana workers are more likely to suffer pesticide intoxications than other workers.

• *Matching driving forces.* If institutions are not sufficiently well established and implemented at the level where the driving forces for the problems originate, the governance measures at other levels will merely target symptoms. Moreover, from an ethical perspective, it would be preferable if those who are explicitly responsible for the problems would be more often targeted in governance.[65] There are layers of direct and indirect driving forces for pesticide problems and they differ depending on what risk factor is in focus. The strongest driving forces for using pesticides emerge at the global level, where the agrochemical industry, along with governments, promote incentives for modern high-input agriculture, and at the national level, where agricultural policy, research, and extension advice and marketing strategies of the agrochemical companies create similar incentives. However, these are the drivers least addressed in pesticide governance.

The strongest driving forces that determine the types of pesticides used are also located at the global and national levels. At the global level, multinational corporations develop and choose which of their products to market in developing countries. The Stockholm and Rotterdam Conventions target specific pesticides to regulate or ban; if implemented, these rules can reduce risks in developing countries. However, the number of substances included is very small, making this process a weak match of institutions and driving forces. At the national level, the governments in Kenya and

Table 5a. Matching institutions

		Global		National		Local	
		Negative effects	Driving forces	Driving forces	Capacity to act	Negative effects	Driving forces
Example		Global transboundary pollution of pesticides	Institutions giving incentives for high-input agriculture, resulting in increased pesticide use	Regulations prescribing how toxic pesticides are allowed to be used in countries	Governmental authority to regulate pesticide storage, sales, and marketing	Local health and environmental effects	Influence that farmers and workers who handle pesticides have on how safely they are used
Degree example is matched in pesticide risk reduction		High: The Stockholm Convention addresses many persistent organic pollutant pesticides.	Low: Regulation is absent and there are few incentives for alternative agricultural systems.	Moderate: Many developing countries allow many highly toxic pesticides.	Moderate: These institutions are weak or very weakly enforced in many developing countries.	Low: Many places have poor healthcare facilities and no institutions that address potential environmental effects.	Low: Most countries have limited and/or ineffective safe-use education and training.

Table 5b. Changing institutions

Types of institution	Example	Proneness to change
Operational	Guidelines for safe use	• Relatively easy to change with few resources and within a short time span. • Difficult to implement/enforce, particularly at global scale.
Collective-choice	Regulations prescribing which pesticides are banned at national and global levels	• Moderately difficult to change with large variances in time and resources required.
Constitutional-choice	Favored type of agricultural system (including system of pest management)	• Very difficult to change, requiring significant political will and involving multiple sectors.

SOURCE: S. I. Karlsson, 2004.

Costa Rica control which pesticides may be used in their respective countries—although this also depends on which products the pesticide industry chooses to market. Overall, developing countries have no influence on pesticide development: Most research in this regard centers around pests and crops in the temperate regions.[66] And while Kenya and Costa Rica have the authority to ban pesticides, the overwhelming majority are approved.[67] In terms of how pesticides are actually used, there are a number of driving forces, including images presented in marketing campaigns. Ultimately, however, individual farmers or workers are the ones who determine how they are used. This is where the major mismatch between institutions and driving forces lies. Even if many efforts are made at national and higher levels to make farmers use pesticides more safely, such measures have a long way to go before they actually reach individuals who spray—or to a sufficient degree effect a change in their behavior.

• *Matching capacity.* If there is no capacity to act at the level where effects or driving forces originate, then there is not much one can expect in terms of institution building.[68] Stakeholders at a governance level where there is capacity to act may not necessarily have contributed to the problem but could take on responsibility for governance because of a sense of concern and moral obligation.[69] The whole focus at the global level to assist developing countries with scientific and technical information on pesticide risk is largely due to the fact that IGOs have the capacity to assess those risks, while many developing countries have neither the expertise nor resources needed. The governments in Kenya and Costa Rica focus on deciding which products are allowed for use because they have the capacity to make those decisions, while local actors are considered not to have such capacity. Moreover, for the most part, farmers do not have the capacity to adopt risk-reduction strategies—either because they are not aware of risks or because they believe it is impossible to farm without pesticides and have no alternative pest-control strategies to adopt. Those stakeholders who do have the capacity to influence this—such as the pesticide industry and the international community—have not fully utilized their resources to this end.[70]

Changing Institutions

Another key set of criteria relates to the cost (in monetary or human-resource terms, for example) required to create a new institutional arrangement or to enforce or change an existing one. Three types of institutions are considered here: operational, collective-choice, and constitutional-choice institutions.[71]

- *Operational institutions.* These institutions provide structures for making day-to-day decisions in a wide diversity of operational situations, which means a large number of individual actors are involved.[72] In a local context, it is assumed that these are the institutions that can be most quickly changed. However, this may not always be the case on the global scale, where the sheer number and diversity of multiple localities make the picture more complex. When pesticides continue to be the primary pest-management tool, and risk reduction follows the mode strategy of ensuring safe use, an effective approach to effect change must include a strong focus on operational institutions at the local level. A global code of conduct is not enough nor are national laws that make unsafe use illegal. Any institutions established at higher levels have to be implemented by a large

number of stakeholders: farmers and workers—female and male alike; their families; and all others who handle pesticides throughout their life cycle. Even if changes can be initiated quickly, they are costly and time consuming to implement on a large scale.

Government officials in Costa Rica recognized herbicides had negative impacts, including erosion. Many local officials later discouraged their use.

- *Collective-choice institutions.* This category describes those institutions that indirectly affect the options for operational rules.[73] The number of stakeholders involved in designing these institutions are fewer but, depending on the governance level, can range from a single local NGO to all the member states of the United Nations. The cost of changing such institutions, and the time it takes to do so, will vary accordingly. When risk-reduction efforts target the types of pesticides used, the focus is on collective-choice institutions at the national level. Here, decisions on which pesticides farmers will have access to involve only very few individuals—and in Costa Rica, for example, the pesticide registration process takes 6-12 months.[74] Implementation of these governmental decisions involves a smaller number of stakeholders, such as customs officers and pesticide retailers, who are charged with ensuring that the unwanted pesticides do not reach the farmers. This strategy also involves measures at the global level, through, for example, sharing toxicity information about certain chemicals or banning specific substances. Such global processes can take a very long time: For instance, the process to establish the Stockholm Convention began in 1997 and will not have entered into force until mid-May 2004. Change to collective-choice institutions can thus in

some cases be made at relatively low cost within a short time frame, but in other cases the process is lengthy and cumbersome.

- *Constitutional-choice institutions.* These determine the specific institutions that create the collective-choice institutions.[75] These types of institutions may involve the smallest number of the most powerful and knowledgeable stakeholders when crafting formal institutions. Or, when institutions are part of deeply rooted structures in society, the number of stakeholders involved may be innumerable and not easy to pinpoint. These types of institutions are usually the slowest to change. The risk-reduction strategy to reduce or eliminate the use of pesticides requires changes in constitutional-choice institutions. These are institutions that favor one type of agricultural system: one dependent on pesticides, one less dependent on pesticides, or one completely independent of pesticides. Such institutions consist of consumer demands, national and international market and trade structures, government policies, and farmer attitudes (for example). The inter-linkages between sectors, the resistance from prevailing power structures, and the number of decisionmakers involved all present a considerable and time-consuming challenge for anyone attempting to change these institutions.

Correcting Mismatches and Facilitating Change

In very general terms, it appears that institutions must be assigned primarily to those levels where driving forces originate and where stakeholders have the capacity to establish and enforce institutions. Furthermore, if rapid institutional change is desired, the focus should be on enforcing changes in operational institutions at local levels—although this may not be the most cost-effective or long-lasting governance strategy. If slower—but likely more enduring—change is to be achieved it is the constitutional-choice institutions that need to be targeted—not only on global levels but across national and local levels as well. The more pragmatic and manageable approach in time and resources is to target collective-choice institutions.

Many agricultural workers, unaware of the risks or lacking adequate resources, do not always wear protective gear. Effective education programs could change this.

In more specific terms, to address the mismatch between institutions and effects, institutions need to be established that enable developing countries and the international community to incorporate the concerns for local health and environmental effects from pesticides in their specific climatic, ecological, economic, and social context. Currently, developing countries rely on global institutions and knowledge when establishing their own institutions.[76] There is a need to create institutions that facilitate the collection of data locally and nationally—for instance, those that prescribe the monitoring of health and environmental effects after registration is approved.

To address the mismatch between driving forces and institutions, there are specific needs for each risk-reduction strategy. Institutions at national and global levels are needed that discourage the use of pesticides and provide alternative, economically viable farming strategies. Some national and international NGOs are currently involved in small-scale initiatives promoting organic farming, but these are very limited in scope and hard to upscale. The situation is better for IPM initiatives. Such higher-level institutions could include changing or developing new agronomist education programs, reducing hidden pesticide subsidies, increasing the market channels for organic products, or even influencing consumer demand for these

products through education campaigns that raise the general public's awareness. Institutions are also needed at the global level to support those countries that can neither create institutions nor enforce them—a situation that results in a substantial black market of smuggled and substandard pesticides—to prevent the most toxic products from entering their countries.[77] Only global-level phase-outs can, in such cases, keep the most toxic types of pesticides out of the hands of smallholder farmers. This implies that the criteria for the type of products to be banned globally would need to be expanded to include those that are known to produce locally occurring severe health and environmental effects. Finally, to strengthen safer modes of pesticide handling, existing higher-level institutions need to radically upscale or fundamentally change their implementation efforts. A local culture of safe pesticide use can only emerge with long-term, more effective education efforts that reach all groups who come in contact with pesticides. To address the mismatch between capacity and institutions, one step is to look at the pesticide industry—a stakeholder that is a strong driving force and that holds considerable capacity to effect change. They could be charged with greater responsibility to become a stronger player in risk reduction. An example of how this could be implemented is illustrated with the cases in Kenya and Costa Rica: Pesticide companies now pay for field tests of the efficacy of their products on the crops in both countries—the governments require this for registration.[78] National institutions could hold these companies accountable to contribute resources toward national data-gathering programs examining the impacts of their products in each respective country. The international community—the only entity that can regulate global production and trade—could also take on a larger share of responsibility by creating more encompassing institutions, for example by phasing out more substances glo-

bally (see above) and making such institutions more effective with stronger mechanisms of monitoring and enforcement. Combining the two sets of criteria puts in focus the collective-choice institutions targeted at eliminating the pesticides that pose the highest risks. Changing these institutions would take considerable time and effort, particularly in building up a better knowledge base and banning substances on a global scale. Nevertheless, it would be the most accessible "fast track" to reduce risks pending a global-scale mustering of resources for widespread implementation of safe use and changes in agricultural systems.

Conclusions

The final choice of strategies for risk reduction, institution building, and change will be heavily dependent on answers to such questions as

• What are the major contributing factors to risks?

• What are the inherent toxic properties of pesticides?

• What are the exposure patterns under conditions of use?

• What is the acceptable level of risk? and

• Who should be responsible to address the risks?

Some of these questions could be resolved with more monitoring and research, others are more value laden, and these kinds of issues divide stakeholder groups substantially. Stakeholder views range from NGO movements that consider all pesticides inherently toxic to many in the pesticide industry who assert that all pesticides—as long as used as prescribed—are safe. The former are convinced that the conditions in developing countries—especially considering the human and financial input necessary to effect change—make it impossible to change the operational institutions and ensure safe use. The latter fear that changing the constitutional-choice institutions—striving to establish IPM or organic farming on a global scale, for example—would endanger food and economic security. There is no easy way to resolve the debate between the widely diverging knowledge bases and value judgments that underlie these different views. However, some of the recommendations above can help clarify and structure the available options for governance and the institution building and change that they would require.

In addition, the pesticide case and the policy-relevant conclusions drawn from it have much to teach us in other policy areas. Pesticides are one of the first groups of toxic chemicals introduced on a large scale in developing countries, and the complexities involved with their use illustrate many of the challenges and possibilities for the management of other groups of chemicals in these regions. As one of the environmental issues that exhibits a number of local-global linkages, the impacts of pesticide use illuminate directions that future research and policy discussions need to take. Governance needs to be analyzed and addressed with a much more holistic approach, viewing the efforts at all levels—local, national, regional, and global—as elements of one system of governance. Only then can we evaluate how individual policies operate in the context of a large, interconnected system. Only then can research start identifying the most important elements of establishing multi-layered governance, with a nested hierarchy of mutually supportive policies and institutions initiated at all governance levels.[79]

NOTES

1. The term "governance" has emerged as one of the most-used concepts when discussing measures taken in society to address a particular issue—specifically when stressing that there are many more actors than just governments involved. It is particularly useful for the global level where there is no world government

but still a lot of governance. The Commission on Global Governance defined governance as "the sum of the many ways individuals and institutions, public and private, manage their common affairs." See Commission on Global Governance, *Our Global Neighbourhood* (Oxford, UK: Oxford University Press), 2.

2. This figure is for 2001, a year in which the market suffered a 7.4 percent decline. See Phillipps McDougall (2002) quoted in CropLife International, *Facts and Figures*, accessed via http://www.gcpf.org on 21 January 2004. In 1999, the market was more than US$30 billion. See "World Agrochemical Market Held Back by Currency Factors," *Agrow*, 11 June 1999, 19-20.

3. See, for example, World Health Organization (WHO), *Public Health Impact of Pesticides Used in Agriculture* (Geneva: WHO, 1990); P. N. Viswanathan and V. Misra, "Occupational and Environmental Toxicological Problems of Developing Countries," *Journal of Environmental Management* 28 (1989): 381-86; L. A. Thrupp, "Exporting Risk Analysis to Developing Countries," *Global Pesticide Campaigner* 4, no. 1 (1994): 3-5; Health Council of the Netherlands, *Risks of Dangerous Substances Exported to Developing Countries* (Den Haag: Health Council of the Netherlands, 1992); and C. Wesseling, R. McConnell, T. Partanen, and C. Hogstedt, "Agricultural Pesticide Use in Developing Countries: Health Effects and Research Needs," *International Journal of Health Services* 27, no. 2 (1997): 273-308. For a special analysis of the impact on women, see, M. Jacobs and B. Dinham, eds., *Silent Invaders: Pesticides, Livelihoods and Women's Health* (New York: Zed Books, 2003).

4. Food and Agriculture Organization of the United Nations (FAO), *Analysis of Government Responses to the Second Questionnaire on the State of Implementation of the International Code of Conduct on the Distribution and Use of Pesticides* (Rome: FAO, 1996). For example, highly toxic insecticides is the main pesticide category in use in many less developed countries. Wes-

seling, McConnell, Partanen, and Hogstedt, note 3 above, page 276.

5. WHO, note 3 above, pages 85-86. This estimate is calculated by using a 6:1 ratio between nonhospitalized (unreported) and hospitalized (reported) cases.

6. WHO, note 3 above, page 86. The background for this is that pesticides in rural areas are among the most accessible types of toxic substances. Because these data are based on hospital registers, they probably overestimate the proportion of suicides. Wesseling, McConnell, Partanen, and Hogstedt note 3 above, page 283.

7. The figures from the WHO 1990 report have been strongly challenged by the pesticide industry. Anonymous official, International Programme on Chemical Safety (IPCS), interview by author, Geneva, 25 June 1998. IPCS has a project to support developing countries to gather data on pesticide intoxications more systematically. IPCS, *Pesticide Project: Collection of Human Case Data on Exposure to Pesticides* (Geneva), accessed via http://www.intox.org/pagesource/intox%20area/other/pesticid.htm on 22 January 2004. However, the project has not yet resulted in new global estimates. In the first stage, studies were carried out in India, Indonesia, Myanmar, Nepal, and Thailand, based on hospital records (some results of these are available at http://www.nihsgo.jp/GINC/meeting/7th/profile.html), but the results were not satisfactory. In a second phase, they will use community-based studies in pilot countries. Dr. Nida Besbelli, IPCS, e-mail message to author, 10 February 2004. While there are some developing countries where intoxications have to be reported to the authorities, overall there is limited data on pesticide health impacts in developing countries. Wesseling, McConnell, Partanen, and Hogstedt, note 3 above, page 284.

8. The lack of data provides a significant obstacle for global estimates of the number of people suffering from chronic effects. WHO, note 3 above, page 87. The few studies that have been done have demonstrated

neurotoxic, reproductive, and dermatologic effects. Wesseling, McConnell, Partanen, and Hogstedt, note 3 above, page 273.

9. See, for example, P. Bourdeau, J. A. Haines, W. Klein and C. R. K. Murti, eds., *Ecotoxicology and Climate With Special Reference to Hot and Cold Climates* (Chichester, UK: John Wiley and Sons Ltd., 1989); and T. E. Lacher and M. I. Goldstein, "Tropical Ecotoxicology: Status and Needs," *Environmental Toxicology and Chemistry* 16, no. 1 (1997): 100-11.

10. For example, 89 percent of the 60 developing countries who responded to an FAO questionnaire in 1993 reported that they are not studying the effects of pesticides on the environment. FAO, note 4 above, page 61. Research in disciplines that are essential for detecting and understanding environmental degradation—such as biology, ecology, and ecotoxicology—is very limited in sub-tropical and tropical regions compared to that in nontropical latitudes. Bourdeau, Haines, Klein and Murti, note 9 above; and Lacher and Goldstein, note 9 above. This situation reflects a general knowledge divide in the environmental field between developed and developing countries, which in turn reflects the generic divide in resources (human and financial) available for monitoring and research. S. Karlsson, "The North-South Knowledge Divide: Consequences for Global Environmental Governance," in D. C. Esty and M. H. Ivanova, eds., *Global Environmental Governance: Options & Opportunities* (New Haven, CT: Yale School of Forestry & Environmental Studies, 2002), 53-76. It is estimated that about 5 percent of the world's scientific production comes from developing countries. International Development Research Centre, "The Global Research Agenda: A South-North Perspective," *Interdisciplinary Science Reviews* 16, no. 4 (1991): 337-4. The number of scientists/engineers per million inhabitants in developed countries is 2,800 on average; in developing countries it is 200. T. H. I. Serageldin, "The Social-Natural Science Gap in Educating for Sustain-

able Development," in T. H. I. Serageldin, J. Martin-Brown, G. López Ospina, and J. Dalmatian, eds., *Organizing Knowledge for Environmentally Sustainable Development* (Washington, DC: The World Bank, 1998).

11. See B. Dinham, *The Pesticide Trail: The Impact of Trade Controls on Reducing Pesticide Hazards in Developing Countries* (London: The Pesticide Trust, 1995), which summarizes case studies from a number of countries. One of the few larger studies on environmental impact is the Locustox project studying the impact from large-scale locust sprayings in Africa. See J. W. Everts, D. Mbaye, and O. Barry, *Environmental Side-Effects of Locust and Grasshopper Control, Volume I and II*, (Senegal: FAO and the Plant Protection Directorate, Ministry of Agriculture 1997, 1998). As part of that study ponds were treated with deltamethrin (a synthetic pyrethroid) and bendiocarb (a carbamate). Delthamethrin had considerable acute effects on most macroinvertebrates, and bendiocarb affected a number of zooplankton.

12. This does not mean that organochlorines have not caused environmental effects in the tropics. For some examples, see F. Bro-Rasmussen, "Contamination by Persistent Chemicals in Food Chain and Human Health," *The Science of the Total Environment* 188 Suppl. 1 (1996): S45-60.

13. The proposed process of long-range transport in the atmosphere of certain organic compounds is called "global distillation," or "global fractionation." F. Wania and D. Mackay, "Global Fractionation and Cold Condensation of Low Volatility Organochlorine Compounds in Polar Regions" *Ambio* 22, no.1 (1993): 10-18. The theory has received support by modeling and monitoring data. H. W. Vallack et al, "Controlling Persistent Organic Pollutants—What Next?" *Environmental Toxicology and Pharmacology* 6 (1998): 143-75; and S. N. Meijer, W. A. Ockenden, E. Steinnes, H. P. Corrigan, and K. C. Jones, "Spatial and Temporal Trends of POPs in Norwegian and UK Background Air: Implications for Global Cycling;" *En-

vironmental Science and Technology* 37, no. 3 (2003): 454-61. Atmospheric deposition is considered to be the major source of POPs in the Arctic. A. Godduhn and L. K. Duffy, "Multigeneration Health Risks of Persistent Organic Pollution in the Far North: Use of the Precautionary Approach in the Stockholm Convention," *Environmental Science & Policy* 6 (2003): 341-53. There is thus a growing scientific consensus for the global distillation/fractionation hypothesis. Vallack et al, this note. This has also been reflected in policy where the Stockholm Convention includes in its screening criteria for adding further substances to the convention the potential for long-range transport through air, for example (one of the criteria for such substances is that their half-life in air must be greater than two days). Stockholm Convention on Persistent Organic Pollutants (POPs): Texts and Annexes (Geneva: Interim Secretariat for the Stockholm Convention on Persistent Organic Pollutants, 2001), 46-47.

14. An alternative concept for what has been referred to here as governance levels is levels of social organization. O. Young, *Institutional Dimensions of Global Environmental Change (IDGEC) Science Plan* (Bonn: International Human Dimensions Programme on Global Environmental Change, 1999).

15. Several methods were combined to solicit the perspectives and approaches of these stakeholders, including semistructured interviews and policy document analyses. A total number of 204 interviews were carried out during 8.5 months of fieldwork in the years 1997-1999. For further details on methodology, theoretical framework, and results of the study that are associated with this article, see S. Karlsson, *Multilayered Governance. Pesticides in the South: Environmental Concerns in a Globalised World* (Linköping, Sweden: Linköping University, 2000).

16. In Costa Rica, the average quantity of imported formulated pesticides rose from 8,100 tons to 15,300 tons between the second half of the

1980s and the first half of the 1990s. A. C. Rodríguez, R. van der Haar, D. Antich, and C. Jarquín, *Desarollo e Implementacion de un Sistema de Vigilancia de Intoxicaciones con Plaguicidas, Experenica en Costa Rica, Informe Tecnico Proyecto Plagsalud Costa Rica, Fase 1* (Development and Implementation of a Pesticide Intoxication Monitoring System, Costa Rica's Experience, Technical Report, Plagsalud Costa Rica Project, Phase I), (San José, Costa Rica: Ministerio de Salud Departamento de Sustancias Toxicas y Medicina del Trabajo (Ministry of Health, Department of Toxic Substances and Occupational Medicine), 1997). The nominal value of pesticide imports rose approximately 50 percent between 1990 and 1994. In 1994, the value of pesticide imports reached US$84.2 million and in the same year on average more than US$170 were spent on pesticides per ha agricultural land. S. Agne, *Economic Analysis of Crop Protection Policy in Costa Rica, Publication Series No. 4* (Hannover, Germany: Pesticide Policy Project, 1996): 6, 12. Because of the substantive formulating industry, the trade figures on the value of pesticide purchases are gross underestimates of the amount spent on pesticides in Costa Rican agriculture. Agne, this note, page 12. Statistics are limited on the use of pesticides in the African continent. J. J. Ondieki, "The Current State of Pesticide Management in Sub-Saharan Africa" *The Science of the Total Environment* 188, Suppl. no. I (1996): S30-34; and S. Williamson, *Pesticide Provision in Liberalised Africa: Out of Control? Network Paper No. 126* (London: Overseas Development Institute Agricultural Research & Extension Network, 2003). As many as 47 percent of African countries responding to an FAO questionnaire do not collect any statistics on pesticide import and use. FAO, note 4 above, page 71. One study reported that in Kenya the average annual import of pesticides for 1989-1993 was just over 5,000 tons at a value of US$28 million. H. Partow, *Pesticide Use and Management in Kenya* (Geneva: Institut Universitaire D'Etudes du

Développement (Graduate Institute of Development Studies), 1995): 205-6. The data reported for Kenya and Costa Rica only goes to the mid-1990s. (See note 2 above on falling pesticide sales in 2001.) The increasing trend is not taking place in all developing countries. Furthermore, studies in Africa have shown significant variation in impacts of, for example, liberalization on pesticide prices, access, and use. See A. W. Shepherd and S. Farfoli, "Export Crop Liberalisation in Africa: A Review," *FAO Agricultural Services Bulletin No. 135* (Rome: FAO, 1999); and Williamson, this note.

17. Agne, ibid., page 8. One source listed the main groups of pesticides used in the Central American countries: the insecticides organophosphates, carbamates, and pyrethroids; fungicides, mainly dithiocarbamics; and the herbicides phenoxyacids, dipyridyls, and more recently, triazines. L. E. Castillo, E. de la Cruz, and C. Rupert, "Ecotoxicology and Pesticides in Tropical Aquatic Ecosystems of Central America," *Environmental Toxicology and Chemistry* 16, no. 1 (1997): 41-51.

18. Partow, note 16 above, page vii. Inorganic pesticides accounted for 21 percent of imported pesticides, organophosphates accounted for 15 percent, organochlorines accounted for 11 percent, thiocarbamates accounted for 7 percent, and phtalimides accounted for 7 percent. These figures do not distinguish between products imported as technical grade or formulated product. For example, around 25 percent of the organophosphates are imported as technical grade material. Partow, note 16 above, pages vii, 39.

19. Agne, note 16 above and C. Conejo, R. Díaz, E. Furst, E. Gitli, and L. Vargas, *Comercio y Medio Ambiente: El Caso de Costa Rica* (Trade and the Environment: The Case of Costa Pica) (San José, Costa Rica: Centro Internacional en Política Económica Para el Desarollo Sostenible (International Center of Political Economy for Sustainable Development), 1996). The banana sector uses 45 percent of all

pesticides and at the end of the 1970s the disease commonly known in Spanish as *sigatoka negra* (the scientific name is *Mycosphaerella fijiensis*), arrived in the country, which affected production severely and led to significant pesticide use.

20. Castillo, de la Cruz, and Rupert note 17 above. Permanent crops like coffee, oilpalm, and cacao are sprayed less often (1-5 times/year) compared to annual crops such as tobacco, potatoes, or vegetables and products like banana, melon, watermelon, or flowers—extreme cases of which can be sprayed up to 39 times per cycle. J. E. García, *Introducción a los Plaguicidas* (Introduction to Pesticides) (San José, Costa Rica: Editorial Universidad Estatal a Distancia (Publishing Trust of the State University for Distance Education), 1997).

21. For a long time, coffee was the main export earner in agriculture, but tea has taken over the lead position. Since 1985, horticulture has grown substantially. In 1990 it came in third place as an export earner. General Agreement on Tariffs and Trade (GATT, *Trade Policy Review Kenya Volume 1* (Geneva: GATT, 1994). Between 1991 and 1994 the value of horticultural exports increased by 113 percent, the principle horticultural crops being cut flowers, French beans, mangoes, avocados, pineapples, and Asian vegetables. Republic of Kenya, *National Development Plan 1997-2001* (Nairobi: Government Printer, 1996).

22. While most pesticides are used on export crops, studies have shown that in some areas where the coffee economy has introduced a modernized agricultural system—for example, by using pesticides, their use on food crops has increased. A. Goldman, "Tradition and Change in Postharvest Pest Management in Kenya" *Agriculture and Human Values* 8, no. 1-2 (1991): 91-113. This is a phenomenon found in other countries as well. See, for example, Williamson, note 16 above.

23. Karlsson, note 15 above, pages 209-13 and 238—40.

24. In Naranjo, coffee farmers apply fungicides on average three times a year, nematicides a maximum of once per year and herbicides once or twice a year. Karlsson, note 15 above, page 262. About 7 percent of all pesticide purchases in Costa Rica are used on the 20 percent of agricultural land that is under coffee production. See Agne, note 16 above, page 13. In Meru, on the other hand, those who still spray their coffee apply fungicides either 2–4 times or 8–12 times a year. Herbicides are sprayed occasionally there. Karlsson, note 15 above, page 262. It is important to note that the comparisons here are only made from the self-reported numbers of sprayings per year. The dose in each application is not addressed.

25. This result is based on careful analysis of a large number of interviews with stakeholders and study of policy documents. Thus, the summary conclusions cannot be attributed to a single source. Karlsson, note 15 above. The problem categorization is not clear-cut since several of the areas are closely interrelated. The environment serves as a medium for transport of pesticides and their metabolites, which may expose humans to these substances via air and water (for example) and potentially affect human health. Pesticides, as a trade issue, emerge in the regulations established to address the concern of long-term, low-level exposure of pesticide residues in food. Pesticides that exert effects on non-target organisms—those on the farm and surrounding environment—can disrupt populations of natural enemies of the original pest, leading to increased and different pest attacks and thus production problems.

26. In addition to evidence from the interviews (see Karlsson, note 15 above, pages 144-45), references to environmental considerations were absent in pesticide management plans. Partow, note 16 above, page xii. A study from the early 1990s on the legislation and institutional framework for environmental protection and natural resource management in Kenya concluded that there was no

stress on environmental problems in the agricultural areas. S. H. Bragdon, *Kenya's Legal and Institutional Structure for Environmental Protection and Natural Resource Management—An Analysis and Agenda for the Future* (Washington, DC: Economic Development Institute of the World Bank, 1992). However, the Kenya National Environment Action Plan urges the adoption of as many nonchemical measures as possible and it urges the use of the least toxic chemicals as a last resort. Ministry of Environment and Natural Resources, *The Kenya National Environment Action Plan* (NEAP) (Nairobi, 1994).

27. For example, in Kenya, stricter European Union pesticide residue limits on agricultural products had caused significant concern both in government and in the pesticide industry. Standing Committee on the Use of Pesticides, *Interim Report of the Standing Committee on the Use of Pesticides* (Nairobi, 1996); and "Editorial," *Newsletter for the Pesticide Chemicals Association of Kenya*, February 1995. This was also one reason the existence of 100 tons of pesticide wastes (including many from banned substances) raised concern that these may find their way on to horticultural produce. Standing Committee on the Use of Pesticides, this note. It should be emphasized that the trade issue is very sensitive for national governments because of the high economic stakes involved, and they are likely to be very reluctant to discuss possible problems openly.

28. Republic of Kenya, "Chemical Safety Aspects in Kenya" (A country paper presented by the Kenyan delegation during IPCS Intensive Briefing Session on Toxic Chemicals, Environment and Health for Developing Countries, held in Arusha, United Republic of Tanzania, 1997).

29. These figures are quoted in M. A. Mwanthi and V. N. Kimani, "Patterns of Agrochemical Handling and Community Response in Central Kenya," *Journal of Environmental Health* 55, no. 7 (1993): 11-16. This figure—7 percent of the agricultural population suffering poisonings an-

nually—lies within the range between 2 and 9 percent reported in various studies in developing countries. Wesseling, McConnell, Partanen, and Hogstedt, note 3 above, page 283.

30. V. W. Kimani, "Studies of Exposure to Pesticides in Kibirigwi Irrigation Scheme, Kirinyaga District" (Submitted Ph.D. thesis, Department of Crop Science, University of Nairobi, 1996).

31. Ondieki, note 16 above, page S32.

32. Kimani, note 30 above, page 23.

33. A study in the 1980s of organochlorine residues in domestic fowl eggs in Central Kenya showed levels of dichlorodiphenyltrichloroethane (DDT) and dieldrin especially high, exceeding the acceptable daily intake (ADI) for children. (This ADI standard was developed by a panel of experts linked to WHO as part of the FAO/WHO Joint Meeting on Pesticide Residues (JMPR). Residues of dieldrin exceeded ADI for adults. J. M. Mugambi, L. Kanja, T. E. Maitho, J. U. Skaare, and P. Lökken, "Organochlorine Pesticide Residues in Domestic Fowl (*Gallus domesticus*) Eggs from Central Kenya," *Journal of the Science of Food and Agriculture* 48, no. 2 (1989): 165-76. Another study found organochlorines in mothers' milk exceeding the ADI for infants; except for lindane the exposure occurred long ago. Kimani note 30 above, pages 224-25. Because these substances have been banned, the levels will decline.

34. Regulation No. 20345-S in E. Wo-Ching Sancho and R. Castro Córdoba, *Compendio de Legislacion Sobre Plaguicidas* (Compendium of Pesticide Legislation) (San José, Costa Rica: Organizacion Panamericana de la Salud Proyecto Plag-Salud, Centro de Derecho Ambiental y de los Recursos Naturales (CEDARENA) (Pan American Health Organization Plag-Salud Project, Center for Environmental Rights and Natural Resources), 1996). A project was started in 1993 to improve reporting and help to prevent intoxications. R. Castro Córdoba, N. Morera González, and C.

Jarquín Nuñez, *Sistema de Vigilancia Epidemiologica de Intoxicaciones con Plaguicidas, la Experiencia de Costa Rica, 1994-1996* (Epidemiological Monitoring System of Pesticide Intoxications, Costa Rica's Experience, 1994-1996) (San José, Costa Rica: Departamento de Registro y Control de Sustancias y Medicina del Trabajo del Ministerio de Salud (Ministry of Health Department of Registration and Control of Occupational Substances and Medicine), 1998).

35. Rodriguez, van der Haar, Antich, and Jarquín, note 16 above.

36. Ministerio de Salud, División de Saneamiento Ambiental, Departamento de Registro y Control de Sustancias Tóxicas y Medicina del Trabajo (Ministry of Health, Division of Environmental Health, Department of Registration and Control of Toxic Substances and Occupational Medicine), *Reporte Oficial intoxicaciones con Plaguicidas 1996* (Official Report of Pesticide Intoxications 1996) (Costa Rica, 1997).

37. Castro Córdoba, Morera González, and Jarquín Núñez, note 34 above. A study of the percentage of underreporting of intoxications in the system of monitoring gave an underreporting of 43 percent of symptoms that should have been linked to pesticides. Most of these were either dermal lesions or intoxications occurring outside the working environment, Rodríguez, van der Haar, Antich, and Jarquín, note 16 above, page 19.

38. Ministerio de Salud, note 36 above, page 13.

39. In the 1980s, even workers younger than 15 had very high occupational incidence rates. Today such instances are less frequent due to better enforcement of legislation that prohibits children under 18 to work with pesticides. Catharina Wesseling, Instituto Regional de Estudios en Sustancias Tóxicas (Central American Institute for Studies on Toxic Substances), e-mail message to author, 22 March 2004.

40. C. Wesseling, *Health Effects From Pesticide Use in Costa Rica—An Epidemiological Approach* (Stockholm:

Institute of Environmental Medicine, Karolinska Institute, 1997).

41. C. Wesseling, L. Castillo, and C.F. Elinder, "Pesticide Poisoning in Costa Rica," *Scandinavian Journal of Work and Environmental Health* 19 (1993): 227-35. The data on hospitalizations and occupational accidents included a significant number of cases where the pesticide substance had not been identified.

42. Wesseling, note 40 above, page 42

43. Wesseling, note 40 above page 50.

44. Wesseling, note 40 above, page 50

45. For example, a study of organochlorine pesticides along the Kenyan coast only revealed very low levels compared to other areas, including tropical areas, J. M. Everaarts, E. M. van Weerlee, C. V. Fischerm, and T J. Hillebrand, "Polychlorinated Byphenyls and Cyclic Pesticides in Sediments and Macro-invertebrates from the Coastal Zone and Continental Slope of Kenya," *Marine Pollution Bulletin* 36, no. 6 (1998): 492-500. In a study of the concentration in water of organochlorine pesticides in coffee- and tea-growing areas that was made in 1994-1995, the mean pesticide levels did not exceed WHO or U.S. Environmental Protection Agency limits for drinking water. M. A. Mwanthi, "Occurrence of Three Pesticides in Community Water Supplies, Kenya" *Bulletin of Environmental Contamination and Toxicology* 60, no. 4 (1998): 601-8. However, the Development Plan of 1989-1993 reported that agrochemicals had led to severe pollution effects in the Tana and Athi Rivers regimes. B. D. Ogolla, "Environmental Management Policy and Law" *Environmental Policy and Law* 22, no. 3 (1992): 164-75. There had also been a case of paraquat contamination in the water supply for one town (Anonymous official, Ministry of Land Reclamation, Regional Development and Water, interview by author, Nairobi, Kenya, 15 October 1997).

46. L. Corrales and A. Salas, *Diagnóstico Ambiental de la Actividad Bananera en Sarapiquí Tortuguero y*

Talamanca, Costa Rica 1990-1992 (Con Actualizaciones Parciales a 1996) (Environmental Assessment of Banana Cultivation in Sarapiquí, Tortuguero, and Talamanca, Costa Rica 1990-1992 (with Partial Updates until 1996) (San José, Costa Rica: Oficina Regional para Mesoamérica, Unión Mundial para la Naturaleza (Regional Office for Mesoamerica, International Union for the Conservation of Nature (IUCN)), 1997). The environmental pollution reported included: high concentrations of heavy metals in coral reef along the coast that could partly be due to pesticides; detection of DDT residues in some fish species in the late 1980s; detection of hexachlorobenzene (HCB), dieldrin, DDT, 1, 1-dichloro-2,2-bis(p-chlorophenyl)ethylene (DDE), paraquat and lindane in soil; detection of organochlorines and organophosphates in rivers and along the coast of the Atlantic; residues of various pesticides in sediments primarily chlorothalonil; the detection of chlorothalonil in ground water around banana plantations.

47. Karlsson, note 15 above, page 264. The study was not a quantitative survey with the rigidity that is implied in random farmer selection. The more than 30 farmers who were interviewed in each district were selected with purposeful sampling, with the goal of seeking the broadest range of views rather than averages. It is thus not possible to give figures on the number of experienced intoxications and pesticide products used (for example), but qualitative judgments on differences between districts can be made.

48. When looking at how individuals and organizations address pesticide-associated problems, the study focuses on those measures that aim to reduce primarily the health and environmental risks. Evidently, such measures can be of relevance for reducing trade, production, and economic problems, and conversely measures taken to address these may have positive effects on health and environment as well. Karlsson, note 15 above.

49. In addition to the Rotterdam and Stockholm Conventions described in Table 4, there is, for example, the Basel Convention on the Control of Transboundary Movements of Hazardous Waste and Their Disposal (www.basel.int) and the ILO Convention on Safety in the Use of Chemicals (www.ilo.org/public/english/protection/safework/cis/products/safetytm/c170.htm). Furthermore, the Codex Alimentarius (linked to WHO and FAO) establishes recommended maximum residue limits (MRLs) of pesticides in traded food products, but with the Agreement of Sanitary and Phytosanitary Measures these standards have indirectly become legally binding on the member countries of the WTO. Karlsson, note 15 above, page 96.

50. The FAO Code was first adopted by the FAO General Assembly in 1985 and has since been amended twice, in 1989 and 2002. FAO, International Code of Conduct on the Distribution and Use of Pesticides, Revised Version (Rome: FAO, 2003), accessed via http://www.fao.org/ag/agp/agpp/pesticid/ on 19 January 2004.

51. Examples of guidelines include FAO, *Guidelines for Legislation on the Control of Pesticides* (Rome: FAO, 1989); and FAO, *Revised Guidelines on Environmental Criteria for the Registration of Pesticides* (Rome: FAO, 1989).

52. Support for agricultural pesticide use has been a commonplace element in development projects, supported by various multi- and bilateral donors, but data on these are often missing from government import data. Williamson, note 16 above, page 3. While this practice is now less common, it is still taking place. Williamson, note 16 above, page 11. The IPM concept had been present in IGO discussions for many decades. Originally it was focused on controlling pest populations through a combination of all suitable techniques below thresholds that would cause economic damage. FAO, *International Code of Conduct on the Distribution and Use of Pesticides, Amended Version*, (Rome:

FAO, 1990). In the 1990s, however, the IPM concept came to be understood as a means to minimize the use of pesticides and increase reliance on alternative pest management technologies. See "Agenda 21," Rio de Janeiro, 2002, in Report of the United Nations Conference on Environment and Development, Rio de Janeiro, 3-14 June 1992, Volume I Resolutions Adopted by the Conference, A/CONF.151/26.Rev.1.

53. For example, it took until 1998 for FAO to have a first meeting with the International Federation of Organic Agriculture Movements (IFOAM), whose member organizations around the world are involved in research on alternatives to pesticides and in the design and manufacture of technology for controlling weeds, Karlsson, note 15 above, pages 107-8.

54. Organizations and IGOs left out of Table 4, but which were part of the study, include the International Labour Organization (ILO), United Nations Institute for Training and Research (UNITAR), World Health Organization (WHO), and the World Trade Organization (WTO).

55. Kenya's Pest Control Products Act came into force in 1984 and is often referred to as one of the most comprehensive laws in Africa. Republic of Kenya, The Pest Control Products Act Chapter 346 (Nairobi: Government Printer, 1985). In many respects it conforms to the guidelines for pesticide legislation from FAO. In 1997, 241 pesticide products had been registered and the rest were being screened. Republic of Kenya, note 28 above. The first effort to regulate pesticides in Costa Rica was made in 1954 and a modern pesticide registration process was established in 1976. R. Castro Córdoba, *Estudio Diagnostico Sobre la Legislación de Plaguicidas en Costa Rica* (Diagnostic Study of Pesticide Legislation in Costa Rica) (San José, Costa Rica: CEDARENA, 1995). The most recent revision was made in 1995 when law No. 24337MAG-S was published. Wo-Ching Sancho and Castro Córdoba, note 34 above. In 1993, there were 1,213 pesticides registered in Costa Rica, 347 generics,

and 38 mixtures. Garcia, note 20 above, page 235.

56. As a basis for the government agency's decision, the companies need to submit data that supports the efficacy of the product to control pests on the crops it was intended to be used as well as a long list of physical, chemical, toxicological and ecotoxicological data. Pest Control Products Board (PCPB), *Data Requirements for Registration of Pest Control Products, Legal Notice N46* (Nairobi, 1994); and Ministerio de Salud (Ministry of Health), *Pesticide Registration in Costa Rica* (Costa Rica, n.d.).

57. Kenya has banned the following pesticides: in 1986, dibromochloropropane, ethylene dibromide, 2,4,5 Trichlorophenoxyacetic acid (2,4,5-T), chlordimeform, hexachlorocychlohexane (HCH), chlordane, heptachlor, endrin, toxaphene; in 1988, parathion; and in 1989, captafol. In addition, in 1986, lindane was restricted use for seed dressing only; aldrin and diledrin were restricted for termite control in the building industry and DDT was restricted for use in public health. PCPB, Banned/Restricted Pesticides in Kenya (Nairobi, n.d.). Costa Rica has banned the following pesticides: in 1987, 2,4,5-T; in 1988, alchin, captafol, chlordecone, chlordimeform, DDT, dibromochloropropane (DBCP), dinoseb, ethylendibromide (EDB), nitrofen, toxafen; in 1990, lead arsenate, cyhexatin, endrin, pentachlorophenol; in 1991, chlordane, heptachlor; and in 1995, lindane. Ministerio de Salud, *Lista de Plaguicidas Prohibidos y Restringidos en Costa Rica* (List of Banned and Restricted Pesticides in Costa Rica) (Costa Rica, n.d.)

58. In Kenya, the pesticide industry, through the Global Crop Protection Federation (GCPF), cooperated with the government's extension system in a safe-use project, and between 1991 and 1993, about 280,000 people were trained, including 2,800 retailers. Anonymous official, GCPF-Kenya, interview by author, Nairobi, Kenya, 23 September 1997. In Costa Rica, cooperation between the government and the pesticide industry

association has taken place throughout the 1990s, reaching 110,000 people. Anonymous official, Cámara Insumos Agropecuarios (Chamber of Agricultural and Livestock Inputs), interview by author, San Jose, Costa Rica, 4 February 1998. In 1991, the industry association and two ministries initiated the project "Teach," which was geared at training teachers in the rural areas so that they can teach children about pesticide issues. Conejo, Díaz, Furst, Gitli, and Vargas, note 19 above. Despite these efforts, there were a number of stakeholders in both countries who were concerned about the low effectiveness of the training measures. Karlsson, note 15 above. In Kenya, for example, results in the form of increased understanding of the toxic effects of pesticides were noted, but less than 30 percent of the farmers trained adopted the safety measures prescribed. Kimani, note 30 above, page 49. The Kenya Safe Use Project was also criticized for neglecting to train pesticide workers in the plantation sector. Partow, note 16 above, page xiv.

59. The Pest Control Products (Labeling, Advertising and Packaging) Regulation 3 (2) (n) requires each pesticide label to state that it is against the law to "use or store pest control products under unsafe conditions." Republic of Kenya, note 55 above. In Costa Rica, the banana sector is regulated by a special law (No. 7147), which obliges employers to train workers in appropriate use of pesticides and their associated risks. Castro Córdoba, note 55 above, page 21.

60. The Kenyan government has been involved in research on IPM and there were some pilot projects (often supported by donors). P. C. Matteson and M. I. Meltzer, *Environmental and Economic Implications of Agricultural Trade and Promotion Policies in Kenya: Pest and Pesticide Management* (Arlington, VA: Winrock International Environmental Alliance, 1995); O. Zethner, "Practice of Integrated Pest Management in Sub-Tropical Africa: An Overview of Two Decades (1970-1990)," in A. N. Mengech, K. N. Saxena and H. N. B. Gopalan, eds., *Inte-*

grated Pest Management in the Tropics, Current Status and Future Prospects (New York: John Wiley & Sons, 1995); and Organisation for Economic Co-operation and Development (OECD), *Report of the OECD/FAO Workshop on Integrated Pest Management and Pesticide Risk Reduction, Neuchâtel, Switzerland, 28 June—2 July 1998. ENV/JM/ MONO (99)* 7 (Paris: OECD, 1999). In Costa Rica, the official extension service promotes IPM. Ague, note 16 above, page 27. Several programs have been introduced in the country to reduce the volume of pesticide use. A. Faber, *Study on Investigations on Pesticides and the Search for Alternatives in Costa Rica* (Guápiles, Costa Rica: Wageningen Agricultural University, 1997). However, these initiatives remained isolated in a dominating agricultural system where pesticide use has become considered essential to increase productivity in most crops Organización Panamericana de la Salud, Programa Medío Ambiente y Salud en el Istmo CentroAmericano (Pan American Health Organization, Program of Environment and Health in the Central American Isthmus), *Aspectos Ocupacionales y Ambientales de la Exposición a Plaguicidas en el Istmo Centroamericano PLAGSALUD-Fase II* (Environmental and Occupational Aspects of Pesticide Exposure in the Central American Isthmus PLAGSALUD-Phase II), PLG97ESP.POR (San José, Costa Rica: 1997).

61. In Kenya, the Kenya Institute of Organic Farming (KIOF) began in 1986, and in the first years they met resistance from the government. Anonymous official, KIOF, interview by author, Nairobi, Kenya, 26 September 1997. KIOF trains farmers' groups in the field and arrange exchange visits among groups. KIOF, *Organic Farming, A Sustainable Method of Agriculture* (Nairobi, n.d.). Organic farming has been through a period of significant growth in Costa Rica with estimates of about 3,000 ha being under organic agriculture. Faber, ibid., page 14. There was high demand for organic products in the export market but virtually no demand in the domestic market. Karlsson, note 15

above, page 180. The Costa Rican government supported organic farming, for instance, it created a law for organic agriculture included in the environmental law in 1995, and it established a special office in the Ministry of Agriculture. The Asociación Nacional de Agricultura Organica (ANAO) (National Association for Organic Agriculture) had received funds to establish a nationally based organic certification system in the second half of the 1990s. Karlsson, note 15 above, page 181.

62. This definition of institutions is often used in social science and economics and differs somewhat from the common-language use that defines institutions as organizations. For a discussion on the role of institutions in environmental governance, see, for example, Young, note 14 above.

63. The issue of fit is extensively explored in relation to common property resource management. See, for example, R. J. Oakerson, "Analyzing the Commons: A Framework," in D. W. Bromley, ed., *Making the Commons Work: Theory, Practice and Policy* (San Francisco: ICS Press, 1992); E. Ostrom, "Designing Complexity to Govern Complexity," in S. Hanna and M. Munasinghe, eds., *Property Rights and the Environment, Social and Ecological Issues* (Washington, DC: Beijer International Institute of Ecological Economics and the World Bank, 1995); M. McGinnis and E. Ostrom "Design Principles for Local and Global Commons," in O. R. Young, ed., *The International Political Economy and International Institutions Volume II* (Cheltenham, UK: Edward Elgar Publishing Ltd., 1996). For a general discussion of the concept of fit, see C. Folke, L. Jr. Pritchard, F. Berkes, J. Coiling and U. Svedin, *The Problem of Fit between Ecosystems and Institutions* (Bonn, Germany: IHDP, 1998) accessed via www.uni-bonn.de/ihdp/ wp02main.htm, on 27 June 2000.

64. Anonymous official, CEDARENA, interview by author, San José, Costa Rica, 10 March 1998. There have been efforts to harmonize pesticide registration in Central America

under the umbrella of Organismo Internacional Regional de Sanidad Agropecuaria (OIRSA) (International Regional Organization for Plant and Animal Health) with some support from FAO, but at the time of the study (1998/1999) efforts seemed to have halted. See Karlsson, note 15 above, page 199.

65. This strategy is not straightforward, however, as there are usually layers of driving forces and responsible stakeholders, often at different levels. It can be difficult to identify the original causes (or ultimate drivers), and this confuses the allocation of responsibility between levels, J. Saurin, "Global Environmental Degradations, Modernity and Environmental Knowledge," in C. Thomas, ed., *Rio Unravelling the Consequences* (Essex, UK: Frank Cass, 1994).

66. Karlsson, note 15 above, page 87.

67. Karlsson, note 15 above, pages 156-60 and 184-89.

68. It has been argued that it is difficult to be held responsible for a problem if one does not have the means to respond to it. T. Princen, "From Property Regime to International Regime: An Ecosystems Perspective," *Global Governance* 4, no. 4 (1998): 395-413.

69. For a discussion on altruistic motivations for behavior, see, for example, J. J. Mansbridge, ed. *Beyond Self-Interest* (London: University of Chicago Press, 1990).

70. In an FAO questionnaire, 46 percent of the responding developing countries felt that the pesticide industry acted only partly responsibly or not responsibly in adhering to the provisions of the FAO Code of Conduct as a standard for the manufacture, distribution, and advertising of pesticides. FAO, note 4 above, page 9.

71. L. L. Kiser and E. Ostrom, "The Three Worlds of Action: A Metatheoretical Synthesis of Institutional Approaches," in E. Ostrom, ed., *Strategies of Political Inquiry* (London: Sage Publications, 1982); and E. Ostrom, *Governing the Commons, The Evolution of Institutions for Collective Action* (New York: Cambridge University Press, 1990). While these authors refer to them as rules, they fall within the definition of institutions and that term is used here for clarity.

72. C. C. Gibson, E. Ostrom, and T K. Ahn "The Concept of Scale and the Human Dimensions of Global Change: A Survey," *Ecological Economics* 32, no. 2 (2000): 217-39.

73. Ostrom, note 71 above, page 52.

74. Karlsson, note 15 above, page 185.

75. Ostrom, note 71 above, page 52.

76. Karlsson, note 15 above; and S. Karlsson, "Institutionalized Knowledge Challenges in Pesticide Governance—The End of Knowledge and Beginning of Values in Governing Globalized Environmental Issues," *International Environmental Agreements* (forthcoming in 2004).

77. Williamsson, note 16 above.

78. In Kenya, the efficacy tests were made under some kind of cost-sharing arrangement with the government, while in Costa Rica the companies had to cover the full costs. Karlsson, note 15 above, pages 157, 186.

79. The term "multilayered governance" was constructed and defined as a system of coordinated and collective governance across levels in Karlsson, note 15 above, page 40. See also P. Hirst and G. Thompson, *Globalization in Question* (Cambridge, UK: Polity Press, 1996), 184, who argue for more 'sutured' governance across levels when discussing economic aspects of globalization, and Young, note 14 above, page 34, who argues for the need to influence behavior at all levels of governance.

Sylvia I. Karlsson is a postdoctoral fellow at Yale University's Center for Environmental Law and Policy and is a research fellow with the Institutional Dimensions of Global Environmental Change (IDGEC) project. Her current research focuses on cross-level aspects of global sustainable development governance. In 2001-2003 she worked as International Science Project Coordinator at the International Human Dimensions Programme on Global Environmental Change (IHDP) in Bonn, Germany. Sylvia I. Karlsson has worked for a short time at UNEP Chemicals in Geneva and the Economic Development Institute of the World Bank and as program officer for an action research project in Eastern Africa. Parallel to her studies and research, she has been actively engaged in the NGO processes of the Rio Conference in 1992, the World Summit for Social Development in 1995, and most recently, the World Summit on Sustainable Development, where she headed the delegation of the International Environment Forum, a scientific NGO accredited to the summit. The author wishes to thank Dr. Arthur L. Dahl and Ms. Agneta SundénBýléhn and several anonymous reviewers for their helpful comments on earlier drafts. She further wishes to express her gratitude to all the people around the world who patiently answered questions during interviews—from the UN halls of Geneva to the soft grass of a Kenyan *shamba* (farm). The research was made possible through a grant from the Swedish International Development Agency (Sida). Karlsson's work has appeared in the peer-reviewed journals *International Environmental Agreements* and *The Common Property Resource Digest* and in a book edited by D.C. Esty and M. Ivanova, *Strengthening Global Environmental Governance: Options and Opportunities* (New Haven, CT: Yale School of Forestry & Environmental Studies, 2002). Karlsson can be contacted via e-mail at sylvia.karlsson@yale.edu.

The Quest for Clean Water

As water pollution threatens our health and environment, we need to implement an expanding array of techniques for its assessment, prevention, and remediation.

Joseph Orlins and Anne Wehrly

In the 1890s, entrepreneur William Love sought to establish a model industrial community in the La Salle district of Niagara Falls, New York. The plan included building a canal that tapped water from the Niagara River for a navigable waterway and a hydroelectric power plant. Although work on the canal was begun, a nationwide economic depression and other factors forced abandonment of the project.

By 1920, the land adjacent to the canal was sold and used as a landfill for municipal and industrial wastes. Later purchased by Hooker Chemicals and Plastics Corp., the landfill became a dumping ground for nearly 21,000 tons of mixed chemical wastes before being closed and covered over in the early 1950s. Shortly thereafter, the property was acquired by the Niagara Falls Board of Education, and schools and residences were built on and around the site.

In the ensuing decades, groundwater levels in the area rose, parts of the landfill subsided, large metal drums of waste were uncovered, and toxic chemicals oozed out. All this led to the contamination of surface waters, oily residues in residential basements, corrosion of sump pumps, and noxious odors. Residents began to question if these problems were at the root of an apparent prevalence of birth defects and miscarriages in the neighborhood.

Eventually, in 1978, the area was declared unsafe by the New York State Department of Health, and President Jimmy Carter approved emergency federal assistance. The school located on the landfill site was closed and nearby houses were condemned. State and federal agencies worked together to relocate hundreds of residents and contain or destroy the chemical wastes.

That was the bitter story of Love Canal. Although not the worst environmental disaster in U.S. history, it illustrates the tragic consequences of water pollution.

Water quality standards

In addition to toxic chemical wastes, water pollutants occur in many other forms, including pathogenic microbes (harmful bacteria and viruses), excess fertilizers (containing compounds of phosphorus and nitrogen), and trash floating on streams, lakes, and beaches. Water pollution can also take the form of sediment eroded from stream banks, large blooms of algae, low levels of dissolved oxygen, or abnormally high temperatures (from the discharge of coolant water at power plants).

The United States has seen a growing concern about water pollution since the middle of the twentieth century, as the public recognized that pollutants were adversely affecting human health and rendering lakes unswimmable, streams unfishable, and rivers flammable. In response, in 1972, Congress passed the Federal Water Pollution Control Act Amendments, later modified and referred to as the Clean Water Act. Its purpose was to "restore and maintain the chemical, physical, and biological integrity of the nation's waters."

The Clean Water Act set the ambitious national goal of completely eliminating the discharge of pollutants into navigable waters by 1985, as well as the interim goal of making water clean enough to sustain fish and wildlife, while being safe for swimming and boating. To achieve these goals, certain standards for water quality were established.

The "designated uses" of every body of water subject to the act must first be identified. Is it a source for drinking water? Is it used for recreation, such as swimming? Does it supply agriculture or industry? Is it a significant habitat for fish and other aquatic life? Thereafter, the water must be tested for pollutants. If it fails to meet the minimum standards for its designated uses, then steps must be taken to limit pollutants entering it, so that it becomes suitable for those uses.

On the global level, the fundamental importance of clean water has come into the spotlight. In November 2002, the UN Committee on Economic, Cultural and Social Rights declared access to clean

Tragedy at Minamata Bay

The Chisso chemical factory, located on the Japanese island of Kyushu, is believed to have discharged between 70 and 150 tons of methylmercury (an organic form of mercury) into Minamata Bay between 1932 and 1968. The factory, a dominant presence in the region, used the chemical to manufacture acetic acid and vinyl chloride.

Methylmercury is easily absorbed upon ingestion, causing widespread damage to the central nervous system. Symptoms include numbing and unsteadiness of extremities, failure of muscular coordination, and impairment of speech, hearing and vision. Exposure to high levels of the substance can be fatal. In addition, the effects are magnified for infants exposed to methylmercury through their mothers, both before birth and while nursing.

In the 1960s and '70s, it was revealed that thousands of Minamata Bay residents had been exposed to methylmercury. The chemical had been taken up from the bay's waters by its fish and then made its way into the birds, cats, and people who ate the fish. Consequently, methylmercury poisoning came to be called Minamata disease.

Remediation, which took as long as 14 years, involved removing the mercury-filled sediments and containing them on reclaimed land in Minamata Bay. Fish in the bay had such high levels of methylmercury that they had to be prevented from leaving the bay by a huge net, which was in place from 1947 to 1997.

Mercury poisoning has recently appeared in the Amazon basin, where deforestation has led to uncontrolled runoff of natural accumulations of mercury from the soil into rivers and streams. In the United States, testing has revealed that predator fish such as bass and walleye in certain lakes and rivers contain enough mercury to justify warnings against consuming them in large amounts.

—J.O. and A.W.

Minamata Bay residents who were exposed to methylmercury have been suffering from such problems as loss of muscular control, numbing of extremities, and impairment of speech, hearing, and vision.

water a human right. Moreover, the United Nations has designated 2003 to be the International Year of Freshwater, with the aim of encouraging sustainable use of freshwater and integrated water resources management.

Here, there, and everywhere

Implementing the Clean Water Act requires clarifying the sources of pollutants. They are divided into two groups: "point sources" and "nonpoint sources." Point sources correspond to discrete, identifiable locations from which pollutants are emitted. They include factories, wastewater treatment plants, landfills, and underground storage tanks. Water pollution that originates at point sources is usually what is associated with headline-grabbing stories such as those about Love Canal.

Nonpoint sources of pollution are diffuse and therefore harder to control. For instance, rain washes oil, grease, and solid pollutants from streets and parking lots into storm drains that carry them into bays and rivers. Likewise, irrigation and rainwater leach fertilizers, herbicides, and insecticides from farms and lawns and into streams and lakes.

In the United States, the Clean Water Act requires that industrial wastes be neutralized or broken down before being released into rivers and lakes.

The direct discharge of wastes from point sources into lakes, rivers, and streams is regulated by a permit program known as the National Pollutant Discharge Elimination System (NPDES). This program, established through the Clean Water Act, is administered by the Environmental Protection Agency (EPA) and authorized states. By regulating the wastes discharged, NPDES has helped reduce point-source pollution dramatically. On the other hand, water pollution in the United States is now mainly from nonpoint sources, as reported by the EPA.

In 1991, the U.S. Geological Survey (USGS, part of the Department of the Interior) began a systematic, long-term program to monitor watersheds. The National Water-Quality Assessment Program (NAWQA), established to help manage surface and groundwater supplies, has involved the collection and analysis of water quality data in over 50 major river basins and aquifer systems in nearly all 50 states.

The program has encompassed three principal categories of investigation: (1) the current conditions of surface water and groundwater; (2) changes in those conditions over time; and (3) major factors—such as climate, geography, and land use—that affect water quality. For each of these categories, the water and sediment have been tested for such pollutants as pesticides, plant nutrients, volatile organic compounds, and heavy metals.

The NAWQA findings were disturbing. Water quality is most affected in watersheds with highest population density and urban development. In agricultural areas, 95 percent of tested streams and 60 percent of shallow wells contained herbicides, insecticides, or both. In urban areas, 99 percent of tested streams and 50 percent of shallow wells had herbicides, especially those used on lawns and golf courses. Insecticides were found more frequently in urban streams than in agricultural ones.

The study also found large amounts of plant nutrients in water supplies. For instance, 80 percent of agricultural streams and 70 percent of urban streams were found to contain phosphorus at concentrations that exceeded EPA guidelines.

Moreover, in agricultural areas, one out of five well-water samples had nitrate concentrations higher than EPA standards for drinking water. Nitrate contamination can result from nitrogen fertilizers or material from defective septic systems leaching into the groundwater, or it may reflect defects in the wells.

Effects of pollution

According to the UN World Water Assessment Programme, about 2.3 billion people suffer from diseases associated with polluted water, and more than 5 million people die from these illnesses each year. Dysentery, typhoid, cholera, and hepatitis A are some of the ailments that result from ingesting water contaminated with harmful microbes. Other illnesses—such as malaria, filariasis, yellow fever, and sleeping sickness—are transmitted by vector organisms (such as mosquitoes and tsetse flies) that breed in or live near stagnant, unclean water.

A number of chemical contaminants—including DDT, dioxins, polychlorinated biphenyls (PCBs), and heavy metals—are associated with conditions ranging from skin rashes to various cancers and birth defects. Excess nitrate in an infant's drinking water can lead to the "blue baby syndrome" (methemoglobinemia)—a condition in which the child's digestive system cannot process the nitrate, diminishing the blood's ability to carry adequate concentrations of oxygen.

Besides affecting human health, water pollution has adverse effects on ecosystems. For instance, while moderate amounts of nutrients in surface water are generally not problematic, large quantities of phosphorus and nitrogen compounds can lead to excessive growth of algae and other nuisance species. Known as *eutrophication*, this phenomenon reduces the penetration of sunlight through the water; when the plants die and decompose, the body of water is left with odors, bad taste, and reduced levels of dissolved oxygen.

Low levels of dissolved oxygen can kill fish and shellfish. In addition, aquatic weeds can interfere with recreational activities (such as boating and swimming) and can clog intake by industry and municipal systems.

Some pollutants settle to the bottom of streams, lakes, and harbors, where they may remain for many years. For instance, although DDT and PCBs were banned years ago, they are still found in sediments in many urban and rural streams. They occur at levels harmful to wildlife at more than two-thirds of the urban sites tested.

Prevention and remediation

As the old saying goes, an ounce of prevention is worth a pound of cure. This is especially true when it comes to controlling water pollution. Several important steps taken since the passage of the Clean Water Act have made surface waters today cleaner in many ways than they were 30 years ago.

For example, industrial wastes are mandated to be neutralized or broken down before being discharged to streams, lakes, and harbors. Moreover, the U.S. government has banned the production and use of certain dangerous pollutants such as DDT and PCBs.

In addition, two major changes have been introduced in the handling of sewage. First, smaller, less efficient sewage treatment plants are being replaced with modern, regional plants that include biological treatment, in which microorganisms are used to break down organic matter in the sewage. The newer plants are releasing much cleaner discharges into the receiving bodies of water (rivers, lakes, and ocean).

Second, many jurisdictions throughout the United States are building separate sewer lines for storm water and sanitary wastes. These upgrades are needed because excess water in the older, "combined" sewer systems would simply bypass the treatment process, and untreated sewage would be discharged directly into receiving bodies of water.

To minimize pollutants from nonpoint sources, the EPA is requiring all municipalities to address the problem of runoff from roads and parking lots. At the same time, the use of fertilizers and pesticides needs to be reduced. Toward this end, county extension agents are educating farmers and homeowners about their proper application and the availability of nutrient testing.

To curtail the use of expensive and potentially harmful pesticides, the approach known as *integrated pest management* can be implemented [see "Safer Modes of Pest Control," THE WORLD & I, May 2000, p. 164]. It involves the identification of specific pest problems and the use of nontoxic chemicals and chemical-free alternatives whenever possible. For instance, aphids can be held in check by ladybug beetles and caterpillars can be controlled by applying neem oil to the leaves on which they feed.

Moreover, new urban development projects in many areas are required to implement storm-water management practices. They include such features as: oil and grease traps in storm drains; swales to slow down runoff, allowing it to infiltrate back into groundwater; "wet" detention basins (essentially artificial ponds) that allow solids to settle out of runoff; and artificial wetlands that help break down contaminants in runoff. While such additions may be costly, they significantly improve water quality. They are of course much more expensive to install after those areas have been developed.

Once a waterway is polluted, cleanup is often expensive and time consuming. For instance, to increase the concentration of dissolved oxygen in a lake that has undergone eutrophication, fountains and aerators may be necessary. Specially designed boats may be needed to harvest nuisance weeds.

At times, it is costly just to identify the source of a problem. For example, if a body of water contains high levels of coliform bacteria, expensive DNA testing may be needed to determine whether the bacteria came from leakage of human sewage, pet waste, or the feces of waterfowl or other wildlife.

Testing the effect of simulated rainfall on fresh manure, scientists with the Agricultural Research Service have found that grass strips are highly effective at preventing manure-borne microbes from being washed down a slope and contaminating surface waters.

Contaminated sediments are sometimes difficult to treat. Available techniques range from dredging the sediments to "capping" them in place, to limit their potential exposure. Given that they act as reservoirs of pollutants, it is often best to remove the sediments and burn off the contaminants. Alternatively, the extracted sediments may be placed in confined disposal areas that prevent the pollutants from leaching back into groundwater. Dredging, however, may create additional problems by releasing pollutants back into the water column when the sediment is stirred up.

The future of clean water

The EPA reports that as a result of the Clean Water Act, millions of tons of sewage and industrial waste are being treated before they are discharged into U.S. coastal waters. In addition, the majority of lakes and rivers now meet mandated water quality goals.

Yet the future of federal regulation under the Clean Water Act is unclear. In 2001, a Supreme Court decision (*Solid Waste Agency of Northern Cook County v. United States Army Corps of Engineers, et al.*) brought into question the power of federal agencies to regulate activities affecting water quality in smaller, nonnavigable bodies of water. This and related court decisions have set the stage for the EPA and other federal agencies to redefine which bodies of water can be protected from unregulated dumping and discharges under the Clean Water Act. As a result, individual states may soon be faced with much greater responsibility for the protection of water resources.

Worldwide, more than one billion people presently lack access to clean water sources, and over two billion live without basic sanitation facilities. A large proportion of those who die from water-related diseases are infants. We would hope that by raising awareness of these issues on an international level, the newly recognized right to clean water will become a reality for a much larger percentage of the world's population.

Joseph Orlins, professor of civil engineering at Rowan University in Glassboro, New Jersey, specializes in water resources and environmental engineering. Anne Wehrly is an attorney and freelance writer.

On the Internet

2003: INTERNATIONAL YEAR OF FRESHWATER (UN)
www.wateryear2003.org

CLEAN WATER ACT (EPA)
www.epa.gov/watertrain/cwa

DO'S AND DON'T'S AROUND THE HOME (EPA)
www.epa.gov/owow/nps/dosdont.html

LOVE CANAL COLLECTION (SUNY BUFFALO)
ublib.buffalo.edu/libraries/projects/lovecanal

NATIONAL WATER-QUALITY ASSESSMENT PROGRAM (USGS)
water.usgs.gov/nawqa

OFFICE OF WATER (EPA)
www.epa.gov/ow

THE WORLD'S WATER
www.worldwater.org

This article appeared in the May 2003 issue and is reprinted with permission from *The World & I*, www.WorldandIJournal.com

A LITTLE ROCKET FUEL WITH YOUR SALAD?

By Gene Ayres

I SPENT TWO DECADES living in Southern California, during which time I married, fathered a child, got divorced, and became a single parent. I finally left the Golden State in 1989, taking my young son with me because I was concerned about his health. It wasn't just the smog, although that was bad enough around our house in the San Fernando Valley, where chronic inversions turned the sky over the San Gabriel Mountains the color of dried blood. My son had been born with myriad allergies, and had experienced a terrifying reaction to the DPT vaccine (screaming fits suggestive of being tortured).

Then came the medfly. It was 1988, and I can still recall the sight of the State agricultural commissioner going on TV and dramatically drinking a glass of malathion to reassure a panicky public that, despite the conspicuous presence of a skull and crossbones on all malathion containers (the pesticide was readily available at most garden centers), it was really quite harmless. Of course, we only had the man's word on that, as well as on what was really in his glass, but it was a classic case of the "as-seen-on-TV" school of credibility.

What had prompted this extraordinary demonstration was the discovery of a single medfly in the Port of Long Beach, some 20 miles south of Los Angeles. The big citrus producers had few, if any, commercial groves south of Ventura County, which was two mountain ranges north of Los Angeles. No matter. The medfly was capable of destroying entire groves of citrus, which would cost state growers untold millions. Governor George Deukmejian declared a state of emergency. The Air National Guard was called out to begin spraying with malathion, despite howls of protest from the 11 million inhabitants of Los Angeles. Within a matter of weeks, as the panic in Sacramento spread south along the Central Valley, scores of choppers swarmed the skies over southern California, coming out at night like bats, to saturation-spray the entire urban area. It soon became apparent that malathion could take the paint off your car, but we were assured it was harmless to humans.

At the time I was recently divorced, and my son's mother and I had separate households a few miles apart.

We would trade phone calls, warning, "They're coming!" "Move your car, they're headed your way!" And then a nervous, "They're here!" Our son was acting up by then, having increasingly bad attacks of asthma, and developing behavioral problems. By age four, to make matters worse, he'd been diagnosed with attention deficit disorder (ADD).

We decided we'd had enough. It was agreed I would pack up, take the boy, and head for Florida, where my parents lived.

AN INTRODUCTION TO ROCKET FUEL

Florida, however, proved to be anything but a safe haven. Within a few years of our move, a medfly turned up in the Port of Miami. Florida's response, while it didn't include drinking pesticide on TV, was if anything even more draconian. Florida, like California, is the turf of large agribusiness concerns, especially citrus growers. Like their West Coast counterparts before them, they flew into a panic. This, of course, led to the now no-longer-unprecedented decision to mount a preemptive strike. The state legislature quickly passed a new law, written (as since has become common practice) by the affected industry. The Florida law required that all citrus trees within 1,800 feet of an infected tree (the evident roving range of the average medfly) be destroyed. In the years since, the flies have continued to creep northward, and commercial groves and private yard trees alike have fallen before them. Lawsuits, protests, and petitions have done nothing to slow this preemptive juggernaut.

Not long after my move to Florida, my thyroid gland ceased functioning. I learned later that my ex-wife, as well as my brother Ed, both of whom had lived in California in those years, had also suffered hypothyroid disease. Doctors still don't know the cause, although—as we'd eventually learn—they have some suspects. In my case they blamed an unknown virus.

It wasn't until years later that we found that malathion wasn't the only toxic chemical to which we'd been exposed during our time in California. On December 16,

2003, *The Wall Street Journal* published an article by investigative reporter Peter Waldman on the history of California's experience with a chemical called perchlorate, a component of rocket fuel dating back to the first solid-fuel rockets of World War II. Perchlorates are actually a group of salts—ammonium perchlorate, potassium perchlorate, sodium perchlorate, cobalt perchlorate, and a score of others. They were developed mainly as oxidizer components for propellants and other explosive materials (including flares and fireworks) in the 1940s, emerging into a full-bore industry during the Cold War buildup of the 1950s. They have more recently turned up in such diverse products as automobile airbags and certain fertilizers, particularly those produced in Chile.

Despite repeated efforts by California water managers and regulators to stop the dumping of rocket fuel and related toxic chemicals into the state's groundwater and wastewater systems, defense contractors such as Lockheed Martin and Aerojet General pumped and dumped millions of gallons of these chemicals into unlined pits or holding ponds, or injected them deep into the ground. They did this with impunity, considering themselves answerable only to the U.S. Department of Defense (DOD), whose view on the subject, according to Waldman, was that "its job is national security, not environmental safety." That view seems to persist today, as reflected in the recent push by DOD to attain wide-ranging exemption from environmental regulation and restriction.

The U.S. Environmental Protection Agency (EPA) has a very different view than DOD, although no one in either organization denies that perchlorates are highly toxic. No one from DOD has volunteered to drink a glassful of the stuff, to my knowledge. Nineteen recent studies tracked by the Washington-based Environmental Working Group (EWG) between 1997 and 2002 have associated perchlorates with thyroid damage ranging from metabolic and hormone disruption to cancer in adults, and with impaired neurological and bone development in fetuses. They have linked the chemical to reduced IQs, mental retardation, loss of hearing or speech, deficits in motor skills, and (surprise, surprise) learning disorders and ADD in children. Studies in rats have found tumors developing at extremely early stages. And these studies only focus on the apparent effects of perchlorates in water. EPA scientists now believe that the levels of this chemical to which Californians and others are being exposed are far too high. And they wonder what happens when huge quantities of this water pass through the roots of irrigated vegetable crops and end up concentrating in someone's salad.

WHO WORRIES ABOUT MISSILE FUEL IN OUR FOOD, WHEN THERE'S A MISSILE CRISIS IN OUR BACK YARD?

Perchlorates were first developed by a group of aeronautics engineers at the California Institute of Technology in Pasadena, led by a Hungarian immigrant professor named Theodore von Karman. He and a group of colleagues from the university founded Aerojet, which pioneered "jet-assisted" takeoff rockets that enabled the new generation of military jets to take off from the decks of aircraft carriers. They also developed the Minuteman missile. The developers dubbed perchlorates "powdered oxygen" for their rapid and intense combustibility.

As it happens, these chemicals break down over time, and require replacement—hence the large-scale dumping over the decades. Aerojet was first warned to stop dumping as early as 1949, at its Azusa manufacturing plant east of Pasadena, by Los Angeles County engineers who even then were aware of the likely dangers to groundwater. Aerojet ignored those warnings and many others that followed, and received no sanctions. At one point, to further facilitate its perchlorates disposal, the company hooked itself up to a public sewer line.

In 1951, Aerojet moved north to Sacramento and the suburb of Rancho Cordova. According to Peter Waldman, the company's unlined holding ponds and pits leached up to 1,000 gallons of liquid waste and 300 pounds of ammonium perchlorate into the local aquifer every day. Today, a number of families there are suffering from cancer and other ailments alleged to be attributable to perchlorates. With Southern California left eating Aerojet's dust, other authorities took up the pursuit. In 1952, the California Central Valley Regional Water Pollution Control Board issued a resolution specifically intended to block further dumping of perchlorates into local groundwater or the nearby American River. Nothing changed. Aerojet's defense was that, according to guidelines issued by the DOD, its unlined holding ponds and pits were quite adequate methods of disposal.

By 1957, an underground toxic plume had spread across several square miles east of Sacramento. According to a national task force and *The Wall Street Journal*, the plume ranged in perchlorates concentration from 3.5 to 5 parts per billion (ppb). Surely, scientists began to hypothesize, this had to be bad for you. That year, a study at Harvard University found that perchlorates passed through the placenta of guinea pigs and affected the development of the thyroid and its hormones that regulate growth and development.

Still, Aerojet continued its stonewalling, even refusing to disclose exactly what chemicals it was using. In 1962, the board tried again, passing a resolution prohibiting Aerojet from disposing of anything "deleterious to human, animal, plant, or aquatic life."

At this point, national security once again assumed priority—in the form of the Cuban Missile Crisis. Nothing I have seen in the accounts of that time suggests that anyone saw any irony in the possibility that American missiles might be poisoning the American people in order to protect them from a theoretical attack by missiles from another country.

And so, again, nothing changed. Over the ensuing decades, thousands of tons of these toxic chemicals were deposited into open ditches, canals, holding ponds, and pits. Only in 1985 did perchlorates finally become a "drinking water problem," when they were detected by the EPA in wells serving 42,000 households in the vicinity of Aerojet's original plant in Azusa, back in Southern California.

In 1992, the EPA turned to the Centers for Disease Control in Atlanta for help. CDC declared that "the effects of low level perchlorate ingestion need to be described as soon as possible." So the EPA went back to the 1952 Harvard study linking the chemicals to thyroid damage, and issued its first health assessment, recommending an "initial reference dose" of no more than 4 ppb in drinking water. In response, DOD insisted that the reference dose should be 42,000 ppb. The dispute remains unresolved, although DOD has since shifted its estimate of acceptable levels sharply downward. While EPA holds firmly to its recommended level of 4 ppb, the people who brought us Agent Orange and Gulf War Syndrome now tell us that 200 ppb is safe.

THE UNLUCKY EMPLOYEES OF LUCKY FARMS

The concept of bioaccumulation has become chillingly familiar to those who have followed the stories of PCBs, mercury, and other contaminants that may be found in low levels in insects or algae, but that concentrate as they rise through the food chain. The same kind of concentration can occur as polluted ground water is consumed by irrigated plants.

The first clues that perchlorates were accumulating in vegetables emerged around October 1996, when a commercial grocers' farm, Lucky Farms of San Bernardino, California, began handing a release form to its employees—"todos los empleados"—which it required them to sign. The disclaimer stated, in English and Spanish:

> I have been informed of the dangers if I drink irrigation water from the sprinklers, valves, or faucets that are marked red. This water may cause cancer or birth defects. I know that I am only to drink water from the orange coolers. And drink only from the water specified as "good" water, which is the faucet located by the shop.

Each of the predominately Mexican farm workers was requested to sign that form, presumably to protect Lucky Farms from bad luck in the form of lost labor due to illness, or in the form of lawsuits.

Someone at the farm had been savvy enough to become suspicious of the water supply there and had had some water tested. Apparently, there was a growing awareness in the Redlands area of San Bernardino County that there was something wrong with the water. Maybe it was not mere coincidence that Lockheed Martin had operated missile testing facilities in that area for many years, using a large amount of rocket fuel. The facilities had been long since closed, but they'd left a stain of perchlorates in the ground.

On December 19, 1997, a test report was delivered to farm manager Robert Liso of Lucky Farms by Weck Laboratories of Industry, California. The preliminary report indicated that Week had tested a "vegetable" (later reported as lettuce) and found 110 micrograms (mcg) of perchlorate in the sample. While perchlorates were (and are still) an unregulated commodity, and this was not an astronomical quantity, it was well in excess of EPA's recommended level for a lettuce leaf. It was the first indication that perchlorates had migrated from the water supply to the food chain.

Then, on February 9, 1998, Weck delivered a bombshell. A second study, analyzing four lettuce samples, revealed the presence of massive amounts of perchlorates:

Leafy vegetable no. 401635: 3,260 mcg

Leafy vegetable no. 401636: 6,590 mcg

Leafy vegetable no. 401637: 6,900 mcg

Leafy vegetable no. 401638: 3,210 mcg

But none of this information reached the public. Both Weck Laboratories and Lucky Farms refused to comment on these findings when I called for clarification. Unless decimal points migrated along with the perchlorates, these were highly alarming numbers. According to Kevin Mayer of the EPA's San Francisco office, if water were the only source of perchlorate intake (which evidently it is not), the EPA's 4 ppb standard would translate to about 1 microgram per liter, or 2.1 micrograms per day for a 150-pound (70 kilogram) person drinking 2 liters of water a day. One of the Weck samples, it seemed, would deliver several thousand times that amount.

A flurry of behind-the-scenes legal activity quickly followed. By May 7, 1998, Lockheed Martin had engaged the services of the Los Angeles law firm of Gibson, Dunn & Crutcher to represent it in a secret settlement agreement with Lucky Farms. Evidently intimidated by the defense contractor's heavy guns being brought to bear, Lucky Farms backed off making any threat of a lawsuit. Lockheed Martin, in turn, generously offered to pay for the cost of Weck Laboratories' testing—an offer that was apparently accepted, because no further action was taken. Instead, Lucky Farms continued to require signed releases from its employees and continued with business as usual, at least until 1999. Lettuce continued to be shipped to markets throughout the country.

In 1999, the EPA swung ponderously into action, ordering a study by its National Environmental Research Laboratory in Athens, Georgia. The new study found that perchlorate accumulated in leaves by factors of 100 or more times the agency's recommended levels. The researchers also found that lettuce leaves were capable of absorbing and storing up to 95 percent of the perchlorates in the water supply. This meant that even small levels of

HAVE YOU HEARD THIS ONE BEFORE?

The perchlorate story brings together three all-too-familiar themes of the modern military-industrial saga:

First, there's the technological hubris that so often surrounds a major new invention. Immediate beneficiaries become so focused on the invention's primary intended use that they overlook the secondary, often slower-acting and less visible, effects on health or the environment. The dramatic Cold War takeoff of the rocket industry, with little thought to its eventual, insidious effects on the health of millions of people, parallels the histories of many other lucrative technologies—from thalidomide to DDT to breast implants—that have been rushed into use. Not coincidentally, the fiery power of rockets, and the apparent lack of thought about what longer-term effects might follow, was replicated on a grand scale in the "shock and awe" with which the U.S. launched its rocket-raining war on Iraq.

Second, there's the secrecy and denial—and attempted coverup—that repeatedly accompanies the rise or perpetuation of highly risky but lucrative industries. Think of the decades-long coverup of the health effects of tobacco smoke, or the chemical industry effort to destroy the credibility of Rachel Carson after the publication of her book Silent Spring, or the Cheney-Bush administration's collusion with fossil-fuel industries to distract public attention from the dangers of climate change, which far exceed those of terrorism but pose an unwanted challenge to those industries. In the case of perchlorates, it's likely that a large part of the coverup has yet to be exposed. But already, there's a telling pattern—in the Air Force's dog-ate-my-homework story that someone stole its study results; in the refusals of Lucky Farms or Weck Labs to speak; in the Department of Agriculture's unexplained cancellation of a perchlorate study for other irrigated crops (what about our tomatoes, chard, and grapes?); and most of all, in the White House gag order that has made both the EPA and the Pentagon clam up. We talked with a top official of EPA's San Francisco office, who said his agency had worked long and hard to determine its recommended level of 4 ppb, and was prepared to make those recommendations final, when "someone" stepped in and requested that the data be turned over to the National Academy of Science for review. And there, so far, it has remained.

Third, and perhaps most dangerous of all, there is the familiar presumption—once unthinkable but now pervasive—that scientific findings are not matters of public knowledge but proprietary industrial or government information that the holders need disclose only if it is to their financial or political advantage to do so. While the rocket-fuel contamination is mainly an American problem (it's mainly Americans who make and shoot off rockets), the water that flows through our lettuce—and through the Earth's hydrological cycle—is a global commons. The knowledge of what happens to it is an essential part of the global public domain. So, why are all the unanswered questions about perchlorates being treated like a national security secret? The need to keep our water clean is common sense, not rocket science.

perchlorate in the water could concentrate into extremely high levels in the lettuce leaves.

Curiously, the EPA discounted this study due to the fact that the water used in the study had excessive amounts of perchlorates in it and was presumably not representative of typical California irrigation water, though these doses weren't anywhere near the concentrations of those found in the Rancho Cordova and Azusa plumes. Even more curiously, EPA also concluded that "foods do not contribute to" perchlorate accumulation in the human body—an assertion that appears to be directly contradicted by the 1952 Harvard study and the 19 studies that followed.

SHADES OF SILKWOOD

By 2002, sources of perchlorates in irrigation water were being traced to the Colorado River, which was found to be seriously contaminated from Las Vegas to the Mexican border. This was especially troublesome news, since Colorado River water was the sole source of irrigation water for the entire Coachella and Imperial Valley regions, which produce 90 percent of America's lettuce. On July 14, 2001, The Sacramento Bee reported that 20 million people in California, Arizona, and Nevada had some level of perchlorates in their drinking water, averaging between 5 and 10 ppb. According to The Wall Street Journal, scientists have traced perchlorates found in the Los Angeles water supply 400 miles up the Colorado River to Lake Mead, above Hoover Dam. From there, according to the Journal's Waldman, they tracked the plume 10 miles west up a desert riverbed called the Las Vegas Wash, to a giant ammonium perchlorate plant in Henderson, Nevada, owned and operated by Oklahoma City-based Kerr-McGee Corporation. The Las Vegas Wash is the main drain leading into Lake Mead, the primary source of water for Las Vegas and the lower Colorado River.

Kerr-McGee is the company that was featured in the movie *Silkwood*, based on the story of Karen Silkwood, a chemical technician at the company who claimed that her employer was exposing its unwitting employees to plutonium radiation. Silkwood was killed in an unwitnessed one-car crash while gathering evidence implicating the company. This time, rather than trying to hide its involvement and providing the plot for another movie, Kerr-McGee has sued the Pentagon for reimbursement of cleanup costs. Its plant is now closed but continues to leak 900 pounds of perchlorate a day into the Las Vegas Wash.

Other lawsuits have been filed as well, including a class action suit by the residents of Rancho Cordova, where Aerojet operated with impunity all those years. Many residents there have developed thyroid and other cancers. These suits, against Lockheed Martin and others, have gone nowhere, except to the extent that the State of California agreed to pay the cost of a suit by a Rancho Cordova water company that, according to *The Sacramento Bee*, "accuses pollution enforcers of having willfully allowed Aerojet Corp. to contaminate the ground water with rocket fuel." So, Aerojet continues to escape unscathed, while the people who tried for decades to stop it from perchlorate dumping end up taking the hit.

In April 1999 the EPA convened an "eco-summit" of representatives from the Air Force (the prime perchlorates consumer, other than NASA), a coalition of perchlorate manufacturers and users called the Perchlorate Study Group, and members of five Indian tribes whose livelihoods are based on produce farming along the lower Colorado River. The DOD promised a grant of $650,000, or less than one-fourth the cost of a single cruise missile, according to *The Los Angeles Times*, for a so-called "real world" study that would test a variety of crops through the auspices of the U.S. Department of Agriculture. Then, after consultations with the Food and Drug Administration, the project was indefinitely postponed.

Instead, the Air Force, according to records obtained through the Freedom of Information Act by *The Riverside (CA) Press-Enterprise*, obtained a $500,000 grant from DOD earmarked for two studies: perchlorates in crops, and perchlorates in wild plants and animals. Yet the first study, it seems, was never done—or if it was, the findings were never disclosed. Other documents, obtained by the Environmental Working Group, indicate that the Air Force did in fact conduct a second study of greenhouse-grown lettuce. In October 2002, at an industry-sponsored perchlorate conference in Ontario, California (not far from Aerojet's original plant in Azusa), EWG questioned Air Force spokesman David Mattie about the second lettuce study and was told that the study had in fact been completed, but "someone walked away with the data."

When prodded for the results under the Freedom of Information Act, the Air Force then claimed that any findings are "fully exempt from disclosure until the formally sponsored EPA peer review is complete."

In the spring of 2003, the lettuce finally hit the fan. On April 2, California's U.S. Senator Diane Feinstein accused the Defense Department of "dragging its feet in the cleanup of rocket fuel from old military facilities, which state officials say has contaminated hundreds of wells in California." On April 27, reporters David Danelski and Douglas Beeman of *The Riverside Press-Enterprise* released the results of a study commissioned by the newspaper of 18 winter lettuce samples and one mustard greens sample harvested in the Imperial and Coachella Valleys, both irrigated by the Colorado River. All 19 samples tested were found to be contaminated with perchlorates. A spokesman for the U.S. Department of Agriculture insisted that the levels found by *The Press-Enterprise* were too low to pose a health risk, but nevertheless expressed concern that other crops might also be contaminated.

The day after the *Press-Enterprise* story, the EWG released its own study of 22 samples of lettuce purchased at Northern California supermarkets. EWG had found four to be contaminated with perchlorates. This study received more national attention than the *Press-Enterprise* one, possibly because some of the EWG findings, based on tests conducted at Texas Tech University, showed levels of contamination "as much as 20 times as high as the amount California considers safe for drinking water."

In the meantime, renewed attention was focused on the EPA, which while talking tough about recommended levels, has yet to make any firm decision pending a peer review by the National Academy of Sciences. EPA said it needed to see the review before it could issue any regulations, reports, or guidelines on the subject of perchlorates.

So what has become of that review? EPA has been less than forthcoming on this subject. Sometime in the weeks since the most recent revelations of the presence of perchlorates in the U.S. food supply, the White House issued a gag order to the EPA prohibiting its researchers or scientists from discussing perchlorates with the press. Interestingly, during the unfolding of these events, EPA chief Christine Todd Whitman resigned. Calls requesting information on this matter have gone unanswered, but Whitman's home state of New Jersey is one of many states threatened by the discovery of perchlorates in its water. Furthermore, New Jersey—"The Garden State"—is one of the country's biggest sources both of vegetable produce and of perchlorate manufacturing.

California Congresswoman Lois Capps, in quick response to the two lettuce studies, wrote to the White House—together with 57 other members of the House of Representatives—demanding explanation and rescinding of the reported EPA gag order, noting that "perchlorate is known or suspected to be a contaminant in hundreds of locations in 43 states. It has been confirmed in more than 100 drinking water sources in 19 states including Texas, California, Arizona, Nebraska, Iowa, New York, Maryland, and Massachusetts.... It is highly disturbing to think that the agency charged with protecting our environmental health and safety could be barred from discussing an increasingly prevalent and potentially dangerous chemical contaminant like perchlorate. Americans deserve to get information from the scientists who work on their behalf. President Bush should see to it that this 'gag order' is lifted immediately." The White House did not respond.

Meanwhile, while the EPA has since shown a willingness to talk with *World Watch* about the issue and openly acknowledges the gravity of the situation, it has still not established any firm safety standards for perchlorates, which remain completely unregulated at both the state and Federal levels. At a time when the secretary of defense often seems to hold as much sway as the Congress, which makes the laws, the Pentagon doesn't want stan-

dards. So far, Kerr-McGee is the only defense contractor to have voluntarily embarked on cleaning up its mess, which in its case was inherited from an old Navy lab at the same site.

The DOD's stonewalling harks back half a century to its 1950s positions about matters of munitions and pollution: the Army, Navy, and Air Force are in the business of national security, not environmental protection. The military hasn't yet come around to the radical idea that when it comes to national security—or even global security, for that matter—protecting our food and water may be the first line of defense.

Gene Ayres is a writer based in St. Petersburg, Florida.

From *World Watch Magazine*, November/December 2003, pp. 12-20. Copyright © 2003 by Worldwatch Institute, www.worldwatch.org. Reprinted by permission.

HOW DID HUMANS FIRST ALTER GLOBAL CLIMATE?

A bold new hypothesis suggests, that our ancestors' farming practices kicked off global warming thousands of years before we started burning coal and driving cars

William E. Ruddiman

The scientific consensus that human actions first began to have a warming effect on the earth's climate within the past century has become part of the public perception as well. With the advent of coal-burning factories and power plants, industrial societies began releasing carbon dioxide (CO_2) and other greenhouse gases into the air. Later, motor vehicles added to such emissions. In this scenario, those of us who have lived during the industrial era are responsible not only for the gas buildup in the atmosphere but also for at least part of the accompanying global warming trend. Now, though, it seems our ancient agrarian ancestors may have begun adding these gases to the atmosphere many millennia ago, thereby altering the earth's climate long before anyone thought.

New evidence suggests that concentrations of CO_2 started rising about 8,000 years ago, even though natural trends indicate they should have been dropping. Some 3,000 years later the same thing happened to methane, another heat-trapping gas. The consequences of these surprising rises have been profound. Without them, current temperatures in northern parts of North America and Europe would be cooler by three to four degrees Celsius—enough to make agriculture difficult. In addition, an incipient ice age—marked by the appearance of small ice caps—would probably have begun several thousand years ago in parts of northeastern Canada. Instead the earth's climate has remained relatively warm and stable in recent millennia.

Until a few years ago, these anomalous reversals in greenhouse gas trends and their resulting effects on climate had escaped notice. But after studying the problem for some time, I realized that about 8,000 years ago the gas trends stopped following the pattern that would be predicted from their past long-term behavior, which had been marked by regular cycles. I concluded that human activities tied to farming—primarily agricultural deforestation and crop irrigation—must have added the extra CO_2 and methane to the atmosphere. These activities explained both the reversals in gas trends and the ongoing increases right up to the start of the industrial era. Since then, modern technological innovations have brought about even faster rises in greenhouse gas concentrations.

> My claim that human contributions have been ALTERING THE EARTH'S CLIMATE FOR MILLENNIA is provocative and controversial.

My claim that human contributions have been altering the earth's climate for millennia is provocative and controversial. Other scientists have reacted to this proposal with the mixture of enthusiasm and skepticism that is typical when novel ideas are put forward, and testing of this hypothesis is now under way.

The Current View

THIS NEW IDEA builds on decades of advances in understanding long-term climate change. Scientists have known since the 1970s that three predictable variations in the earth's orbit around the sun have exerted the dominant control over long-term global climate for millions of years.

Overview/Early Global Warming

- A new hypothesis challenges the conventional assumption that greenhouse gases released by human activities have perturbed the earth's delicate climate only within the past 200 years.
- New evidence suggests instead that our human ancestors began contributing significant quantities of greenhouse gases to the atmosphere thousands of years earlier by clearing forests and irrigating fields to grow crops.
- As a result, human beings kept the planet notably warmer than it would have been otherwise—and possibly even averted the start of a new ice age.

As a consequence of these orbital cycles (which operate over 100,000, 41,000 and 22,000 years), the amount of solar radiation reaching various parts of the globe during a given season can differ by more than 10 percent. Over the past three million years, these regular changes in the amount of sunlight reaching the planet's surface have produced a long sequence of ice ages (when great areas of Northern Hemisphere continents were covered with ice) separated by short, warm interglacial periods.

Dozens of these climatic sequences occurred over the millions of years when hominids were slowly evolving toward anatomically modern humans. At the end of the most recent glacial period, the ice sheets that had blanketed northern Europe and North America for the previous 100,000 years shrank and, by 6,000 years ago, had disappeared. Soon after, our ancestors built cities, invented writing and founded religions. Many scientists credit much of the progress of civilization to this naturally warm gap between less favorable glacial intervals, but in my opinion this view is far from the full story.

In recent years, cores of ice drilled in the Antarctic and Greenland ice sheets have provided extremely valuable evidence about the earth's past climate, including changes in the concentrations of the greenhouse gases. A three-kilometer-long ice core retrieved from Vostok Station in Antarctica during the 1990s contained trapped bubbles of ancient air that revealed the composition of the atmosphere (and the gases) at the time the ice layers formed. The Vostok ice confirmed that concentrations of CO_2 and methane rose and fell in a regular pattern during virtually all of the past 400,000 years.

Particularly noteworthy was that these increases and decreases in greenhouse gases occurred at the same intervals as variations in the intensity of solar radiation and the size of the ice sheets. For example, methane concentrations fluctuate mainly at the 22,000-year tempo of an orbital cycle called precession. As the earth spins on its rotation axis, it wobbles like a top, slowly swinging the Northern Hemisphere closer to and then farther from the sun. When this precessional wobble brings the northern continents nearest the sun during the summertime, the at-

mosphere gets a notable boost of methane from its primary natural source—the decomposition of plant matter in wetlands.

After wetland vegetation flourishes in late summer, it then dies, decays and emits carbon in the form of methane, sometimes called swamp gas. Periods of maximum summertime heating enhance methane production in two primary ways: In southern Asia, the warmth draws additional moisture-laden air in from the Indian Ocean, driving strong tropical monsoons that flood regions that might otherwise stay dry. In far northern Asia and Europe, hot summers thaw boreal wetlands for longer periods of the year. Both processes enable more vegetation to grow, decompose and emit methane every 22,000 years. When the Northern Hemisphere veers farther from the sun, methane emissions start to decline. They bottom out 11,000 years later—the point in the cycle when Northern Hemisphere summers receive the least solar radiation.

Unexpected Reversals

EXAMINING RECORDS from the Vostok ice core closely, I spotted something odd about the recent part of the record. Early in previous interglacial intervals, the methane concentration typically reached a peak of almost 700 parts per billion (ppb) as precession brought summer radiation to a maximum. The same thing happened 11,000 years ago, just as the current interglacial period began. Also in agreement with prior cycles, the methane concentration then declined by 100 ppb as summer sunshine subsequently waned. Had the recent trend continued to mimic older interglacial intervals, it would have fallen to a value near 450 ppb during the current minimum in summer heating. Instead the trend reversed direction 5,000 years ago and rose gradually back to almost 700 ppb just before the start of the industrial era. In short, the methane concentration rose when it should have fallen, and it ended up 250 ppb higher than the equivalent point in earlier cycles.

Like methane, CO_2 has behaved unexpectedly over the past several thousand years. Although a complex combination of all three orbital cycles controls CO_2 variations, the trends during previous interglacial intervals were all surprisingly similar to one another. Concentrations peaked at 275 to 300 parts per million (ppm) early in each warm period, even before the last remnants of the great ice sheets finished melting. The CO_2 levels then fell steadily over the next 15,000 years to an average of about 245 ppm. During the current interglacial interval, CO_2 concentrations reached the expected peak around 10,500 years ago and, just as anticipated, began a similar decline. But instead of continuing to drop steadily through modern times, the trend reversed direction 8,000 years ago. By the start of the industrial era, the concentration had risen to 285 ppm—roughly 40 ppm higher than expected from the earlier behavior.

What could explain these unexpected reversals in the natural trends of both methane and CO_2? Other investigators suggested that natural factors in the climate system provided the answer. The methane increase has been ascribed to expansion of wetlands in Arctic regions and the CO_2 rise to natural losses of carbon-rich vegetation on the continents, as well as to changes in the chemistry of the ocean. Yet it struck me that these explanations were doomed to fail for a simple reason. During the four preceding interglaciations, the major factors thought to influence greenhouse gas concentrations in the atmosphere were nearly the same as in recent millennia. The northern ice sheets had melted, northern forests had reoccupied the land uncovered by ice, meltwater from the ice had returned sea level to its high interglacial position, and solar radiation driven by the earth's orbit had increased and then began to decrease in the same way.

Why, then, would the gas concentrations have fallen during the last four interglaciations yet risen only during the current one? I concluded that something new to the natural workings of the climate system must have been operating during the past several thousand years.

The Human Connection

THE MOST PLAUSIBLE "new factor" operating in the climate system during the present interglaciation is farming. The basic timeline of agricultural innovations is well known. Agriculture originated in the Fertile Crescent region of the eastern Mediterranean around 11,000 years ago, shortly thereafter in northern China, and several thousand years later in the Americas. Through subsequent millennia it spread to other regions and increased in sophistication. By 2,000 years ago, every crop food eaten today was being cultivated somewhere in the world.

Several farming activities generate methane. Rice paddies flooded by irrigation generate methane for the same reason that natural wetlands do—vegetation decomposes in the stagnant standing water. Methane is also released as farmers burn grasslands to attract game and promote growth of berries. In addition, people and their domesticated animals emit methane with feces and belches. All these factors probably contributed to a gradual rise in methane as human populations grew slowly, but only one process seems likely to have accounted for the abruptness of the reversal from a natural methane decline to an unexpected rise around 5,000 years ago—the onset of rice irrigation in southern Asia.

Farmers began flooding lowlands near rivers to grow wet-adapted strains of rice around 5,000 years ago in the south of China. With extensive floodplains lying within easy reach of several large rivers, it makes sense that broad swaths of land could have been flooded soon after the technique was discovered, thus explaining the quick shift in the methane trend. Historical records also indicate a steady expansion in rice irrigation throughout the interval when methane values were rising. By 3,000 years ago the tech-

nique had spread south into Indochina and west to the Ganges River Valley in India, further increasing methane emissions. After 2,000 years, farmers began to construct rice paddies on the steep hillsides of Southeast Asia.

Future research may provide quantitative estimates of the amount of land irrigated and methane generated through this 5,000-year interval. Such estimates will be probably difficult to come by, however, because repeated irrigation of the same areas into modern times has probably disturbed much of the earlier evidence. For now, my case rests mainly on the basic fact that the methane trend went the "wrong way" and that farmers began to irrigate wetlands at just the right time to explain this wrong-way trend.

Another common practice tied to farming—deforestation—provides a plausible explanation for the start of the anomalous CO_2 trend. Growing crops in naturally forested areas requires cutting trees, and farmers began to clear forests for this purpose in Europe and China by 8,000 years ago, initially with axes made of stone and later from bronze and then iron. Whether the fallen trees were burned or left to rot, their carbon would have soon oxidized and ended up in the atmosphere as CO_2.

Scientists have precisely dated evidence that Europeans began growing nonindigenous crop plants such as wheat, barley and peas in naturally forested areas just as the CO_2 trend reversed 8,000 years ago. Remains of these plants, initially cultivated in the Near East, first appear in lake sediments in southeastern Europe and then spread to the west and north over the next several thousand years. During this interval, silt and clay began to wash into rivers and lakes from denuded hillsides at increasing rates, further attesting to ongoing forest clearance.

The most unequivocal evidence of early and extensive deforestation lies in a unique historical document—the Doomsday Book. This survey of England, ordered by William the Conqueror, reported that 90 percent of the natural forest in lowland, agricultural regions was cleared as of A.D. 1086. The survey also counted 1.5 million people living in England at the time, indicating that an average density of 10 people per square kilometer was sufficient to eliminate the forests. Because the advanced civilizations of the major river valleys of China and India had reached much higher population densities several thousand years prior, many historical ecologists have concluded that these regions were heavily deforested some two or even three thousand years ago. In summary, Europe and southern Asia had been heavily deforested long before the start of the industrial era, and the clearance process was well under way throughout the time of the unusual CO_2 rise.

An Ice Age Prevented?

IF FARMERS WERE responsible for greenhouse gas anomalies this large—250 ppb for methane and 40 ppm for CO_2 by the 1700s—the effect of their practices on the earth's climate would have been substantial. Based on the average sensitivity shown by a range of climate models, the

Human Disease and Global Cooling

Concentrations of CO_2 in the atmosphere have been climbing since about 8,000 years ago. During the past two millennia, however, that steady increase at times reversed direction, and the CO_2 levels fell for decades or more. Scientists usually attribute such CO_2 drops—and the accompanying dips in global temperature—to natural reductions in the sun's energy output or to volcanic eruptions. These factors have been regarded as major drivers of climate change over decades or centuries, but for the CO_2 patterns, such explanations fall short—which implies that an additional factor forced CO_2 levels downward. Because I had already concluded that our human ancestors had caused the slow rise in CO_2 for thousands of years by clearing forests for agriculture [*see main article*], this new finding made me wonder whether some kind of reversal of the ongoing clearance could explain the brief CO_2 drops.

The most likely root cause turns out to be disease—the massive human mortality accompanying pandemics. Two severe outbreaks of bubonic plague, the single most devastating killer in human history, correlate well with large CO_2 drops at approximately A.D. 540 and 1350 [*graph*]. Plague first erupted during the Roman era, with the most virulent pandemic, the Plague of Justinian, in A.D. 540 to 542. The infamous "Black Death" struck between 1347 and 1352, followed by lesser outbreaks for more than a century. Each of these pandemics killed some 25 to 40 percent of the population of Europe. An even worse catastrophe followed in the Americas after 1492 when Europeans introduced smallpox and a host of other diseases that killed around 50 million people, or about 90 percent of the pre-Columbian population. The American pandemic coincides with the largest CO_2 drop of all, from 1550 to 1800.

Observers at the time noted that the massive mortality rates produced by these pandemics caused widespread abandonment of rural villages and farms, leaving untended farmland to revert to the wild. Ecologists have shown that forests will reoccupy abandoned land in just 50 years. Coupled with estimates of human population and the acreage cultivated by each farmer, calculations of forest regrowth in pandemic-stricken regions indicate that renewed forest's could have sequestered enough carbon to reduce concentrations of CO_2 in the atmosphere by the amounts observed. Global climate would have cooled as a result, until each pandemic passed and rebounding populations began cutting and burning forests anew.

—W.F.R.

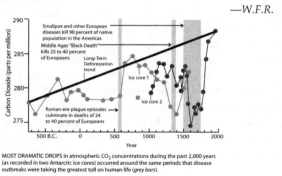

MOST DRAMATIC DROPS in atmospheric CO_2 concentrations during the past 2,000 years (as recorded in two Antarctic ice cores) occurred around the same periods that disease outbreaks were taking the greatest toll on human life (*grey bars*).

exceeds the combined changes registered during the time of rapid industrialization.

How did this dramatic warming effect escape recognition for so long? The main reason is that it was masked by natural climatic changes in the opposite direction. The earth's orbital cycles were driving a simultaneous natural cooling trend, especially at high northern latitudes. The net temperature change was a gradual summer cooling trend lasting until the 1800s.

Had greenhouse gases been allowed to follow their natural tendency to decline, the resulting cooling would have augmented the one being driven by the drop in summer radiation, and this planet would have become considerably cooler than it is now. To explore this possibility, I joined with Stephen J. Vavrus and John E. Kutzbach of the University of Wisconsin-Madison to use a climate model to predict modern-day temperature in the absence of all human-generated greenhouse gases. The model simulates the average state of the earth's climate—including temperature and precipitation—in response to different initial conditions.

For our experiment, we reduced the greenhouse gas levels in the atmosphere to the values they would have reached today without early farming or industrial emissions. The resulting simulation showed that our planet would be almost two degrees C cooler than it is now—a significant difference. In comparison, the global mean temperature at the last glacial maximum 20,000 years ago was only five to six degrees C colder than it is today. In effect, current temperatures would be well on the way toward typical glacial temperatures had it not been for the greenhouse gas contributions from early farming practices and later industrialization.

I had also initially proposed that new ice sheets might have begun to form in the far north if this natural cooling had been allowed to proceed. Other researchers had shown previously that parts of far northeastern Canada might be ice covered today if the world were cooler by just 1.5 to two degrees C—the same amount of cooling that our experiment suggested has been offset by the greenhouse gas anomalies. The later modeling effort with my Wisconsin colleagues showed that snow would now persist into late summer in two areas of northeastern Canada: Baffin Island, just east of the mainland, and Labrador, farther south. Because any snow that survives throughout the summer will accumulate in thicker piles year by year and eventually become glacial ice, these results suggest that a new ice age would have begun in northeast Canada several millennia ago, at least on a small scale.

This conclusion is startlingly different from the traditional view that human civilization blossomed within a period of warmth that nature provided. As I see it, nature would have cooled the earth's climate, but our ancestors kept it warm by discovering agriculture.

combined effect from these anomalies would have been an average warming of almost 0.8 degree C just before the industrial era. That amount is larger than the 0.6 degree C warming measured during the past century—implying that the effect of early farming on climate rivals or even

Implications for the Future

THE CONCLUSION THAT humans prevented a cooling and arguably stopped the initial stage of a glacial cycle bears directly on a long-running dispute over what global climate has in store for us in the near future. Part of the reason that policymakers had trouble embracing the initial predictions of global warming in the 1980s was that a number of scientists had spent the previous decade telling everyone almost exactly the opposite—that an ice age was on its way. Based on the new confirmation that orbital variations control the growth and decay of ice sheets, some scientists studying these longer-scale changes had reasonably concluded that the next ice age might be only a few hundred or at most a few thousand years away.

In subsequent years, however, investigators found that greenhouse gas concentrations were rising rapidly and that the earth's climate was warming, at least in part because of the gas increases. This evidence convinced most scientists that the relatively near-term future (the next century or two) would be dominated by global warming rather than by global cooling. This revised prediction, based on an improved understanding of the climate system, led some policymakers to discount all forecasts—whether of global warming or an impending ice age—as untrustworthy.

My findings add a new wrinkle to each scenario. If anything, such forecasts of an "impending" ice age were actually understated: new ice sheets should have begun to grow several millennia ago. The ice failed to grow because human-induced global warming actually began far earlier than previously thought—well before the industrial era.

In these kinds of hotly contested topics that touch on public policy, scientific results are often used for opposing ends. Global-warming skeptics could cite my work as evidence that human-generated greenhouse gases played a beneficial role for several thousand years by keeping the earth's climate more hospitable than it would otherwise have been. Others might counter that if so few humans with relatively primitive technologies were able to alter the course of climate so significantly, then we have reason to be concerned about the current rise of greenhouse gases to unparalleled concentrations at unprecedented rates.

The rapid warming of the past century is probably destined to persist for at least 200 years, until the economically accessible fossil fuels become scarce. Once that happens, the earth's climate should begin to cool gradually as the

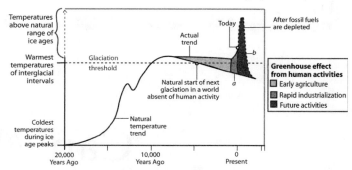

GREENHOUSE EFFECT from human activities has warded off a glaciation that otherwise would have begun about 5,000 years ago. Early human agricultural activities produced enough greenhouse gases to offset most of the natural cooling trend during preindustrial times, warming the planet by an average of almost 0.8 degrees Celsius. That early warming effect (*a*) rivals the 0.6 degree Celsius (*b*) warming measured in the past century of rapid industrialization. Once most fossil fuels are depleted and the temperature rise caused by greenhouse gases peaks, the earth will cool toward the next glaciation—now thousands of years overdue.

deep ocean slowly absorbs the pulse of excess CO_2 from human activities. Whether global climate will cool enough to produce the long-overdue glaciation or remain warm enough to avoid that fate is impossible to predict.

MORE TO EXPLORE

Plagues and Peoples. William McNeill. Doubleday, 1976.

Ice Ages: Solving the Mystery. John Imbrie and Katherine Palmer Imbrie. Enslow, 1979.

Guns, Germs, and Steel: The Fates of Human Societies. Jared Diamond. W. W. Norton, 1999.

Earth's Climate: Past and Future. William F. Ruddiman. W. H. Freeman, 2001.

The Anthropogenic Greenhouse Era Began Thousands of Years Ago. William F. Ruddiman in *Climatic Change*, Vol. 61, No. 3, pages 261–293; 2003.

Deforesting the Earth: From Prehistory to Global Crisis. Michael A. Williams. University of Chicago Press, 2003.

Plows, Plagues, and Petroleum: How Humans Took Control of Climate. William F. Ruddiman. Princeton University Press [in press].

WILLIAM F. RUDDIMAN is a marine geologist and professor emeritus of environmental sciences at the University of Virginia. He joined the faculty there in 1991 and served as department chair from 1993 to 1996. Ruddiman first began studying records of climate change in ocean sediments as a graduate student at Columbia University, where he received his doctorate in 1969. He then worked as a senior scientist and oceanographer with the U.S. Naval Oceanographic Office in Maryland and later as a senior research scientist at Columbia's Lamont-Doherty Earth Observatory.

Can We Bury
GLOBAL WARMING?

*Pumping carbon dioxide underground to avoid warming the atmosphere is feasible,
but only if several key challenges can be met*

Robert H. Socolow

When William Shakespeare took a breath, 280 molecules out of every million entering his lungs were carbon dioxide. Each time you draw breath today, 380 molecules per million are carbon dioxide. That portion climbs about two molecules every year.

No one knows the exact consequences of this upsurge in the atmosphere's carbon dioxide (CO_2) concentration nor the effects that lie ahead as more and more of the gas enters the air in the coming decades—humankind is running an uncontrolled experiment on the world. Scientists know that carbon dioxide is warming the atmosphere, which in turn is causing sea level to rise, and that the CO_2 absorbed by the oceans is acidifying the water. But they are unsure of exactly how climate could alter across the globe, how fast sea level might rise, what a more acidic ocean could mean, which ecological systems on land and in the sea would be most vulnerable to climate change and how these developments might affect human health and well-being. Our current course is bringing climate change upon ourselves faster than we can learn how severe the changes will be.

If slowing the rate of carbon dioxide buildup were easy, the world would be getting on with the job. If it were impossible, humanity would be working to adapt to the consequences. But reality lies in between. The task can be done with tools already at hand, albeit not necessarily easily, inexpensively or without controversy.

Were society to make reducing carbon dioxide emissions a priority—as I think it should to reduce the risks of environmental havoc in the future—we would need to pursue several strategies at once. We would concentrate on using energy more efficiently and on substituting noncarbon renewable or nuclear energy sources for fossil fuel (coal, oil and natural gas—the primary sources of manmade atmospheric carbon dioxide). And we would employ a method that is receiving increasing attention: capturing carbon dioxide and storing, or sequestering, it underground rather than releasing it into the atmosphere. Nothing says that CO_2 must be emitted into the air. The atmosphere has been our prime waste repository, because discharging exhaust up through smokestacks, tailpipes and chimneys is the simplest and least (immediately)

costly thing to do. The good news is that the technology for capture and storage already exists and that the obstacles hindering implementation seem to be surmountable.

Carbon Dioxide Capture

THE COMBUSTION of fossil fuels produces huge quantities of carbon dioxide. In principle, equipment could be installed to capture this gas wherever these hydrocarbons are burned, but some locations are better suited than others.

If you drive a car that gets 30 miles to the gallon and go 10,000 miles next year, you will need to buy 330 gallons—about a ton—of gasoline. Burning that much gasoline sends around three tons of carbon dioxide out the tailpipe. Although CO_2 could conceivably be caught before leaving the car and returned to the refueling station, no practical method seems likely to accomplish this task. On the other hand, it is easier to envision trapping the CO_2 output of a stationary coal-burning power plant.

It is little wonder, then, that today's capture-and-storage efforts focus on those power plants, the source

Overview/*Entombing CO₂*

- A strategy that combines the capture of carbon dioxide emissions from coal power plants and their subsequent injection into geologic formations for long-term storage could contribute significantly to slowing the rise of the atmospheric CO_2 concentration.
- Low-cost technologies for securing carbon dioxide at power plants and greater experience with CO_2 injection to avoid leakage to the surface are key to the success of large-scale CO_2 capture and storage projects.
- Fortunately, opportunities for affordable storage and capture efforts are plentiful. Carbon dioxide has economic value when it is used to boost crude oil recovery at mature fields. Natural gas purification and industrial hydrogen production yield CO_2 at low cost. Early projects that link these industries will enhance the practitioners' technical capabilities and will stimulate the development of regulations to govern CO_2 storage procedures.

of one quarter of the world's carbon dioxide emissions. A new, large (1,000-megawatt-generating) coal-fired power plant produces six million tons of the gas annually (equivalent to the emissions of two million cars). The world's total output (roughly equivalent to the production of 1,000 large plants) could double during the next few decades as the U.S., China, India and many other countries construct new power-generating stations and replace old ones. As new coal facilities come online in the coming quarter of a century, they could be engineered to filter out the carbon dioxide that would otherwise fly up the smokestacks.

Today a power company planning to invest in a new coal plant can choose from two types of power systems, and a third is under development but not yet available. All three can be modified for carbon capture. Traditional coal-fired steam power plants burn coal fully in one step in air: the heat that is released converts water into high-pressure steam, which turns a steam turbine that generates electricity. In an unmodified version of this system—the workhorse of the coal power industry for the past century—a mixture of exhaust (or flue) gases exits a tall stack at atmospheric pressure after having its sulfur removed. Only about 15 percent of the flue gas is carbon dioxide; most of the remainder is nitrogen and water vapor. To adapt this technology for CO_2 capture, engineers could replace the smokestack with an absorption tower, in which the flue gases would come in contact with droplets of chemicals called amines that selectively absorb CO_2. In a second reaction column, known as a stripper tower, the amine liquid would be heated to release concentrated CO_2 and to regenerate the chemical absorber.

The other available coal power system, known as a coal gasification combined-cycle unit, first burns coal partially in the presence of oxygen in a gasification chamber to produce a "synthetic" gas, or syngas—primarily pressurized hydrogen and carbon monoxide. After removing sulfur compounds and other impurities, the plant combusts the syngas in air in a gas turbine—a modified jet engine—to make electricity. The heat in the exhaust gases leaving the gas turbine turns water into steam, which is piped into a steam turbine to generate additional power, and then the gas turbine exhaust flows out the stack. To capture carbon from such a facility, technicians add steam to the syngas to convert (or "shift") most of the carbon monoxide into carbon dioxide and hydrogen. The combined cycle system next filters out the CO_2 before burning the remaining gas, now mostly hydrogen, to generate electricity in a gas turbine and a steam turbine.

The third coal power approach, called oxyfuel combustion, would perform all the burning in oxygen instead of air. One version would modify single-step combustion by burning coal in oxygen, yielding a fuel gas with no nitrogen, only CO_2 and water vapor, which are easy to separate. A second version would modify the coal gasification combined-cycle system by using oxygen, rather than air, at the gas turbine to burn the carbon monoxide and hydrogen mixture that has exited the gasifier. This arrangement skips the shift reaction and would again produce only CO_2 and water vapor. Structural materials do not yet exist, though, that can withstand the higher temperatures that are created by combustion in oxygen rather than in air. Engineers are exploring whether reducing the process temperature by recirculating the combustion exhaust will provide a way around these materials constraints.

Tough Decisions

MODIFICATION FOR carbon dioxide capture not only adds complexity and expense directly but also cuts the efficiency of extracting energy from the fuel. In other words, safely securing the carbon by-products means mining and burning more coal. These costs may be partially offset if the plant can filter out gaseous sulfur simultaneously and store it with the CO_2, thus avoiding some of the considerable expense of sulfur treatment.

Utility executives want to maximize profits over the entire life of the plant, probably 60 years or more, so they must estimate the expense of complying not only with today's environmental rules but also with future regulations. The managers know that the extra costs for CO_2 capture are likely to be substantially lower for coal gasification combined-cycle plants than for traditional plants. Removing carbon dioxide at high pressures, as occurs in a syngas operation, costs less because smaller equipment can be employed. But they also know that only a few demonstration gasification plants are running today, so that opting for gasification will require spending extra on backup equipment to ensure reliability. Hence, if the management bets on not having to pay for CO_2 emissions until late in the life of its new plant, it will probably choose a traditional coal plant, although perhaps one with the potential to be modified later for carbon capture. If, however, it believes that government directives to capture CO_2 are on their way within a decade or so, it may select a coal gasification plant.

To get a feel for the economic pressures the extra cost of carbon sequestration would place on the coal producer, the power plant operator and the home owner who consumes the electricity, it helps to choose a reasonable cost estimate and then gauge the effects. Experts calculate that the total additional expense of capturing and storing a ton of carbon dioxide at a coal gasification combined-cycle plant will be about $25. (In fact, it may be twice that much for a traditional steam plant using today's technology. In both cases, it will cost less when new technology is available.)

The coal producer, the power plant operator and the home owner will perceive that $25 cost increase quite differently. A coal producer would see a charge of about $60 per ton of coal for capturing and storing the coal's carbon, roughly tripling the cost of coal delivered to an electric utility customer. The owner of a new coal power plant would face a 50 percent rise in the cost of power the coal plant puts on the grid, about two cents per kilowatt-hour (kWh) on top of a base cost of around four cents per kWh. The home owner buying only coal-based electricity, who now pays an average of about 10 cents per kWh, would experience one-fifth higher electricity costs (provided that the extra two cents per kWh cost for capture and storage is passed on without increases in the charges for transmission and distribution).

First and Future Steps

RATHER THAN WAITING for the construction of new coal-fired power plants to begin carbon dioxide capture and storage, business leaders are starting the process at existing facilities that produce hydrogen for industry or purify natural gas (methane) for heating and power generation. These operations currently generate concentrated streams of CO_2. Industrial hydrogen production processes, located at oil refineries and ammonia plants, remove carbon dioxide from a high-pressure mix of CO_2 and hydrogen, leaving behind carbon dioxide that is released skyward. Natural gas purification plants must remove CO_2 because the methane is heading for a liquefied natural gas tanker and must be kept free of cold, solid carbon dioxide (dry ice) that could clog the system or because the CO_2 concentration is too high (above 3 percent) to be allowed on the natural gas distribution grid.

Many carbon dioxide capture projects using these sources are now under consideration throughout the oil and gas industry. Hydrogen production and natural gas purification are the initial stepping-stones to full-scale carbon capture at power plants; worldwide about 5 percent as much carbon dioxide is produced in these two industries as in electric power generation.

In response to the growing demand for imported oil to fuel vehicles, some nations, such as China, are turning to coal to serve as a feedstock for synthetic fuels that substitute for gasoline and diesel fuel. From a climate change perspective, this is a step backward. Burning a coal-based synthetic fuel rather than gasoline to drive a set distance releases approximately double the carbon dioxide, when one takes into account both tailpipe and synfuels plant emissions. In synthetic fuels production from coal, only about half the carbon in the coal ends up in the fuel, and the other half is emitted at the plant. Engineers could modify the design of a coal synfuels plant to capture the plant's CO_2 emissions. At some point in the future, cars could run on electricity or carbon-free hydrogen extracted from coal at facilities where CO_2 is captured.

Electricity can also be made from biomass fuels, a term for commercial fuels derived from plant-based materials: agricultural crops and residues, timber and paper industry waste, and landfill gas. If the fossil fuels used in harvesting and processing are ignored, the exchanges between the atmosphere and the land balance because the quantity of carbon dioxide released by a traditional biomass power plant nearly equals that removed from the atmosphere by photosynthesis when the plants grew. But biomass power can do better: if carbon capture equipment were added to these facilities and the harvested biomass vegetation were replanted, the net result would be to scrub the air of CO_2. Unfortunately, the low efficiency of photosynthesis limits the opportunity for atmospheric scrubbing because of the need for large land areas to grow the trees or crops. Future technologies may change that, however. More efficient carbon dioxide removal by green plants and direct capture of CO_2 from the air (accomplished, for example, by flowing air over a chemical absorber) may become feasible at some point.

Carbon Dioxide Storage

CARBON CAPTURE is just half the job, of course. When an electric utility builds a 1,000-megawatt coal plant designed to trap CO_2, it needs to have somewhere to stash securely the six million tons of the gas the facility will generate every year for its entire life. Researchers believe that the best destinations in most cases will be underground formations of sedimentary rock loaded with pores now filled with brine (salty water). To be suitable, the sites typically would lie far below any source of drinking water, at least 800 meters under the surface. At 800 meters, the ambient pressure is 80 times that of the atmosphere, high enough that the pressurized injected CO_2 is in a "supercritical" phase—one that is nearly as dense as the brine it replaces in geologic formations. Sometimes crude oil or natural gas will also be found in the brine formations, having invaded the brine millions of years ago.

The quantities of carbon dioxide sent belowground can be expressed in "barrels," the standard 42-gallon unit of volume employed by the petroleum industry. Each year at a 1,000-megawatt coal plant modified for carbon capture, about 50 million barrels of supercritical carbon dioxide would be secured—about 100,000 barrels a day. After 60 years of operation, about three billion barrels (half a cubic kilometer) would be sequestered below the surface. An oil field with a capacity to produce three billion barrels is six times the size of the smallest of what the industry calls "giant" fields, of which some 500 exist. This means that each large modified coal plant would need to be associated with a "giant" CO_2 storage reservoir. About two thirds of the 1,000 billion barrels of oil the world has produced to date has come from these giant oil fields, so the industry already has a good deal of experience with the scale of the operations needed for carbon storage.

Many of the first sequestration sites will be those that are established because they can turn a profit. Among these are old oil fields into which carbon dioxide can be injected to boost the production of crude. This so-called enhanced oil recovery process takes advantage of the fact that pressurized CO_2 is chemically and physically suited to displacing hard-to-get oil left behind in the pores of the geologic strata after the first stages of production. In this process, compressors drive CO_2 into the oil remaining in the deposits, where chemical reactions result in modified crude oil that moves more easily through the porous rock toward production wells. In particular, CO_2 lowers crude oil's interfacial tension—a form of surface tension that determines the amount of friction between the oil and rock. Thus, carbon dioxide injects new life into old fields.

In response to British government encouragement of carbon dioxide capture and storage efforts, oil companies are proposing novel capture projects at natural gas power plants that are coupled with enhanced oil recovery ventures at fields underneath the North Sea. In the U.S., operators of these kinds of fields can make money today while paying about $10 to $20 per ton for carbon dioxide delivered to the well. If oil prices continue to rise, however, the value of injected CO_2 will probably go up because its use enables the production of a more valuable commodity. This market development could lead to a dramatic expansion of carbon dioxide capture projects.

Carbon sequestration in oil and gas fields will most likely proceed side by side with storage in ordinary brine formations, because the latter structures are far more common. Geologists expect to find enough natural storage capacity to accommodate much of the carbon dioxide that could be captured from fossil fuels burned in the 21st century.

Storage Risks

TWO CLASSES of risk must be addressed for every candidate storage reservoir: gradual and sudden leakage. Gradual release of carbon dioxide merely returns some of the greenhouse gas to the air. Rapid escape of large amounts, in contrast, could have worse consequences than not storing it at all. For a storage operation to earn a license, regulators will have to be satisfied that gradual leakage can occur only at a very slow rate and that sudden leakage is extremely unlikely.

Although carbon dioxide is usually harmless, a large, rapid release of the gas is worrisome because high concentrations can kill. Planners are well aware of the terrible natural disaster that occurred in 1986 at Lake Nyos in Cameroon: carbon dioxide of volcanic origin slowly seeped into the bottom of the lake, which sits in a crater. One night an abrupt overturning of the lake bed let loose between 100,000 and 300,000 tons of CO_2 in a few hours. The gas, which is heavier than air, flowed down through two valleys, asphyxiating 1,700 nearby villagers and thou-

sands of cattle. Scientists are studying this tragedy to ensure that no similar man-made event will ever take place. Regulators of storage permits will want assurance that leaks cannot migrate to belowground confined spaces that are vulnerable to sudden release.

Gradual leaks may pose little danger to life, but they could still defeat the climate goals of sequestration. Therefore, researchers are examining the conditions likely to result in slow seepage. Carbon dioxide, which is buoyant in brine, will rise until it hits an impermeable geologic layer (caprock) and can ascend no farther.

Carbon dioxide in a porous formation is like hundreds of helium balloons, and the solid caprock above is like a circus tent. A balloon may escape if the tent has a tear in it or if its surface is tilted to allow a path for the balloon to move sideways and up. Geologists will have to search for faults in the caprock that could allow escape as well as determine the amount of injection pressure that could fracture it. They will also evaluate the very slow horizontal flow of the carbon dioxide outward from the injection locations. Often the sedimentary formations are huge, thin pancakes. If carbon dioxide is injected near the middle of a pancake with a slight tilt, it may not reach the edge for tens of thousands of years. By then, researchers believe, most of the gas will have dissolved in the brine or have been trapped in the pores.

Even if the geology is favorable, using storage formations where there are old wells may be problematic. More than a million wells have been drilled in Texas, for example, and many of them were filled with cement and abandoned. Engineers are worried that CO_2-laden brine, which is acidic, could find its way from an injection well to an abandoned well and thereupon corrode the cement plug and leak to the surface. To find out, some researchers are now exposing cement to brine in the laboratory and sampling old cements from wells. This kind of failure is less likely in carbonate formations than in sandstone ones; the former reduce the destructive potency of the brine.

The world's governments must soon decide how long storage should be maintained. Environmental ethics and traditional economics give different answers. Following a strict environmental ethic that seeks to minimize the impact of today's activities on future generations, authorities might, for instance, refuse to certify a storage project estimated to retain carbon dioxide for only 200 years. Guided instead by traditional economics, they might approve the same project on the grounds that two centuries from now a smarter world will have invented superior carbon disposal technology.

The next few years will be critical for the development of carbon dioxide capture-and-storage methods, as policies evolve that help to make CO_2-emission reduction profitable and as licensing of storage sites gets under way. In conjunction with significant investments in improved energy efficiency, renewable energy sources and, possibly, nuclear energy, commitments to capture and storage can reduce the risks of global warming.

ROBERT H. SOCOLOW *is professor of mechanical and aerospace engineering at Princeton University. He teaches in both the School of Engineering and Applied Science and the Woodrow Wilson School of Public and International Affairs. A physicist by training, Socolow is currently co-principal investigator (with ecologist Stephen Pacala) of the university's Carbon Mitigation Initiative, supported by BP and Ford, which focuses on global carbon management, the hydrogen economy and fossil-carbon sequestration. In 2003 he was awarded the Leo Szilard Lectureship Award by the American Physical Society.*

Dozens of Words for Snow, None for Pollution

Perched atop the Arctic food chain, the people of the Far North face an impossible choice: abandon their traditional foods, or ingest the rest of the world's poisons with every bite.

Marla Cone

ON A SHEET OF ICE where the Arctic Ocean meets the North Atlantic in the territorial waters of Greenland, Mamarut Kristiansen kneels beside the carcass of a narwhal, the elusive animal sometimes known as "the unicorn of the sea" for its spiraled ivory tusk. He slices off a piece of *mattak*, the whale's raw pink blubber and mottled gray skin, and bites into it. "*Peqqinnartoq*," he says in Greenlandic. Healthy food. Nearby, Mamarut's wife, Tukummeq Peary, a descendant of North Pole explorer Robert Peary, is boiling the main entrée on a camp stove. She, Mamarut, and his brother Gedion dip their hunting knives into the kettle and pull out steaming ribs of ringed seal.

From their home in Qaanaaq, a village in Greenland's Thule region, the Kristiansens have traveled here, to the edge of the world, by dog sledge. It took six hours to journey the 30 miles across a rugged glacier to this sapphire-hued fjord, where every summer they camp on the precarious ice awaiting their prey. The family lives much as their ancestors did thousands of years ago, relying on the bounty of the sea and skills honed by generations. Their lifestyle isn't quaint; it is a necessity in this hostile and isolated expanse. Survival here, in the northernmost civilization on earth, means living the way marine mammals live, hunting as they do, wearing their skins. No factory-engineered fleece compares to the warmth of a sealskin parka. No motorboat can sneak up on a whale like a handmade kayak lashed together with strips of hide. And no imported food nourishes the people's bodies and warms their spirits like the meat they slice from the flanks of a whale or seal.

Traditionally, this marine diet has made the people of the Arctic Circle among the world's healthiest. Beluga whale, for example, has 10 times the iron of beef, twice the protein, and five times the vitamin A. Omega-3 fatty acids in the seafood protect the indigenous people from heart disease. A 70-year-old Inuit in Greenland has coronary arteries as elastic as those of a 20-year-old Dane eating Western foods, says Dr. Gert Mulvad of the Primary Health Care Clinic in Nuuk, Greenland's capital. Some Arctic clinics do not even keep heart medications like nitroglycerin in stock. Although heart disease has appeared with the introduction of Western foods, it remains "more or less unknown," Mulvad says.

Yet the ocean diet that gives these people life and defines their culture also threatens them. Despite living amid pristine ice and glacier-carved bedrock, people like Mamarut, Tukummeq, and Gedion are more vulnerable to pollution than anyone else on earth. Mercury concentrations in Qaanaaq mothers are the highest ever recorded, 12 times greater than the level that poses neurological risks to fetuses, according to U.S. government standards. A separate study has linked PCBs with slight effects on the intelligence of children in Qaanaaq. Although most of the village's people never leave their hunting grounds, the world travels to them, riding upon wintry winds.

The Arctic has become the planet's chemical trash can, the final destination for toxic waste that originates thousands of miles away.

THE ARCTIC has been transformed into the planet's chemical trash can, the final destination for toxic waste that originates thousands of miles away. Atmospheric and oceanic currents conspire to send industrial chemicals, pesticides, and power-plant emissions on a journey to the Far North. Many airborne chemicals tend to migrate to, and precipitate in, cold climates, where they then endure for decades, perhaps centuries, slow to break down in the frigid temperatures and low sunlight. The Arctic Ocean is a deep-freeze archive, holding the memories of the world's past and present mistakes. Its wildlife, too, are archives, as poisonous chemicals accumulate in the fat that Arctic animals need to survive. Polar bears denning in Norway and Russia near the North Pole carry some of the highest levels of toxic compounds ever found in living animals.

Perched at the top of the Arctic food chain, eating a diet similar to a polar bear's, the Inuit also play unwilling host to some 200 toxic pesticides and industrial compounds. These include all of the "Dirty Dozen"—the 12 pollutants capable of inflicting the most damage—including PCBs and chlorinated pesticides such as chlordane, toxaphene, and DDT, long banned in most of North America and Europe. Other compounds still in use today—flame retardants in furniture and computers, insecticides, and the chemicals used to make Teflon—are growing in concentration as well.

The first evidence of alarming levels of toxic substances in the bodies of Arctic peoples came from the Canadian Inuit. In 1987, Dr. Eric Dewailly, an epidemiologist at Laval University in Quebec, was surveying contaminants in the breast milk of mothers near the industrialized, heavily polluted Gulf of St. Lawrence, when he met a midwife from Nunavik, the Inuit area of Arctic Quebec. (Across the Hudson Bay, the Inuit also have their own self-governing territory, Nunavut, or "our land.") She asked whether he wanted milk samples from Nunavik women. Dewailly reluctantly agreed, thinking they might be useful as "blanks," samples with nondetectable pollution levels.

A few months later, glass vials holding half a cup of milk from each of 24 Nunavik women arrived. Dewailly soon got a phone call from his lab director. Something was wrong with the Arctic milk. The chemical concentrations were off the charts. The peaks overloaded the lab's equipment, running off the page. The technician thought the samples must have been tainted in transit.

Upon testing more breast milk, however, the scientists realized that the readings were accurate: Arctic mothers had seven times more PCBs in their milk than mothers in Canada's biggest cities. Informed of the results, an expert in chemical safety at the World Health Organization told Dewailly that the PCB levels were the highest he had ever seen. Those women, he said, should stop breastfeeding their babies.

Dewailly hung up the phone. "Breast milk is supposed to be a gift," he says. "It isn't supposed to be a poison." And in a place as remote as Nunavik, he knew that mothers often had nothing else to feed their infants. Nearly 18 years have passed since Dewailly tested those first vials of breast milk; subsequent data has emerged to show that people, especially babies, are exposed to dangerous concentrations of contaminants all across the Arctic. The average levels of PCBs and mercury in newborn babies' cord blood and women's breast milk are a staggering 20 to 50 times higher in Greenland than in urban areas of the United States and Europe, according to a 2002 report from the Arctic Monitoring and Assessment Programme (AMAP), a project created by eight governments including the United States. Ninety-five percent of women tested in eastern Greenland, nearly 75 percent of women in Arctic Canada's Baffin Island, and nearly 60 percent in Nunavik exceed Canada's "level of concern" for PCBs. Fewer measurements have been taken in Siberia, but the AMAP says contamination levels are high there as well.

In addition to their potential to cause cancer, many of the compounds found in Arctic inhabitants are capable of altering sex hormones and reproductive systems, suppressing immune systems, and obstructing brain development. Infants are the most vulnerable—subject to exposure both in utero and through breast milk, because contaminants such as PCB and DDT accumulate in the fatty nourishment—and are harmed in subtle but profound ways. Arctic babies with high PCB and DDT exposure suffer greater rates of infectious diseases. A study of such infants in Nunavik found that they have more ear and respiratory infections, a quarter of them severe enough to cause hearing loss. "Nunavik has a cluster of sick babies," says Dewailly. "They fill the waiting rooms of the clinics."

A 2003 study found that, compared to infants in lower Quebec, Nunavik infants had much higher exposure to PCBs, mercury, and lead, which resulted in lower birth weight, impaired memory skills, and difficulty in processing new information.

Excessive levels of contamination are not limited to the Arctic. People throughout the world, especially those in seafood-eating cultures, are at similar risk. In the United States, one of every six babies—about 698,000 a year—is born to a mother carrying more mercury in her body than is considered safe under federal guidelines.

The difference is that Americans and Europeans can make choices in their diets to limit their exposure, avoiding fish such as swordfish that are high on the food chain or from highly contaminated waters. For the 650,000 native people of the circumpolar North—the Inuit of Greenland and Canada, the Aleuts, Yup'ik, and Inupiat of Alaska, the Chukchi and other tribes of Siberia, the Saami of Scandinavia and western Russia—there is no real choice. Spread over three continents and speaking dozens of languages, almost all of them face the same dilemma: whether to eat traditional food and face the health risk—or abandon their food, and with it their culture.

"Our foods do more than nourish our bodies," Inuit rights activist Ingmar Egede said. "When many things in our lives are changing, our foods remain the same. They make us feel the same as they have for generations. When I eat Inuit foods, I know who I am."

Eating traditional food is a way to hold on to a culture under assault. **"When I eat Inuit foods,"** activist Ingmar Egede said, **"I know who I am."**

KNOWN TO NAVIGATORS as the North Water, the ocean off Qaanaaq is a polynya, a spot that remains thawed year-round in an otherwise frozen sea. An upwelling of nutrients draws an array of marine life, and the Kristiansens and the other people of Qaanaaq, an isolated village of 860 on the slope of a granite mountain, come here to hunt seal, beluga, walrus, narwhal, even polar bear. A century ago, the famous Arctic explorers—Peary, Frederick Cook, Knud Rasmussen—learned on their expeditions through the area that eating Inuit food was key to survival.

Greenland has no trees, no grass, no fertile soil, which means no cows, no pigs, no chickens, no grains, no vegetables, no fruit. In fact, there is little need for the word "green" in Greenland. The ocean is its food basket. In the remote villages, people eat marine mammals and seabirds 36 times per month on average, consuming about a pound of seal and whale each week. One-third of their food is the meat of wild animals. The International Whaling Commission has deemed the Inuit "the most hunting-oriented of all humans." Greenland is an independently governed territory of Denmark, but 85 percent—or 48,000—of its people are Inuit, and hunting is essential to everything in their 4,000-year-old culture: their language, their art, their

clothing, their legends, their celebrations, their community ties, their economy, their spirituality.

Today, the Kristiansens are gathered on the edge of the ice, waiting to spot a whale's breath. "If only we could see one, we'd be happy," Mamarut whispers, lifting binoculars and eyeing the mirrorlike water for the pale gray back of *qilalugaq*, or narwhal. "Sometimes they arrive at a certain hour of the day and then the next day, same hour, they come back."

Once, Gedion and Mamarut waited almost a month on the ice before catching a narwhal. During such vigils, hunters must remain alert for cracks or other signs that the ice beneath them is shifting. In an instant, it can break off and carry them out to sea. To Greenlanders, ice is everything—it's danger, it's the source of dinner, it's the water they drink. Their language has several dozen expressions for ice, only one for tree.

Mamarut is big, bawdy, and beefy, the elder brother and joker of the family. He celebrated his 42nd birthday on this hunting trip. Gedion is 10 years younger, lanky, quiet, the expert kayaker, wearing a *National Geographic* cap. The Kristiansen brothers are among the best hunters in a nation of hunters, able to sustain their families without the help of other jobs for their wives or themselves. In a good year, they can eat their fill of whale meat and earn more than $15,000 a year selling the rest to markets. In winter, they sell sealskins to a Greenlandic company marketing them in Europe. The men's hair is black, thick and straight, cut short. Their skin is darkened by the sun, but they have no wrinkles. Their only shelter on the ice is a canvas tarp attached to their dog sledge, a make-shift tent so cramped that one person can't bend a knee or straighten an elbow without disturbing the others. A noxious oil-burning lamp is their only source of heat; the kitchen is a camp stove, used to melt ice for tea and to boil seal meat.

In remote Inuit villages, people eat marine mammals and seabirds 36 times a month, **consuming about a pound of seal and whale each week.**

Hunting narwhal is a dangerous endeavor. When Gedion hears or sees them coming, he quietly climbs into his kayak with his harpoon and sealskin buoy. He must simultaneously judge the ice conditions, the current, the wind, the speed and direction of the whales. If a kayaker makes the slightest noise, a narwhal will hear it. If he throws the harpoon, the whale must be directly in front of his kayak, about 30 feet away, close but not too close—or the animal's powerful dive will submerge him and he will likely drown. Gedion, like most Greenlanders, can't swim. There's not much need to master swimming when one can't survive more than a few moments in the frigid water.

Pollution isn't the first force to disrupt local Inuit culture. A little more than a century ago, the people of Qaanaaq didn't have a written language and had scant contact with the Western world. In the 1950s, during the Cold War, their entire community was moved 70 miles to the north to make way for an American military base. The U.S. and Danish governments built the villagers contemporary prefabricated houses—small red, green, blue, and purple chalets. Qaanaaq's population has since doubled, with people attracted by the good hunting. The move also brought liquor, television, and other distractions of modern life. Alcoholism, violence, domestic abuse, and suicide now exact a heavy toll.

Today, the people of Qaanaaq can smear imported taco sauce on their seal meat, buy dental floss and Danish porn magazines in the small local market, and watch *Nightmare on Elm Street* and *Altered States* in their living rooms on the one TV station that beams into Qaanaaq. When asked how he catches a whale, Gedion jokes that he uses a lasso like American cowboys he's seen on television.

Whatever is not hunted—from tea to bread to cheese—is imported from Denmark. Imported food is expensive, often stale, and not very tasty or nutritious. The average family income is $24,000 in Greenland's capital Nuuk, $13,000 in Qaanaaq, and though food is government subsidized, the price of staples like milk, bread, and beef is still considerably higher than in the United States.

And so Greenland's public health officials are torn between encouraging the Inuit to keep eating their traditional foods and advising them to reduce their consumption. In part, doctors fear the Inuit will switch to processed foods loaded with carbohydrates and sugar. "The level of contamination is very high in Greenland, but there's a lot of Western food that is worse than the poisons," Dr. Mulvad says. Greenland's Home Rule government has issued no advisories, and doctors continue to tell people, even pregnant women, to eat traditional food and nurse their babies without restrictions. Jonathan Motzfeldt, who was Greenland's premier for almost 20 years and is now speaker of the Parliament, says hunting isn't sport for his people; it's survival, and the government will not discourage it. "We eat seal meat as you eat cow in your country," Motzfeldt says. "It's important for Greenlanders to have meat on the table. You don't see many vegetables in Greenland. We integrate imported foods, but hunting and eating seals as well as whales is essential for us to survive as a people."

ACROSS THE BAFFIN BAY, surveys show most Canadian Inuit have not altered their diet either. This is partly the result of a clash of cultures. Inuktitut, the language of Canadian Inuit, has some 50 expressions for snow and ice. *Qanniq* is falling snow. *Maujaq* is deep, soft snow. *Kinirtaq* is wet, compact snow. *Ka-*

takartanaq is crusty snow marked by footsteps. *Uangniut* is a snowdrift made by a northwest wind. *Munnguqtuq* is compressed snow softening in spring. Yet there is no Inuktitut word for "chemical" or "pollution" or "contaminant." Over the millennia that their culture has existed, the Inuit have had no need for such words. Most have never seen soot spew from a factory smokestack, or smelled the stench of truck exhaust, or waded in an oily river. So Canadian health officials have dubbed the toxic chemicals found in native foods *sukkunartuq*—something that destroys or brings about something bad. But use of the word has made the contaminants seem lethal and mysterious, even supernatural, and that—combined with a history of government secrecy and poor communication about health risks—has left the Inuit confused, scared, and sometimes angry.

In 1985, Canadian health officials, concerned that an Arctic radar warning system might be a source of PCBs, decided to study the people of Broughton Island, a tiny hamlet in the Baffin Bay region. Government researchers, led by Dr. David Kinloch, collected blood samples and breast milk. The PCB levels were so high—much higher than what could have come from local military facilities—that the mayor of Broughton Island granted Kinloch permission to test more women. Completed in the summer of 1988, the research confirmed high concentrations of PCBs in breast milk at about the same time that Quebec's Dewailly was finding extraordinary levels of DDT, PCBs, and other toxic chemicals in the women of Nunavik. Before any of this data could be fully analyzed, and before people were notified, the discovery was leaked to the press.

On December 15, 1988, Toronto's *Globe and Mail* published a front-page story, quoting a Canadian environmental official saying that the Inuit were so contaminated that they might have to give up whale, seal, and walrus. The Inuit were terrified; some stopped eating their native foods, or breastfeeding. Overnight,

Arctic contaminants became a crisis for the Canadian government. Health Canada, the nation's public health agency, was paralyzed with indecision. The Nunavik and Baffin data clearly showed that most Inuit were exceeding the agency's "tolerable daily intake levels" for toxic contaminants. If the agency was to adhere to its own policies, it would have to warn the Inuit to stop eating their traditional foods. But public health officials had never encountered a problem like this before, where the contaminated foods were so vital to a society's health, culture, and economy. On the one hand, it seemed irresponsible to advise people not to nurse their babies and eat their foods when the traditional diet had so many health benefits and alternatives were unavailable. On the other hand, if the government ignored its own toxic guidelines when it came to the Inuit, wouldn't that be discriminatory?

Crisis meetings were held in Ottawa; aboriginal leaders begged to be included, but none were allowed to participate. It wasn't until the spring of 1989, more than a year later, that the Broughton Islanders who'd given their blood and breast milk to scientists were allowed to see the results of their own tests. It was a slap in the face that Canada's indigenous people have not forgotten.

A wide chasm has since grown between what scientists say and what native people hear, and health officials have failed to refine their message to resonate with the traditional cultures of the Arctic. As a result, at least three generations of Inuit have had little or no advice from experts on how to reduce their exposure. In the late 1990s, 42 percent of women questioned in Nunavik said they increased their consumption of traditional foods while pregnant. Of the 12 percent who ate less, only 1 of 135 said she did so to avoid contaminants. Among those who ate more native foods during pregnancy, most said they did so because they believed it would be good for their baby.

Inuit Tapiriit Kanatami, an organization that represents the Canadian Inuit, launched a project in the mid-1990s to gauge the success of authorities' efforts to inform nine Arctic communities about contaminants. The researchers found the communication so poorly handled that it caused extreme psychological distress among the Inuit. Fear, they concluded, is as dangerous a threat as the contaminants themselves.

"In every instance, there was a pervasive unease and anxiety about contaminants," the organization wrote in its 1995 report. "Whether or not individuals are exposed to...contaminants, the threat alone leads to anxiety, loss of familiar and staple food, loss of employment or activity, loss of confidence in the basic food source and the environment, and more generally a loss of control over one's destiny and well-being."

Lately, health officials have been doing a better job at informing the Inuit of new data. And in 2003, the Nunavik Nutrition and Health Committee, based in Kuujjuaq and composed of Inuit leaders as well as Quebec medical experts, finally took a different tack, focusing on telling people what they should eat rather than what they should not eat. Women were advised to eat Arctic char, a tasty, popular fish that has low levels of contaminants and high amounts of beneficial fatty acids; a pilot program distributed free char to three communities. The hope is that if the Inuit eat more char they will eat less beluga, the source of two-thirds of the PCBs in Nunavik residents.

THE KRISTIANSENS, like their fellow residents of Qaanaaq, learned about the contaminants from listening to the radio. But like most Greenlandic Inuit, they have not changed their diet. Virtually every day, they eat seal meat and mattak, and with every bite, traces of mercury, PCBs, and other chemicals amass in their bodies. "We can't avoid them," Gedion says with a shrug. "It's our food."

This hunting trip proves to be a short one, only five days, and they reap little reward for their patience. "Sometimes you have to just go back empty-handed and feed your dogs," Mamarut says. Upon returning to their village, hunters share their experiences so that everyone may benefit from them. The Kristiansen brothers learned to hunt narwhal from their father, who, in turn, learned from his relatives. Gedion's seven-year-old son, Rasmus, often comes along on their hunts, pretending to drive the dogs and harpoon narwhals. Soon enough, he will be paddling a kayak beside his father. Since 2500 B.C., when the forebears of the Inuit arrived in Greenland, this legacy has been passed on to generations of boys by generations of men like Gedion and Mamarut. Their ancestors' memories, as vivid as a dream, as ancient as the sea ice, mingle with their own.

"*Qaatuppunga piniartarlunga,*" Mamarut says. As far back as I can remember, I hunted.

Marla Cone is a Los Angeles Times staff writer who has come to know the Arctic as "a paradoxical place of ancient traditions and modern amenities." Her book, Silent Snow: The Slow Poisoning of the Arctic, comes out in May but is already available from Amazon.com.

Glossary

This glossary of environmental terms is included to provide you with a convenient and ready reference as you encounter general terms in your study of environment that are unfamiliar or require a review. It is not intended to be comprehensive, but taken together with the many definitions included in the articles themselves, it should prove to be quite useful.

A

Abiotic Without life; any system characterized by a lack of living organisms.

Absorption Incorporation of a substance into a solid or liquid body.

Acid Any compound capable of reacting with a base to form a salt; a substance containing a high hydrogen ion concentration (low pH).

Acid Rain Precipitation containing a high concentration of acid.

Adaptation Adjustment of an organism to the conditions of its environment, enabling reproduction and survival.

Additive A substance added to another in order to impart or improve desirable properties or suppress undesirable ones.

Adsorption Surface retention of solid, liquid, or gas molecules, atoms, or ions by a solid or liquid.

Aerobic Environmental conditions where oxygen is present; aerobic organisms require oxygen in order to survive.

Aerosols Tiny mineral particles in the atmosphere onto which water droplets, crystals, and other chemical compounds may adhere.

Air Quality Standard A prescribed level of a pollutant in the air that should not be exceeded.

Alcohol Fuels The processing of sugary or starchy products (such as sugar cane, corn, or potatoes) into fuel.

Allergens Substances that activate the immune system and cause an allergic response.

Alpha Particle A positively charged particle given off from the nucleus of some radioactive substances; it is identical to a helium atom that has lost its electrons.

Ammonia A colorless gas comprised of one atom of nitrogen and three atoms of hydrogen; liquefied ammonia is used as a fertilizer.

Anthropocentric Considering humans to be the central or most important part of the universe.

Aquaculture Propagation and/or rearing of any aquatic organism in artificial "wetlands" and/or ponds.

Aquifers Porous, water-saturated layers of sand, gravel, or bedrock that can yield significant amounts of water economically.

Atom The smallest particle of an element, composed of electrons moving around an inner core (nucleus) of protons and neutrons. Atoms of elements combine to form molecules and chemical compounds.

Atomic Reactor A structure fueled by radioactive materials that generates energy usually in the form of electricity; reactors are also utilized for medical and biological research.

Autotrophs Organisms capable of using chemical elements in the synthesis of larger compounds; green plants are autotrophs.

B

Background Radiation The normal radioactivity present; coming principally from outer space and naturally occurring radioactive substances on Earth.

Bacteria One-celled microscopic organisms found in the air, water, and soil. Bacteria cause many diseases of plants and animals; they also are beneficial in agriculture, decay of dead matter, and food and chemical industries.

Benthos Organisms living on the bottom of bodies of water.

Biocentrism Belief that all creatures have rights and values and that humans are not superior to other species.

Biochemical Oxygen Demand (BOD) The oxygen utilized in meeting the metabolic needs of aquatic organisms.

Biodegradable Capable of being reduced to simple compounds through the action of biological processes.

Biodiversity Biological diversity in an environment as indicated by numbers of different species of plants and animals.

Biogeochemical Cycles The cyclical series of transformations of an element through the organisms in a community and their physical environment.

Biological Control The suppression of reproduction of a pest organism utilizing other organisms rather than chemical means.

Biomass The weight of all living tissue in a sample.

Biome A major climax community type covering a specific area on Earth.

Biosphere The overall ecosystem of Earth. It consists of parts of the atmosphere (troposphere), hydrosphere (surface and ground water), and lithosphere (soil, surface rocks, ocean sediments, and other bodies of water).

Biota The flora and fauna in a given region.

Biotic Biological; relating to living elements of an ecosystem.

Biotic Potential Maximum possible growth rate of living systems under ideal conditions.

Birthrate Number of live births in one year per 1,000 midyear population.

Breeder Reactor A nuclear reactor in which the production of fissionable material occurs.

C

Cancer Invasive, out-of-control cell growth that results in malignant tumors.

Carbon Cycle Process by which carbon is incorporated into living systems, released to the atmosphere, and returned to living organisms.

Carbon Monoxide (CO) A gas, poisonous to most living systems, formed when incomplete combustion of fuel occurs.

Carcinogens Substances capable of producing cancer.

Carrying Capacity The population that an area will support without deteriorating.

Chlorinated Hydrocarbon Insecticide Synthetic organic poisons containing hydrogen, carbon, and chlorine. Because they are fat-soluble, they tend to be recycled through food chains, eventually affecting nontarget systems. Damage is normally done to the organism's nervous system. Examples include DDT, Aldrin, Deildrin, and Chlordane.

Chlorofluorocarbons (CFCs) Any of several simple gaseous compounds that contain carbon, chlorine, fluorine, and sometimes hydrogen; they are suspected of being a major cause of stratospheric ozone depletion.

Circle of Poisons Importation of food contaminated with pesticides banned for use in this country but made here and sold abroad.

Clear-Cutting The practice of removing all trees in a specific area.

Climate Description of the long-term pattern of weather in any particular area.

Climax Community Terminal state of ecological succession in an area; the redwoods are a climax community.

Coal Gasification Process of converting coal to gas; the resultant gas, if used for fuel, sharply reduces sulfur oxide emissions and particulates that result from coal burning.

Commensalism Symbiotic relationship between two different species in which one benefits while the other is neither harmed nor benefited.

Community Ecology Study of interactions of all organisms existing in a specific region.

Competitive Exclusion Resulting from competition; one species forced out of part of an available habitat by a more efficient species.

Conservation The planned management of a natural resource to prevent overexploitation, destruction, or neglect.

Conventional Pollutants Seven substances (sulfur dioxide, carbon monoxide, particulates, hydrocarbons, nitrogen oxides, photochemical oxidants, and lead) that make up the largest volume of air quality degradation, as identified by the Clean Air Act.

Core Dense, intensely hot molten metal mass, thousands of kilometers in diameter, at Earth's center.

Cornucopian Theory The belief that nature is limitless in its abundance and that perpetual growth is both possible and essential.

Corridor Connecting strip of natural habitat that allows migration of organisms from one place to another.

Crankcase Smog Devices (PCV System) A system, used principally in automobiles, designed to prevent discharge of combustion emissions into the external environment.

Critical Factor The environmental factor closest to a tolerance limit for a species at a specific time.

Cultural Eutrophication Increase in biological productivity and ecosystem succession resulting from human activities.

D

Death Rate Number of deaths in one year per 1,000 midyear population.

Decarbonization To remove carbon dioxide or carbonic acid from a substance.

Decomposer Any organism that causes the decay of organic matter; bacteria and fungi are two examples.

Deforestation The action or process of clearing forests without adequate replanting.

Degradation (of water resource) Deterioration in water quality caused by contamination or pollution that makes water unsuitable for many purposes.

Demography The statistical study of principally human populations.

Desert An arid biome characterized by little rainfall, high daily temperatures, and low diversity of animal and plant life.

Desertification Converting arid or semiarid lands into deserts by inappropriate farming practices or overgrazing.

Detergent A synthetic soap-like material that emulsifies fats and oils and holds dirt in suspension; some detergents have caused pollution problems because of certain chemicals used in their formulation.

Detrivores Organisms that consume organic litter, debris, and dung.

Dioxin Any of a family of compounds known chemically as dibenzo-p-dioxins. Concern about them arises from their potential toxicity as contaminants in commercial products. Tests on laboratory animals indicate that it is one of the more toxic anthropogenic (man-made) compounds.

Diversity Number of species present in a community (species richness), as well as the relative abundance of each species.

DNA (Deoxyribonucleic Acid) One of two principal nucleic acids, the other being RNA (Ribonucleic Acid). DNA contains information used for the control of a living cell. Specific segments of DNA are now recognized as genes, those agents controlling evolutionary and hereditary processes.

Dominant Species Any species of plant or animal that is particularly abundant or controls a major portion of the energy flow in a community.

Drip Irrigation Pipe or perforated tubing used to deliver water a drop at a time directly to soil around each plant. Conserves water and reduces soil waterlogging and salinization.

E

Ecological Density The number of a singular species in a geographical area, including the highest concentration points within the defined boundaries.

Ecological Succession Process in which organisms occupy a site and gradually change environmental conditions so that other species can replace the original inhabitants.

Ecology Study of the interrelationships between organisms and their environments.

Ecosystem The organisms of a specific area, together with their functionally related environments; considered as a definitive unit.

Ecotourism Wildlife tourism that could damage ecosystems and disrupt species if strict guidelines governing tours to sensitive areas are not enforced.

Edge Effects Change in ecological factors at the boundary between two ecosystems. Some organisms flourish here; others are harmed.

Effluent A liquid discharged as waste.

El Niño Climatic change marked by shifting of a large warm water pool from the western Pacific Ocean toward the East.

Electron Small, negatively charged particle; normally found in orbit around the nucleus of an atom.

Eminent Domain Superior dominion exerted by a governmental state over all property within its boundaries that authorizes it to appropriate all or any part thereof to a necessary public use, with reasonable compensation being made.

Glossary

Endangered Species Species considered to be in imminent danger of extinction.

Endemic Species Plants or animals that belong or are native to a particular ecosystem.

Environment Physical and biological aspects of a specific area.

Environmental Impact Statement (EIS) A study of the probable environmental impact of a development project before federal funding is provided (required by the National Environmental Policy Act of 1968).

Environmental Protection Agency (EPA) Federal agency responsible for control of air and water pollution, radiation and pesticide problems, ecological research, and solid waste disposal.

Erosion Progressive destruction or impairment of a geographical area; wind and water are the principal agents involved.

Estuary Water passage where an ocean tide meets a river current.

Eutrophic Well nourished; refers to aquatic areas rich in dissolved nutrients.

Evolution A change in the gene frequency within a population, sometimes involving a visible change in the population's characteristics.

Exhaustible Resources Earth's geologic endowment of minerals, nonmineral resources, fossil fuels, and other materials present in fixed amounts.

Extinction Irrevocable elimination of species due to either normal processes of the natural world or through changing environmental conditions.

F

Fallow Cropland that is plowed but not replanted and is left idle in order to restore productivity mainly through water accumulation, weed control, and buildup of soil nutrients.

Fauna The animal life of a specified area.

Feral Refers to animals or plants that have reverted to a noncultivated or wild state.

Fission The splitting of an atom into smaller parts.

Floodplain Level land that may be submerged by floodwaters; a plain built up by stream deposition.

Flora The plant life of an area.

Flyway Geographic migration route for birds that includes the breeding and wintering areas that it connects.

Food Additive Substance added to food usually to improve color, flavor, or shelf life.

Food Chain The sequence of organisms in a community, each of which uses the lower source as its energy supply. Green plants are the ultimate basis for the entire sequence.

Fossil Fuels Coal, oil, natural gas, and/or lignite; those fuels derived from former living systems; usually called nonrenewable fuels.

Fuel Cell Manufactured chemical systems capable of producing electrical energy; they usually derive their capabilities via complex reactions involving the sun as the driving energy source.

Fusion The formation of a heavier atomic complex brought about by the addition of atomic nuclei; during the process there is an attendant release of energy.

G

Gaia Hypothesis Theory that Earth's biosphere is a living system whose complex interactions between its living organisms and nonliving processes regulate environmental conditions over millions of years so that life continues.

Gamma Ray A ray given off by the nucleus of some radioactive elements. A form of energy similar to X rays.

Gene Unit of heredity; segment of DNA nucleus of the cell containing information for the synthesis of a specific protein.

Gene Banks Storage of seed varieties for future breeding experiments.

Genetic Diversity Infinite variation of possible genetic combinations among individuals; what enables a species to adapt to ecological change.

Geothermal Energy Heat derived from the Earth's interior. It is the thermal energy contained in the rock and fluid (that fills the fractures and pores within the rock) in the Earth's crust.

Germ Plasm Genetic material that may be preserved for future use (plant seeds, animal eggs, sperm, and embryos).

Global Warming An increase in the near surface temperature of the Earth. Global warming has occurred in the distant past as the result of natural influences, but the term is most often used to refer to the warming predicted to occur as a result of increased emissions of greenhouse gases. Scientists generally agree that the Earth's surface has warmed by about 1 degree Fahrenheit in the past 140 years.

Green Revolution The great increase in production of food grains (as in rice and wheat) due to the introduction of high-yielding varieties, to the use of pesticides, and to better management techniques.

Greenhouse Effect The effect noticed in greenhouses when shortwave solar radiation penetrates glass, is converted to longer wavelengths, and is blocked from escaping by the windows. It results in a temperature increase. Earth's atmosphere acts in a similar manner.

Gross National Product (GNP) The total value of the goods and services produced by the residents of a nation during a specified period (such as a year).

Groundwater Water found in porous rock and soil below the soil moisture zone and, generally, below the root zone of plants. Groundwater that saturates rock is separated from an unsaturated zone by the water table.

H

Habitat The natural environment of a plant or animal.

Habitat Fragmentation Process by which a natural habitat/landscape is broken up into small sections of natural ecosystems, isolated from each other by sections of land dominated by human activities.

Hazardous Waste Waste that poses a risk to human or ecological health and thus requires special disposal techniques.

Herbicide Any substance used to kill plants.

Heterotroph Organism that cannot synthesize its own food and must feed on organic compounds produced by other organisms.

Hydrocarbons Organic compounds containing hydrogen, oxygen, and carbon. Commonly found in petroleum, natural gas, and coal.

Hydrogen Lightest-known gas; major element found in all living systems.

Hydrogen Sulfide Compound of hydrogen and sulfur; a toxic air contaminant that smells like rotten eggs.

Hydropower Electrical energy produced by flowing or falling water.

I

Infiltration Process of water percolation into soil and pores and hollows of permeable rocks.

Intangible Resources Open space, beauty, serenity, genius, information, diversity, and satisfaction are a few of these abstract commodities.

Integrated Pest Management (IPM) Designed to avoid economic loss from pests, this program's methods of pest control strive to minimize the use of environmentally hazardous, synthetic chemicals.

Invasive Refers to those species that have moved into an area and reproduced so aggressively that they have replaced some of the native species.

Ion An atom or group of atoms, possessing a charge; brought about by the loss or gain of electrons.

Ionizing Radiation Energy in the form of rays or particles that have the capacity to dislodge electrons and/or other atomic particles from matter that is irradiated.

Irradiation Exposure to any form of radiation.

Isotopes Two or more forms of an element having the same number of protons in the nucleus of each atom but different numbers of neutrons.

K

Keystone Species Species that are essential to the functioning of many other organisms in an ecosystem.

Kilowatt Unit of power equal to 1,000 watts.

L

Leaching Dissolving out of soluble materials by water percolating through soil.

Limnologist Individual who studies the physical, chemical, and biological conditions of aquatic systems.

M

Malnutrition Faulty or inadequate nutrition.

Malthusian Theory The theory that populations tend to increase by geometric progression (1, 2, 4, 8, 16, etc.) while food supplies increase by arithmetic means (1, 2, 3, 4, 5, etc.).

Metabolism The chemical processes in living tissue through which energy is provided for continuation of the system.

Methane Often called marsh gas (CH_4); an odorless, flammable gas that is the major constituent of natural gas. In nature it develops from decomposing organic matter.

Migration Periodic departure and return of organisms to and from a population area.

Monoculture Cultivation of a single crop, such as wheat or corn, to the exclusion of other land uses.

Mutation Change in genetic material (gene) that determines species characteristics; can be caused by a number of agents, including radiation and chemicals, called mutagens.

N

Natural Selection The agent of evolutionary change by which organisms possessing advantageous adaptations leave more offspring than those lacking such adaptations.

Niche The unique occupation or way of life of a plant or animal species; where it lives and what it does in the community.

Nitrate A salt of nitric acid. Nitrates are the major source of nitrogen for higher plants. Sodium nitrate and potassium nitrate are used as fertilizers.

Nitrite Highly toxic compound; salt of nitrous acid.

Nitrogen Oxides Common air pollutants. Formed by the combination of nitrogen and oxygen; often the products of petroleum combustion in automobiles.

Nonrenewable Resource Any natural resource that cannot be replaced, regenerated, or brought back to its original state once it has been extracted, for example, coal or crude oil.

Nutrient Any nutritive substance that an organism must take in from its environment because it cannot produce it as fast as it needs it or, more likely, at all.

O

Oil Shale Rock impregnated with oil. Regarded as a potential source of future petroleum products.

Oligotrophic Most often refers to those lakes with a low concentration of organic matter. Usually contain considerable oxygen; Lakes Tahoe and Baikal are examples.

Organic Living or once living material; compounds containing carbon formed by living organisms.

Organophosphates A large group of nonpersistent synthetic poisons used in the pesticide industry; include parathion and malathion.

Ozone Molecule of oxygen containing three oxygen atoms; shields much of Earth from ultraviolet radiation.

P

Particulate Existing in the form of small separate particles; various atmospheric pollutants are industrially produced particulates.

Peroxyacyl Nitrate (PAN) Compound making up part of photochemical smog and the major plant toxicant of smog-type injury; levels as low as 0.01 ppm can injure sensitive plants. Also causes eye irritation in people.

Pesticide Any material used to kill rats, mice, bacteria, fungi, or other pests of humans.

Pesticide Treadmill A situation in which the cost of using pesticides increases while the effectiveness decreases (because pest species develop genetic resistance to the pesticides).

Petrochemicals Chemicals derived from petroleum bases.

pH Scale used to designate the degree of acidity or alkalinity; ranges from 1 to 14; a neutral solution has a pH of 7; low pHs are acid in nature, while pHs above 7 are alkaline.

Phosphate A phosphorous compound; used in medicine and as fertilizers.

Glossary

Photochemical Smog Type of air pollution; results from sunlight acting with hydrocarbons and oxides of nitrogen in the atmosphere.

Photosynthesis Formation of carbohydrates from carbon dioxide and hydrogen in plants exposed to sunlight; involves a release of oxygen through the decomposition of water.

Photovoltaic Cells An energy-conversion device that captures solar energy and directly converts it to electrical current.

Physical Half-Life Time required for half of the atoms of a radioactive substance present at some beginning to become disintegrated and transformed.

Phytoplankton That portion of the plankton community comprised of tiny plants, e.g., algae, diatoms.

Pioneer Species Hardy species that are the first to colonize a site in the beginning stage of ecological succession.

Plankton Microscopic organisms that occupy the upper water layers in both freshwater and marine ecosystems.

Plutonium Highly toxic, heavy, radioactive, manmade, metallic element. Possesses a very long physical half-life.

Pollution The process of contaminating air, water, or soil with materials that reduce the quality of the medium.

Polychlorinated Biphenyls (PCBs) Poisonous compounds similar in chemical structure to DDT. PCBs are found in a wide variety of products ranging from lubricants, waxes, asphalt, and transformers to inks and insecticides. Known to cause liver, spleen, kidney, and heart damage.

Population All members of a particular species occupying a specific area.

Predator Any organism that consumes all or part of another system; usually responsible for death of the prey.

Primary Production The energy accumulated and stored by plants through photosynthesis.

R

Rad (Radiation Absorbed Dose) Measurement unit relative to the amount of radiation absorbed by a particular target, biotic or abiotic.

Radioactive Waste Any radioactive by-product of nuclear reactors or nuclear processes.

Radioactivity The emission of electrons, protons (atomic nuclei), and/or rays from elements capable of emitting radiation.

Rain Forest Forest with high humidity, small temperature range, and abundant precipitation; can be tropical or temperate.

Recycle To reuse; usually involves manufactured items, such as aluminum cans, being restructured after use and utilized again.

Red Tide Population explosion or bloom of minute single-celled marine organisms (dinoflagellates), which can accumulate in protected bays and poison other marine life.

Renewable Resources Resources normally replaced or replenished by natural processes; not depleted by moderate use.

Riparian Water Right Legal right of an owner of land bordering a natural lake or stream to remove water from that aquatic system.

S

Salinization An accumulation of salts in the soil that could eventually make the soil too salty for the growth of plants.

Sanitary Landfill Land waste disposal site in which solid waste is spread, compacted, and covered.

Scrubber Antipollution system that uses liquid sprays in removing particulate pollutants from an airstream.

Sediment Soil particles moved from land into aquatic systems as a result of human activities or natural events, such as material deposited by water or wind.

Seepage Movement of water through soil.

Selection The process, either natural or artificial, of selecting or removing the best or less desirable members of a population.

Selective Breeding Process of selecting and breeding organisms containing traits considered most desirable.

Selective Harvesting Process of taking specific individuals from a population; the removal of trees in a specific age class would be an example.

Sewage Any waste material coming from domestic and industrial origins.

Smog A mixture of smoke and air; now applies to any type of air pollution.

Soil Erosion Detachment and movement of soil by the action of wind and moving water.

Solid Waste Unwanted solid materials usually resulting from industrial processes.

Species A population of morphologically similar organisms, capable of interbreeding and producing viable offspring.

Species Diversity The number and relative abundance of species present in a community. An ecosystem is said to be more diverse if species present have equal population sizes and less diverse if many species are rare and some are very common.

Strip Mining Mining in which Earth's surface is removed in order to obtain subsurface materials.

Strontium-90 Radioactive isotope of strontium; it results from nuclear explosions and is dangerous, especially for vertebrates, because it is taken up in the construction of bone.

Succession Change in the structure and function of an ecosystem; replacement of one system with another through time.

Sulfur Dioxide (SO_2) Gas produced by burning coal and as a by-product of smelting and other industrial processes. Very toxic to plants.

Sulfur Oxides (SO_x) Oxides of sulfur produced by the burning of oils and coal that contain small amounts of sulfur. Common air pollutants.

Sulfuric Acid ($H_2 SO_4$) Very corrosive acid produced from sulfur dioxide and found as a component of acid rain.

Sustainability Ability of an ecosystem to maintain ecological processes, functions, biodiversity, and productivity over time.

Sustainable Agriculture Agriculture that maintains the integrity of soil and water resources so that it can continue indefinitely.

T

Technology Applied science; the application of knowledge for practical use.

Tetraethyl Lead Major source of lead found in living tissue; it is produced to reduce engine knock in automobiles.

Thermal Inversion A layer of dense, cool air that is trapped under a layer of less dense warm air (prevents upward flowing air currents from developing).

Thermal Pollution Unwanted heat, the result of ejection of heat from various sources into the environment.

Thermocline The layer of water in a body of water that separates an upper warm layer from a deeper, colder zone.

Threshold Effect The situation in which no effect is noticed, physiologically or psychologically, until a certain level or concentration is reached.

Tolerance Limit The point at which resistance to a poison or drug breaks down.

Total Fertility Rate (TFR) An estimate of the average number of children that would be born alive to a woman during her reproductive years.

Toxic Poisonous; capable of producing harm to a living system.

Tragedy of the Commons Degradation or depletion of a resource to which people have free and unmanaged access.

Trophic Relating to nutrition; often expressed in trophic pyramids in which organisms feeding on other systems are said to be at a higher trophic level; an example would be carnivores feeding on herbivores, which, in turn, feed on vegetation.

Turbidity Usually refers to the amount of sediment suspended in an aquatic system.

U

Uranium 235 An isotope of uranium that when bombarded with neutrons undergoes fission, resulting in radiation and energy. Used in atomic reactors for electrical generation.

Z

Zero Population Growth The condition of a population in which birthrates equal death rates; it results in no growth of the population.

Index

Index

K

L

M

N

O

P

Q

R

Test Your Knowledge Form

We encourage you to photocopy and use this page as a tool to assess how the articles in *Annual Editions* expand on the information in your textbook. By reflecting on the articles you will gain enhanced text information. You can also access this useful form on a product's book support Web site at *http://www.mhcls.com/online/*.

NAME:

DATE:

TITLE AND NUMBER OF ARTICLE:

BRIEFLY STATE THE MAIN IDEA OF THIS ARTICLE:

LIST THREE IMPORTANT FACTS THAT THE AUTHOR USES TO SUPPORT THE MAIN IDEA:

WHAT INFORMATION OR IDEAS DISCUSSED IN THIS ARTICLE ARE ALSO DISCUSSED IN YOUR TEXTBOOK OR OTHER READINGS THAT YOU HAVE DONE? LIST THE TEXTBOOK CHAPTERS AND PAGE NUMBERS:

LIST ANY EXAMPLES OF BIAS OR FAULTY REASONING THAT YOU FOUND IN THE ARTICLE:

LIST ANY NEW TERMS/CONCEPTS THAT WERE DISCUSSED IN THE ARTICLE, AND WRITE A SHORT DEFINITION:

We Want Your Advice

ANNUAL EDITIONS revisions depend on two major opinion sources: one is our Advisory Board, listed in the front of this volume, which works with us in scanning the thousands of articles published in the public press each year; the other is you—the person actually using the book. Please help us and the users of the next edition by completing the prepaid article rating form on this page and returning it to us. Thank you for your help!

ANNUAL EDITIONS: Environment 06/07

ARTICLE RATING FORM

Here is an opportunity for you to have direct input into the next revision of this volume.
We would like you to rate each of the articles listed below, using the following scale:

1. **Excellent: should definitely be retained**
2. **Above average: should probably be retained**
3. **Below average: should probably be deleted**
4. **Poor: should definitely be deleted**

Your ratings will play a vital part in the next revision.
Please mail this prepaid form to us as soon as possible.
Thanks for your help!

RATING	ARTICLE	RATING	ARTICLE
	1. How Many Planets? A Survey of the Global Environment		14. Hydrogen: Waiting for the Revolution
	2. Five Meta-Trends Changing the World		15. Strangers in Our Midst: The Problem of Invasive Alien Species
	3. Globalization's Effects on the Environment		16. Markets for Biodiversity Services: Potential Roles and Challenges
	4. Rescuing a Planet Under Stress		17. Dryland Development: Success Stories from West Africa
	5. Population and Consumption: What We Know, What We Need to Know		18. What's a River For?
	6. A New Security Paradigm		19. How Much Is Clean Water Worth?
	7. Factory Farming in the Developing World		20. A Human Thirst
	8. Where Oil and Water Do Mix: Environmental Scarcity and Future Conflict in the Middle East and North Africa		21. Agricultural Pesticides in Developing Countries
			22. The Quest for Clean Water
	9. The Irony of Climate		23. A Little Rocket Fuel with Your Salad?
	10. World Population, Agriculture, and Malnutrition		24. How Did Humans First Alter Global Climate?
	11. Powder Keg		25. Can We Bury Global Warming?
	12. Personalized Energy: The Next Paradigm		26. Dozens of Words for Snow, None for Pollution
	13. Wind Power: Obstacles and Opportunities		

(Continued on next page)

BUSINESS REPLY MAIL
FIRST CLASS MAIL PERMIT NO. 551 DUBUQUE IA

POSTAGE WILL BE PAID BY ADDRESEE

McGraw-Hill Contemporary Learning Series
2460 KERPER BLVD
DUBUQUE, IA 52001-9902

ABOUT YOU

Name _____ Date _____

Are you a teacher? ❏   A student? ❏
Your school's name _____

Department _____

Address _____ City _____ State _____ Zip _____

School telephone # _____

YOUR COMMENTS ARE IMPORTANT TO US!

Please fill in the following information:
For which course did you use this book?
_____

Did you use a text with this ANNUAL EDITION? ❏ yes ❏ no
What was the title of the text?
_____

What are your general reactions to the *Annual Editions* concept?
_____

Have you read any pertinent articles recently that you think should be included in the next edition? Explain.
_____

Are there any articles that you feel should be replaced in the next edition? Why?
_____

Are there any World Wide Web sites that you feel should be included in the next edition? Please annotate.
_____

May we contact you for editorial input? ❏ yes ❏ no
May we quote your comments? ❏ yes ❏ no